M. R. Lunt.

April 1968.

THE MOLECULAR BIOLOGY OF VIRUSES

Other Publications of the
Society for General Microbiology*

THE JOURNAL OF GENERAL MICROBIOLOGY

SYMPOSIA

* Published by the Cambridge University Press, except for the first Symposium, which was published by Blackwell's Scientific Publications Limited.

THE MOLECULAR
BIOLOGY OF VIRUSES

EIGHTEENTH SYMPOSIUM OF THE
SOCIETY FOR GENERAL MICROBIOLOGY
HELD AT THE
IMPERIAL COLLEGE, LONDON
APRIL 1968

48,871

[Ed. L.V. Crawford & M.G.P. Stoker]

CAMBRIDGE
Published for the Society for General Microbiology
AT THE UNIVERSITY PRESS
1968

Published by the Syndics of the Cambridge University Press
Bentley House, P.O. Box 92, 200 Euston Road, London, N.W. 1
American Branch: 32 East 57th Street, New York, N.Y. 10022

Library of Congress Catalogue Card Number: 54–3996

Standard Book Number: 521 06931 9

Printed in Great Britain
at the University Printing House, Cambridge
(Brooke Crutchley, University Printer)

CONTRIBUTORS

ANDERSON, E. S., Enteric Reference Laboratory, Public Health Laboratory Service, London.

ARBER, W., Laboratoire de Biophysique, Université de Genève, Switzerland.

BANCROFT, J. B., Department of Botany and Plant Pathology, Purdue University, Lafayette, Indiana, U.S.A.

CRAWFORD, L. V., Medical Research Council Experimental Virus Research Unit, Institute of Virology, Glasgow.

DARNELL, J. E., JUN., Department of Biochemistry and Cell Biology, Albert Einstein College of Medicine, Bronx, New York, U.S.A.

HAYES, W., Medical Research Council Microbial Genetics Research Unit, Hammersmith Hospital, London.

KEIR, H. M., Department of Biochemistry, The University, Glasgow.

LEBERMAN, R., Medical Research Council Laboratory of Molecular Biology, Hill's Road, Cambridge.

MARTIN, E. M. AND KERR, I. M., National Institute for Medical Research, Mill Hill, London.

MONTAGNIER, L., Institut du Radium-Biologie, Faculté des Sciences, 91, Orsay, France.

RAPP, F., Department of Virology and Epidemiology, Baylor University College of Medicine, Houston, Texas, U.S.A.

SINSHEIMER, R. L., Division of Biology, California Institute of Technology, Pasadena, California, U.S.A.

SUBAK-SHARPE, H., Medical Research Council Experimental Virus Research Unit, Institute of Virology, Glasgow.

SVOBODA, J., Institute of Experimental Biology and Genetics, Czechoslovak Academy of Sciences, Prague, Czechoslovakia.

THOMAS, R., Laboratoire de Génétique, Faculté des Sciences, Université Libre de Bruxelles, Belgium.

WATSON, D. H., Department of Virology, The University, Birmingham.

CONTENTS

EDITORS' PREFACE

The Society's first Symposium devoted to viruses, held 16 years ago, marked the beginning of a new era, since it coincided with the well-known Hershey-Chase experiment, the results of which were announced during the meeting. This experiment showed how virus nucleic acid and protein could be separated and functions ascribed to each, and it was a major landmark in the understanding of virus growth. The process of separating viral components and analysing their functions has continued, and a second Symposium, in 1959, reported the extension of such studies from bacteriophages to animal and plant viruses.

Since then, the isolation and dissection of component viral molecules has continued, but now there are also attempts to re-assemble viruses and to achieve working systems from component parts. A great deal more is now known of the morphology of virus particles and of the nature of their nucleic acid and protein components. The biochemistry of virus-infected cells and the general outline of the replication of virus nucleic acids have also been elucidated. Our aim here was to correlate virus structure with functions occurring in the virus-infected cell and covering not only the operation of the virus component proteins but also that of the other virus coded molecules, such as the enzymes involved in virus replication.

In this Symposium we have invited contributors from those who are actively engaged in these problems and who will describe the current state of progress, particularly in relation to their own work. They have not been asked to give comprehensive reviews, so the lists of references are not necessarily complete. We have tried to arrange the presentations systematically, but there are inevitably some duplications and even greater omissions.

Our editorial tasks have been considerably eased by the expert assistance of Mrs E. Hendry, of the Institute of Virology, Glasgow, and we are duly grateful.

L. V. CRAWFORD
M. G. P. STOKER

Institute of Virology,
University of Glasgow

TRENDS AND METHODS IN VIRUS RESEARCH

W. HAYES

Medical Research Council, Microbial Genetics Research Unit, Hammersmith Hospital, London, W.12

In a recent essay on the subject of vitalism, Francis Crick has pointed out that a refutation of this philosophy requires more than the *in vitro* synthesis of a living system from chemical elements, for it could be argued that the endowment of the molecular machine with the properties of life was a consequence of its 'colonization' by a soul or vital force. To defeat this line of argument it must be possible to demonstrate that the behaviour of the living system is totally explicable by the laws of physics and chemistry (Crick, 1966), and this, indeed, is what the new science of molecular biology aims to do, and has succeeded to a remarkable degree in the case of bacterial viruses. Our current preoccupation with this approach, which is an index of the progress which molecular biology has made in recent years, may be judged by the frequency with which the words 'structure' and 'function' are hopefully and often meaningfully coupled in the titles of recent biological publications and symposia.

Ever since their existence has been recognized, viruses have attracted an aura of mystery. Initially, their potency to cause disease contrasted strangely with their invisibility. Then the discovery of their nucleoprotein constitution, and the demonstration by Stanley (1935) that tobacco mosaic virus can exist in crystalline form, gave birth to the romantic idea that viruses are a kind of missing link between living and non-living material, and so may provide meaningful clues to the ultimate nature of life. The most crucial events in the evolution of ideas about viruses were the twin discoveries, stemming primarily from the analysis of pneumococcal transforming principle by Avery and his colleagues (Avery, Macleod & McCarty, 1944), that genetic material consists of nucleic acid and that the essence of virus infection is the entry of viral nucleic acid into the cell, as first demonstrated in the case of bacterial DNA viruses (Hershey & Chase, 1952) and the RNA-containing tobacco mosaic virus (Gierer & Schramm, 1956; Fraenkel-Conrat, Singer & Williams, 1957). In fact, research on the molecular biology of viruses can be said to have begun at this point, when a clear understanding that virus infection comprises a genetic re-programming of the cellular factory to produce virus nucleic acid and proteins, was matched by a knowledge of the molecular structure of these basic components of living systems, and of how they are synthesized. Since then, progress has consisted of developing increasingly detailed models of the kinetic interactions of viral nucleic acids and proteins, and of how these are regulated and coordinated in the infected cell. As a result of these developments life no longer seems an insoluble mystery but just very complicated. With this complexity has come an increasing awareness of the beauty and elegance of the mechanisms involved, and an awe of the process of natural selection which led to their evolution.

Despite the great stimulus which the introduction of tissue culture and plaque-counting techniques gave to the study of animal viruses (see Syverton & Scherer, 1953; Dulbecco, 1953), and the importance of early analyses of the composition of tobacco mosaic virus (Bawden & Pirie, 1937 a, b), there is no doubt that most of the molecular models which form the ideological basis of current research emanate from work on bacterial viruses. This is because of the high infectivity of these viruses, the ease with which synchronous infection is initiated, the ability to follow the development of new virus components both chemically and microscopically and, above all, because both host cell and virus are very amenable to detailed genetic analysis. Introducing this symposium, I thought it might prove useful to foreshadow some trends in virus research in general by outlining a few of the more unusual and progressive genetic concepts at present in vogue in the phage world.

A new era in phage genetics began a few years ago with the introduction of conditional lethal mutants. These are either temperature-sensitive, growing normally at 25°–30° but unable to develop at 42° (Edgar & Lielausis, 1964), or belong to the so-called 'amber' group of mutants (see Edgar, 1966) which grow normally in certain 'permissive' hosts but not at all in other, 'restrictive' hosts (Epstein et al. 1963). Amber mutants arise from 'nonsense' mutations which alter a nucleic acid base triplet coding for an amino acid into one which terminates the growing polypeptide chain; however, the permissive host carries a species of transfer-RNA molecule which translates the chain-terminating triplet as a compatible amino acid, so that a functional polypeptide is produced. Three great advantages follow from the fact that conditional lethal mutants are not restricted to modifiable characters such as plaque type or host range, but may be defective in any function, including an indispensable one. The first is that mutations may be isolated in any gene so that it becomes theoretically possible to construct a complete genetic map of the organism. Secondly, mutants involving different functional genes can be distinguished in complementation tests, following mixed infection under non-permissive conditions, so that the physiological genetics of viruses such as RNA phages, in which genetic recombination has not been demonstrated, can nevertheless be studied. Amber mutants are especially suitable for such complementation tests since their failure to produce complete polypeptide chains renders them 'non-leaky'. Thirdly, the ability to isolate mutants which are defective in essential functions permits these functions to be analysed under restrictive conditions by observing, biochemically or microscopically, what synthetic or morphogenetic step is blocked.

A systematic study of morphogenesis, especially in phage T4, by means of conditional lethal mutants has shown that this is a more complicated process than was at first thought. The fact that infectious particles of tobacco mosaic virus reform when separated and purified nucleic acid and protein components of the virus are mixed (Fraenkel-Conrat & Singer, 1957), and that bacterial flagella which have been dissociated into a solution of their constituent protein molecules, re-aggregate to form flagella under the proper physical conditions (Abram & Koffler, 1964), suggested that many protein structures might arise automatically once a sufficient concentration of subunits was produced in the right environment. The phage head, for example, might result from a crystallization of head protein molecules on the condensed DNA which itself folded specifically on a scaffolding of protein 'condensation principle' (Kellenberger, 1961). However, it turns out in the case of phage T4, that although the structure of the head protein subunit is determined by a single gene (no. 23), the function of many other genes is required for the formation of a normal head. Thus a mutation in gene 20, in association with a wild-type gene 23, results in the production of enormously long tubes composed of head protein, called polyheads; some factor required to cap off the growing tube when it reaches the correct length, to form an elongate icosahedron, is missing. Again, a mutation in gene 66 leads to production of abnormally short heads, in gene 21 to heads devoid of DNA, and in gene 31 to the aggregation of head protein into amorphous clumps so that no structure at all is formed. Clearly the head protein subunit itself lacks shape-specifying information, the architecture of the head depending on a number of secondary, specific components. In the same way, mutations in a different gene from that specifying sheath protein structure result in the formation of enormously long sheaths, or polysheaths (Epstein *et al.* 1963; Kellenberger, 1966).

Altogether about 40 genes have now been identified as controlling morphogenesis in phage T4, and it is clear that only a small proportion of these is concerned with directing synthesis of the various major structural proteins of the phage particle, such as head, sheath, tail core, end-plate and tail fibres. Thus, for example, at least 13 genes are involved in tail structure, while at least five cooperate in the production of tail fibres which appear to be made up of a single type of high molecular weight protein molecule. Nevertheless, despite this complexity, certain steps in the assembly of phage particles do appear to be automatic, provided the proper precursors are present. This has been beautifully shown by *in vitro* complementation tests in which purified extracts of pairs of restrictive bacterial cultures, each separately infected with a different amber mutant which itself yields no infective particles, are mixed in a test tube. When 40 such extracts were tested in all possible pairwise combinations, 13 complementation groups were found; mixtures of pairs of extracts from different groups, but not of pairs from the same group, rapidly yield fully mature, infective phage particles (Edgar & Wood, 1966; Wood & Edgar, 1967).

In general, all the mutants within the same complementation group have defects involving the same structure—some make heads and tail fibres but no tails, while others make tails and tail fibres but no heads. It is therefore evident that certain completed components are capable of self-assembly to form normal particles, but that various gene-mediated steps in the maturation of each component can occur only in the cellular environment. For example, mixture of an extract containing active tails and tail fibres with preparations from other mutants which *appear* to contain normal heads, often fails to produce active phage. Further tests succeeded in defining the sequence of assembly reactions, and the role of certain genes of indeterminate phenotype. For example, mutant genes 13 and 14 produce unconnected heads, tails and tail fibres, as do those of some other complementation groups, and might therefore block maturation of any one of these components. When extracts of cells infected with mutants 13 or 14 were mixed with purified and active tails from extracts of another mutant, no active particles were formed; but addition of purified and active heads did yield infective particles. Clearly genes 13 and 14 specify products whose activity results in head completion. By similar methods, a morphogenetic pathway for the assembly of phage T4 has been worked out although, of course, the functions of many genes mediating the pathway remain unknown (see Wood & Edgar, 1967). *In vitro* complementation between extracts of infected restrictive cells has also been demonstrated for amber mutants of the temperate phage λ, by Weigle (1966).

The possibility of an intriguing morphogenetic interrelationship between the head protein and the DNA contained by it has recently come to light. The chromosomes of phages T2 and T4 have two very interesting and unexpected features which were originally deduced from the results of genetic and physical analyses. The first is that the chromosomes of a phage population are circular permutations of one another; that is, the linear chromosomes incorporated into the phage heads behave as if they resulted from breakage at random points of an initial circular structure (Streisinger, Edgar & Denhardt, 1964). This means that any pair of loci which may mark the extremities of the chromosome of any given particle will always be found to be closely linked on the chromosomes of other particles. While this results in a circular linkage map of the genome, it does not necessarily imply that a continuous structure actually exists in the infected bacterium. This circular permutation of the phage T2 chromosome has been beautifully demonstrated by denaturing the double-stranded DNA extracted from a population of particles and then allowing the single strands to re-anneal. If all the Watson and Crick strands were similar, they would renature to form linear molecules as before; in fact, they form *circular* double-stranded molecules with high efficiency due to the bonding of homologous regions of circularly permuted, complementary strands, to leave complementary, single-stranded extremities, $\frac{1-2-3-4-5}{3-4-5-1-2}$, which will unite to form a circle (Thomas & MacHattie, 1964).

The second feature of phages T2 and T4 is that, in every chromosome, the sequence of bases at one extremity is repeated at the other—the so-called 'terminal redundancy'. Put in more genetic terms, every chromosome is diploid over 1–3 % of its extremities but, because the chromosomes are circularly permuted, each has a different diploid region (Streisinger *et al.* 1964). The existence of terminal redundancy has been confirmed in phages T2 and T4, and discovered to be a feature of many other phages (MacHattie, Ritchie, Thomas & Richardson, 1967; Ritchie, Thomas, MacHattie & Wensink, 1967). They employed the elegant device of treating the extracted DNA molecules for a short time with the enzyme exonuclease III which nibbles away the duplexes by attacking the single strands from their 3′ ends only; in the case of terminally redundant molecules this would leave complementary, single-stranded ends, $\frac{1-2-3-4-5}{3'\ \text{end}}\ \frac{3'\ \text{end}}{3-4-5-1-2}$, which should then join up to form circles, and this is what happens experimentally. It is worth noting that the chromosome of the temperate phage λ, although neither circularly permuted nor terminally redundant, is endowed with single-stranded, complementary extremities (Hershey, Burgi & Ingraham, 1963; Strack & Kaiser, 1965); as a result, not only do the two ends of single genomes unite to form circles both *in vitro* and *in vivo*, but pairing of the single-stranded ends of some chromosomes with the complementary single strands of others leads to the formation of long chains made up of chromosome subunits, called concatenates (Ris & Chandler, 1963).

The existence of clonal populations whose chromosomes are both terminally redundant and circularly permuted suggests a mechanism which would account at the same time for both these phenomena, namely, that fixed lengths of DNA longer than a complete genome are chopped off from a concatenate of genomes, thus:

Genomes (10 digits long)

123456789012345678901234567890123456789012345678 90

Chromosomes (12 digits long)

Recombination between homologous regions towards the extremities of circularly permuted chromosomes would yield concatenates and there is, indeed, some evidence that the DNA of phage T4 may replicate as a long polymer (Frankel, 1966*a*, *b*). If this model is the correct one, how is the chromosome actually chopped up? From a study made some years ago by Nomura & Benzer (1961) it is now apparent that the effect of deleting part of the chromosome of phage T4 is to increase the length of the region of terminal redundancy, as follows:

Genomes
(8 digits long, due to deletion of 5,6)

123478901234789012347890

Chromosomes
(12 digits long, as before)

This means that determination of the length of DNA which is to be included in the phage head is not a function of the DNA itself; it is not coded for directly in the genome. An obvious candidate for the role of chopper is the protein head structure which, on completion, could snip off any DNA in excess of a head-full (Streisinger, Emrich & Stahl, 1967). In support of this hypothesis is the fact that, in normal populations of the *Escherichia coli* transducing phage, P1, about one-quarter of the particles are of small size, having heads 650 Å instead of 900 Å in diameter and containing about 67 % of the protein and 40 % of the DNA of the majority particles. As might be expected, these small particles are defective and unproductive in single infections, but their genomes are, presumably, circularly permuted because, in multiple infections, they can complement one another to yield normally infectious particles. Moreover, a proportion of these small particles are transducing, but carry a diminished length of bacterial DNA which is reflected in a reduced probability of co-transduction of bacterial genes. This reduction in both bacterial and phage DNA content clearly implicates a defect in the head structure as the primary lesion (Ikeda & Tomizawa, 1965). A similar, defective variant of phage T4, which has a diminished protein content and carries only 70 % of the normal DNA, but which yields normal particles following multiple infection, has also been described (Mosig, 1963). Some intriguing possibilities might follow this sort of model. For example, the formation of recombinant chromosomes would not be the direct outcome of recombination events between chromosome pairs, but would emerge from the chopping up of concatenates formed by recombination between the ends of parental chromosomes.

A question which arises from the demonstrated ability of certain morphogenetic steps to occur between relatively pure components *in vitro* is the extent to which the intact cell is really necessary for phage development. The formation of infective phage λ particles by addition of λ DNA to mechanically disrupted *E. coli* bacteria, which were, normally, re-sistant to infection by either intact phage λ or its DNA, was first reported by Mackal, Werninghaus & Evans (1964), although the efficiency of the process was rather low. Moreover, no yield was found if the DNA was added to protoplasts unless these were first disrupted mechanically or by dilution in water. More recently, Zgaga (1967) has claimed high titres of infectious particles following addition of phage λ DNA to osmotically ruptured protoplasts of *E. coli* K 12S; the yield is reported to be 100 times greater than when intact protoplasts are used, while the system is sensitive throughout to the action of both DNase and RNase.

With the isolation, and genetic and functional analysis of large numbers of conditional lethal mutants of both phage T4 and phage λ, interest has been aroused in how the expression of these genes is regu-lated. A distinction between 'early' functions needed for chromosome replication, and other 'late' functions was first made by Jacob, Fuerst & Wollman (1957) who observed the synthetic steps blocked in a series of defective λ prophages which were unable to produce infective particles following induction. A rather striking feature of the chromosome maps of phages T4 and λ is the considerable degree of clustering of genes determining related functions, but especially the tendency for 'early' genes, mainly involved in DNA synthesis, and 'late' genes determining head, tail, sheath and tail fibre formation, to lie on different segments of the map. The main feature of the regulation of phage growth which has so far come to light is that, in cells infected with mutants which cannot synthesize DNA, the genes responsible for the synthesis of the structural components of the phage, as well as of lysozyme, fail to be activated, while other genes responsible for the synthesis of early enzymes operate normally. However, under these conditions early enzymes, other than the one blocked in the infecting mutant, continue to be produced instead of stopping early in development, as happens following infection with wild-type phage or with mutants blocked in late functions. Thus there is a mechanism, apparently associated with some step in DNA synthesis, which is needed to switch on the late genes, while some late function turns the early genes off when their operation is no longer needed. There is an obvious correlation between DNA synthesis and the induction of late functions since in the case of a particular amber mutant of phage T4, in which net DNA synthesis begins only after a lag, the late functions are not inhibited but their initiation shows the same lag. Again, another mutant in which DNA synthesis is arrested after its normal initiation, permits the formation of healthy heads, tails and tail fibres, but not of lysozyme. On the other hand, although these correlations show that DNA synthesis is in some way involved, it is not in itself a sufficient trigger for late functions, since mutations in two genes (nos. 33, 55) prevent the onset of all late functions although DNA synthesis is normal (Epstein *et al.* 1963; Wiberg *et al.* 1962; Levinthal, Hosoda & Shub, 1967).

This general picture is borne out by analysis of phage λ, where amber mutations in genes *N*, *O* and *P* not only stop DNA synthesis but also prevent initiation of all the late functions associated with maturation and lysis. On the other hand, mutations in the linked gene *Q* do not interfere with DNA synthesis, but nevertheless prevent late functions from being switched on. Mutants in which any of these early genes are blocked are characterized by a very low level of messenger RNA, so that gene *Q* may be activated by some product of DNA synthesis and may then operate directly in inducing the formation of messenger RNA serving late functions, possibly by determining the synthesis of a new RNA polymerase (Dove, 1966; Joyner, Isaacs, Echols & Sly, 1966; Eisen *et al.* 1966). What experimental evidence there is favours the hypothesis of two transcription groups, and excludes other suggestions based, for example, on the number of gene copies required to support late protein synthesis, or on the relative stabilities of early and late messenger RNA molecules. The positive evidence rests on the ability to break the phage λ DNA molecule transversely into two halves by hydrodynamic shear, and to obtain separate preparations of these halves by density gradient centrifugation, since one half (the 'right' half), is less dense than the other ('left') due to a different base composition (Hershey, Burgi & Davern, 1965). The genes located on these two halves can be identified by 'marker rescue' experiments, because it is possible to introduce the isolated half-molecules into *E. coli* bacteria by a special transformation technique (Kaiser & Hogness, 1960) so that their ability to complement known amber mutants of phage λ may be tested. It happens that the genes determining early functions lie on the right half-molecule and those determining late functions on the left half (Kaiser, 1962; Kaiser & Inman, 1965). Phage-specific messenger RNA can be extracted from infected bacteria, and isolated and identified by its ability to form 'hybrid' molecules with denatured phage DNA. By estimating the degree of hybridization with each half of the phage λ DNA molecule, the pattern of messenger RNA transcription throughout the growth cycle can be assessed. It turns out that during the early stages of infection transcription is restricted to the right half of the molecule which codes for early functions; later, transcription extends to both halves, and this is correlated with a change in the base composition of the messenger RNA (Skalka, 1966). An interesting corollary to this experiment is the finding that *in vitro* transcription of phage λ DNA by *E. coli* RNA polymerase is mainly confined to the right half-molecule, suggesting that some gene product, needed to shift transcription to the left half-molecule determining late functions, is missing from the *in vitro* system (Cohen, Maitra & Hurwitz, 1967). It could well be that the phage utilizes its host's RNA polymerase to initiate early functions, but must make its own for the transcription of late functions.

These results raise the problem of the extent to which viruses depend on their hosts' enzymes, and how one can distinguish those functions which are specifically viral. It is not enough to show that the enzyme is produced only after infection of the host cell, since mechanisms can easily be postulated whereby a virus component (as, for example, a moiety of phage 'internal protein') might specifically induce the synthesis of a host-determined product. Apart from the future possibility of *in vitro* synthesis of functional proteins from viral messenger RNA templates, the most convincing current evidence comes from the use of temperature-sensitive phage mutants. For example, amber mutations in gene 43 of phage T4 block the production of an active T4 DNA polymerase. When a temperature-sensitive mutant involving this gene was used to infect *E. coli* at 25°, and the DNA polymerase then isolated, purified and tested *in vitro*, it was found to be inactivated by heat, unlike the wild-type enzyme. A similar result has been obtained with a single-step mutant of phage T5. Thus the temperature-sensitive product of the phage genome is not some intermediate inducer, but the actual enzyme which mediates the observed function. In addition, these observations reveal that the polymerase induced by phage infection, and demonstrable *in vitro*, is also involved in the replication of the phage DNA; in fact, the level of *E. coli* DNA polymerase remains constant during infection under restrictive conditions, when no phage DNA synthesis occurs (de Waard, Paul & Lehman, 1965).

A recently developed and informative way of following the control of phage protein synthesis involves exposing the infected bacteria to short pulses of radioactive amino acids (or ^{35}S) at intervals after infection, and then immediately extracting the proteins and separating them by gel electrophoresis; the proteins which have incorporated the radioactivity are recognized by autoradiography, their rate of synthesis being indicated by the degree of darkening of the film (Levinthal *et al.* 1967). The particular proteins can be identified by the specific bands which are found to be missing from extracts of restrictive bacteria infected with different amber mutants. By means of this ingenious method three classes of early protein have been discovered in phage T4, distinguished by the rapidity with which their synthesis is initiated after infection, and the duration of synthesis. Interestingly, it turns out that synthesis of all the late proteins begins at the same time and continues until lysis. By an extension of this method, whereby incubation of the culture is continued for some time after the radioactive pulse has been flushed away with a 'chaser' of cold amino acid, before extraction and analysis, the fate of the labelled protein may be followed. Disappearance of the labelled band may indicate incorporation of the protein into a larger structure having a different mobility; thus it is found that radioactivity can be 'chased' from the band corresponding to the head protein when mutants making normal heads are used, but not in the case of mutants which fail to make intact head structures. Again, the finding that the products of certain genes (nos. 22, 24) associated with head formation can be chased from their bands in the same way as the head protein itself (gene 23) provides evidence that these products are minor head components (Levinthal *et al.* 1967).

Finally we should bear in mind a novel category of genetic elements, also to be discussed in this symposium, which have so far been found only in bacteria, but which qualify as viruses by fulfilling Lwoff's (1959) criteria of infectivity, possession of only one type of nucleic acid and dependence on host metabolism for their reproduction. These are the sex (conjugation, transfer) factors which determine the ability of their hosts to conjugate, and their own efficient transmission throughout a bacterial population without the risk of exposure to the environment (Lederberg, Cavalli & Lederberg, 1952; Hayes, 1953). Some of these factors, such as the sex factor of *E. coli*, can become integrated into the host's chromosome and so are strictly analogous to prophage, while others, such as some colicin factors, are potentially pathogenic in so far as they can initiate lethal syntheses. However, unlike most viruses, these factors are usually very stably inherited in the cytoplasmic state, their replication and segregation being synchronized and strictly coordinated with that of the bacterial chromosome. In the case of the *E. coli* sex factor, which has been most thoroughly studied, it has been clearly shown that the independently replicating sex factor and chromosome segregate together and not randomly, implying that both are distributed to daughter cells by the same mechanism; this is very probably an attachment of each to specific sites on the cell membrane, whose outward growth from an equatorial region separates the daughter chromosomes and sex factors with a precision equivalent to that of a mitotic apparatus (Jacob, Ryter & Cuzin, 1966). Thus the *E. coli* sex factor, and probably all factors of this general type, behave more like small, supernumerary chromosomes than like what hitherto have been regarded as viruses. Those factors which are at present known, were recognized only because attention was drawn, often by accident, to some particular function which they happened to express but which, like chromosome transfer, colicin production or transmission of drug resistance, is quite irrelevant to their behaviour as viruses. There is every reason to believe that covert, transmissible elements of this sort are widespread among bacteria (Anderson, 1965; Meynell & Datta, 1966), and it would be surprising if analogous entities did not exist in other forms of life. Speculation about their evolution is a fascinating topic, and who can say whether they represent *alpha* or *omega* in the world of viruses?

REFERENCES

ABRAM, D. & KOFFLER, H. (1964). The *in vitro* formation of flagella-like filaments and other structures from flagellin. *J. molec. Biol.* **9**, 168.

ANDERSON, E. S. (1965). A rapid screening test for transfer factors in drug-sensitive *Enterobacteriaceae*. *Nature, Lond.* **208**, 1016.

AVERY, O. T., MACLEOD, C. M. & McCARTY, M. (1944). Studies on the chemical nature of the substance inducing transformation of pneumococcal types. I. Induction of transformation by a desoxyribonucleic acid fraction isolated from pneumococcus type III. *J. exp. Med.* **79**, 137.

BAWDEN, F. C. & PIRIE, N. W. (1937*a*). The isolation and some properties of liquid crystalline substances from solanaceous plants injected with three strains of tobacco mosaic virus. *Proc. R. Soc.* B, **123**, 274.

BAWDEN, F. C. & PIRIE, N. W. (1937*b*). The relationship between liquid crystalline preparations of cucumber viruses 3 and 4 and strains of tobacco mosaic virus. *Br. J. exp. Path.* **18**, 275.

COHEN, S. N., MAITRA, U. & HURWITZ, J. (1967). Role of DNA in RNA synthesis. XI. Selective transcription of λ DNA segments *in vitro* by RNA polymerase of *Escherichia coli. J. molec. Biol.* **26**, 19.

CRICK, F. (1966). *Of Molecules and Men.* Seattle and London: University of Washington Press.

DOVE, W. F. (1966). Action of the lambda chromosome. I. Control of functions late in bacteriophage development. *J. molec. Biol.* **19**, 187.

DULBECCO, R. (1953). Some problems of animal virology as studied by the plaque technique. *Cold Spring Harb. Symp. quant. Biol.* **18**, 273.

EDGAR, R. S. (1966). Conditional lethals. In *Phage and the Origins of Molecular Biology*, p. 166. Ed. J. Cairns, G. S. Stent and J. D. Watson. Cold Spring Harbor, New York: Cold Spring Harbor Laboratory of Quantitative Biology.

EDGAR, R. S. & LIELAUSIS, I. (1964). Temperature-sensitive mutants of bacteriophage T4D: their isolation and genetic characterisation. *Genetics*, **49**, 649.

EDGAR, R. S. & WOOD, W. B. (1966). Morphogenesis of bacteriophage T4 in extracts of mutant-infected cells. *Proc. natn. Acad. Sci. U.S.A.* **55**, 498.

EISEN, H. A., FUERST, C. R., SIMINOVITCH, L., THOMAS, R., LAMBERT, L., PEREIRA DA SILVA, L. & JACOB, F. (1966). Genetics and physiology of defective lysogeny in K12(λ): studies of early mutants. *Virology*, **30**, 224.

EPSTEIN, R. H., BOLLE, A., STEINBERG, C. M., KELLENBERGER, E., BOY DE LA TOUR, E., CHEVALLEY, R., EDGAR, R. S., SUSSMAN, M., DENHARDT, G. H. & LIELAUSIS, A. (1963). Physiological studies of conditional lethal mutants of bacteriophage T4D. *Cold Spring Harb. Symp. quant. Biol.* **28**, 375.

FRAENKEL-CONRAT, H. & SINGER, B. (1957). Virus reconstitution: Combination of protein and nucleic acid from different strains. *Biochim. biophys. Acta*, **24**, 541.

FRAENKEL-CONRAT, H., SINGER, B. & WILLIAMS, R. C. (1957). Infectivity of viral nucleic acid. *Biochim. biophys. Acta*, **25**, 87.

FRANKEL, F. R. (1966*a*). The absence of mature phage DNA molecules from the replicating pool of T-even-infected *Escherichia coli. J. molec. Biol.* **18**, 109.

FRANKEL, F. R. (1966*b*). Studies on the nature of replicating DNA in T4-infected *Escherichia coli. J. molec. Biol.* **18**, 127.

GIERER, A. & SCHRAMM, G. (1956). Infectivity of ribonucleic acid from tobacco-mosaic virus. *Nature, Lond.* **177**, 702.

HAYES, W. (1953). The mechanism of genetic recombination in *E. coli. Cold Spring Harb. Symp. quant. Biol.* **18**, 75.

HERSHEY, A. D. & CHASE, M. (1952). Independent function of viral protein and nucleic acid in growth of bacteriophage. *J. gen. Physiol.* **36**, 39.

HERSHEY, A. D., BURGI, E. & DAVERN, C. I. (1965). Preparative density-gradient centrifugation of the molecular halves of lambda DNA. *Biochem. biophys. res. Commun.* **18**, 675.

HERSHEY, A. D., BURGI, E. & INGRAHAM, L. (1963). Cohesion of DNA molecules isolated from phage lambda. *Proc. natn. Acad. Sci. U.S.A.* **49**, 748.

IKEDA, H. & TOMIZAWA, J. (1965). Transducing fragments in generalized transduction by phage P1. III. Studies with small phage particles. *J. molec. Biol.* **14**, 120.

JACOB, F., FUERST, C. R. & WOLLMAN, E. L. (1957). Recherches sur les bactéries lysogènes défectives. II. Les types physiologiques liés aux mutations du prophage. *Ann. Inst. Pasteur*, **93**, 724.

JACOB, F., RYTER, A. & CUZIN, F. (1966). On the association between DNA and the membrane in bacteria. *Proc. R. Soc.* B, **164**, 267.

JOYNER, A., ISAACS, L. N., ECHOLS, H. & SLY, W. S. (1966). DNA replication and messenger RNA production after induction of wild-type λ bacteriophage and λ mutants. *J. molec. Biol.* **19**, 174.

KAISER, A. D. (1962). The production of phage chromosome fragments and their capacity for genetic transfer. *J. molec. Biol.* **4**, 275.

KAISER, A. D. & HOGNESS, D. S. (1960). The transformation of *Escherichia coli* with deoxyribonucleic acid isolated from bacteriophage λ*dg*. *J. molec. Biol.* **2**, 392.

KAISER, A. D. & INMAN, R. B. (1965). Cohesion and the biological activity of bacteriophage lambda DNA. *J. molec. Biol.* **13**, 78.

KELLENBERGER, E. (1961). Vegetative bacteriophage and the maturation of the virus particles. *Adv. Virus Res.* **8**, 1.

KELLENBERGER, E. (1966). The genetic control of the shape of a virus. *Scient. Am.* **215**, 32.

LEDERBERG, J., CAVALLI, L. L. & LEDERBERG, E. M. (1952). Sex compatibility in *E. coli. Genetics*, **37**, 720.

LEVINTHAL, C., HOSODA, J. & SHUB, D. (1967). The control of protein synthesis after phage infection. In *Molecular Biology of Viruses*. Ed. J. Colter. New York: Academic Press.

LWOFF, A. (1959). Bacteriophage as a model of host-virus relationship. In *The Viruses*, vol. II, pp. 187–201. Ed. F. M. Burnet and W. M. Stanley. New York: Academic Press.

MACHATTIE, L. A., RITCHIE, D. A., THOMAS, C. A., JUN. & RICHARDSON, C. C. (1967). Terminal repetition in permuted T2 bacteriophage DNA molecules. *J. molec. Biol.* **23**, 355.

MACKAL, R. P., WERNINGHAUS, B. & EVANS, E. A., JUN. (1964). The formation of λ bacteriophage by λ DNA in disrupted cell preparations. *Proc. natn. Acad. Sci. U.S.A.* **51**, 1172.

MEYNELL, E. & DATTA, N. (1966). The nature and incidence of conjugation factors in *Escherichia coli. Genet. Res. Camb.* **7**, 141.

MOSIG, G. (1963). Genetic recombination in bacteriophage T4 during replication of DNA fragments. *Cold Spring Harb. Symp. quant. Biol.* **28**, 35.

NOMURA, M. & BENZER, S. (1961). The nature of the 'deletion' mutants in the rII region of phage T4. *J. molec. Biol.* **3**, 684.

RIS, H. & CHANDLER, B. L. (1963). The ultrastructure of genetic systems in prokaryotes and eukaryotes. *Cold Spring Harb. Symp. quant. Biol.* **28**, 1.

RITCHIE, D. A., THOMAS, C. A., JUN., MACHATTIE, L. A. & WENSINK, P. C. (1967). Terminal repetition in non-permuted T3 and T7 bacteriophage DNA molecules. *J. molec. Biol.* **23**, 365.

SKALKA, A. (1966). Regional and temporal control of genetic transcription in phage lambda. *Proc. natn. Acad. Sci. U.S.A.* **55**, 1190.

STANLEY, W. M. (1935). Isolation of crystalline protein possessing the properties of tobacco mosaic virus. *Science*, **81**, 644.

STRACK, H. B. & KAISER, A. D. (1965). On the structure of the ends of lambda DNA. *J. molec. Biol.* **12**, 36.

STREISINGER, G., EDGAR, R. S. & DENHARDT, G. H. (1964). Chromosome structure in phage T4. I. Circularity of the linkage map. *Proc. natn. Acad. Sci. U.S.A.* **51**, 775.

STREISINGER, G., EMRICH, J. & STAHL, M. M. (1967). Chromosome structure in phage T4. III. Terminal redundancy and length determination. *Proc. natn. Acad. Sci. U.S.A.* **57**, 292.

SYVERTON, J. T. & SCHERER, W. F. (1953). Applications of strains of mammalian cells to the study of animal viruses. *Cold Spring Harb. Symp. quant. Biol.* **18**, 285.

THOMAS, C. A., JUN. & MacHATTIE, L. A. (1964). Circular T2 DNA molecules. *Proc. natn. Acad. Sci., U.S.A.* **52**, 1297.

DE WAARD, A., PAUL, A. V. & LEHMAN, I. R. (1965). The structural gene for deoxyribonucleic acid polymerase in bacteriophages T4 and T5. *Proc. natn. Acad. Sci. U.S.A.* **54**, 1241.

WEIGLE, J. (1966). Assembly of phage lambda *in vitro*. *Proc. natn. Acad. Sci. U.S.A.* **55**, 1462.

WIBERG, J. S., DIRKSEN, M., EPSTEIN, R. H., LURIA, S. E. & BUCHANAN, J. M. (1962). Early enzyme synthesis and its control in *E. coli* infected with some amber mutants of bacteriophage T4. *Proc. natn. Acad. Sci. U.S.A.* **48**, 293.

WOOD, W. B. & EDGAR, R. S. (1967). Building a bacterial virus. *Scient. Am.* **217**, 60.

ZGAGA, V. (1967). Formation of bacteriophage lambda infective particles from lambda DNA in the presence of the crude extract of *Escherichia coli* K12S. *Virology*, **31**, 559.

VIRUS-INDUCED CHANGES IN HOST-CELL MACROMOLECULAR SYNTHESIS

E. M. MARTIN AND I. M. KERR

National Institute for Medical Research, Mill Hill, London, N.W.7

INTRODUCTION

Viruses are obligate parasites and in nature many of them give rise to latent infections causing little inconvenience to the host cell; only occasionally do they break out to produce symptoms of disease. In the laboratory, however, we have concentrated on virus-host systems in which infections cause profound changes in the host cell, frequently resulting in its death, as this is the only means by which we are able conveniently to detect and measure the infective process. Investigations on the nature and cause of these changes were begun as soon as systems amenable to biochemical studies became available. One of the first of these was the poliovirus-infected HeLa cell, and this was the subject of the earliest report by Ackermann and his colleagues of changes in patterns of macromolecular synthesis in infected cells; they found that infection resulted in an increase in the rates of synthesis of RNA and protein (Ackermann, 1958; Ackermann, Loh & Payne, 1959). Shortly thereafter, however, Salzman, Lockart & Sebring (1959) showed that poliovirus infection of HeLa cells could cause a marked inhibition of cellular RNA, DNA and protein synthesis, and suggested that this resulted from an inhibition of cell growth. Later studies on cells infected with encephalomyocarditis (EMC) (Martin, Malec, Sved & Work, 1961), mengo- (Baltimore & Franklin, 1962) and polio- (Holland, 1962) viruses confirmed the observations of Salzman *et al.*, in that infection with these small RNA-containing picornaviruses resulted in a rapid and profound inhibition of cellular RNA and protein synthesis. This occurred before the appearance of mature virus and was frequently accompanied by a somewhat later inhibition of DNA synthesis.

Subsequent studies with other members of the picornavirus group, and with other RNA- and DNA-containing animal viruses, have shown that a virus-induced inhibition of host-cell macromolecular synthesis is a relatively common feature of virus infection, and a number of examples have been listed in Tables 1–3.

Despite the widespread occurrence of these inhibitory phenomena, and some intensive investigations of their mechanism, we are still in

Table 1. *Early inhibition of host-cell RNA synthesis during infection with RNA-containing viruses*

Virus	Host cell	Timing of cut-off (hr. p.i.)			Factors modifying RNA cut-off	References
		Start	50% inhibition	Relation to protein cut-off		
A. Picornaviruses						
Polio 1	HeLa	2	3·5	After	—	Zimmerman et al. (1963)
Polio 1	ERK	—	1·5	—	Cut-off unaffected by incubation at 40°*	Fenwick (1963)
Polio 1	HeLa	—	—	—	Guanidine* blocked cut-off when added at 0 hr. p.i., but not at 2 hr. p.i. Blocking of cut-off by guanidine reversed by very high multiplicities	Holland (1963, 1964)
Polio 2	HeLa	Before 1	3·5	After	Cut-off delayed in HEL cells by guanidine* (50% inhibition at 7·5 hr. p.i.)	Bablanian et al. (1965a)
	Human embryo lung	Before 1	3·0	Before		
	ERK	Before 1	2·7	Concurrent		
Mengo	L	0	0·5	Concurrent	Cut-off abolished by puromycin* and p-FPA*	Baltimore & Franklin (1962); Baltimore, Franklin & Callender (1963)
Mengo	L	0	1-1·5	Before	Cut-off delayed 1 hr. by crude interferon*	Levy (1964)
Mengo	Hamster ovary	0	1·25	Before	—	Tobey & Campbell (1965)
Mengo	Novikoff-63	0	1	Before	—	Plagemann & Swim (1966)
	Novikoff-67	No inhibition of RNA synthesis				
Mengo	L	0	1	Before	—	McCormick & Penman (1967)
	HeLa	2·5	3·5	After		

Virus	Cell				Conditions	Reference
ME	L	0·5	1·5	Before	Cut-off abolished by p-FPA*	Verwoerd & Hausen (1963)
ME	Ehrlich ascites	Before 1·0	2	Before	—	Holoubek & Rueckert (1964)
Encephalomyo-carditis	Krebs II	0	1·5–2·5	Before	—	Martin et al. (1961); Martin (1967)
Foot-and-mouth disease	BHK 21	Before 0·5	1·5	After	—	Brown et al. (1966)
B. Other RNA-containing viruses						
Sindbis	Chick fibroblasts (in suspension)	—	1·5	—	Cut-off absent in monolayer cultures. Cut-off abolished by crude chick interferon*	Levy et al. (1966)
Venezuelan equine encephalitis	HeLa	Before 1·0	2 (66 % of control)			Lust (1966)
Vesicular stomatitis	Krebs II ascites	0	1·5		Cut-off present in cells infected with u.v.-irradiated virus*	Wagner & Huang (1966)
Semliki forest	Chick fibroblasts	No cut-off of RNA synthesis				Taylor (1965)
Newcastle disease	HeLa	No cut-off of RNA synthesis				Wheelock & Tamm (1961)
Reovirus	L	No cut-off of RNA synthesis				Gomatos & Tamm (1963); Kudo & Graham (1965)

* Conditions under which viral RNA synthesis is inhibited.

Table 2. *Early inhibition of host-cell protein synthesis during infection with RNA-containing viruses*

Virus	Host cell	Time of 50% inhibition (hr. p.i.)	Factors modifying protein cut-off	References
A. Picornaviruses				
Polio 1	HeLa	1·5	—	Levy (1961); Zimmerman et al. (1963)
Polio 1	HeLa	90% by 2·75 hr.	Breakdown of host-cell polysomes stimulated by actinomycin	Penman et al. (1963)
Polio 1	HeLa	2·5	Cut-off stimulated by actinomycin. Ribosomes in cell-free system still active. Cut-off abolished by guanidine at low multiplicity (10)* but not at high multiplicities (10^4/cell)	Holland & Peterson (1964); Holland (1964)
Polio 1	HeLa	90% by 1·5 hr.	Synchronous infection. Cut-off abolished by infection with u.v.-inactivated virus.* Cut-off only slightly affected by guanidine*	Penman & Summers (1965)
Polio 2	HeLa Human embryo lung ERK	1·5 3 2·7	Cut-off delayed (50% at 7 hr.) in HEL cells by guanidine*	Bablanian et al. (1965a)
Mengo	L	0·5	Cut-off unaffected by actinomycin. Cut-off abolished by p-FPA*	Baltimore, Franklin & Callender (1963)
Mengo	L	3	Cut-off unaffected by crude interferon	Levy (1964)
Mengo	Hamster ovary	3	—	Tobey & Campbell (1965)
Mengo	(a) Novikoff-63 (b) Novikoff-67	(a) Minimum (63% of control) at 3 hr. (b) No cut-off	Cut-off mimicked by actinomycin	Plagemann & Swim (1966)

Virus	Cell	Rapid polysome breakdown		Reference
Mengo	L		Unaffected by treatment with purified interferon*	Joklik & Merigan (1966)
Mengo	L HeLa	1·5-2·0 3·5-4·0	Cut-off unaffected by actinomycin Cut-off more rapid in actinomycin-treated cells	McCormick & Penman (1967)
ME	L	2·5	Cut-off stimulated by actinomycin (50 % at 1·75 hr.); abolished by p-FPA* and starvation for phenylalanine	Hausen & Verwoerd (1963); Verwoerd & Hausen (1963)
Encephalomyocarditis	Krebs II ascites	4·5-5·0	Stimulated by actinomycin	Martin et al. (1961); Martin (1967)
Foot-and-mouth disease	BHK 21	0·5	Cut-off unaffected by guanidine*	Brown et al. (1966)
B. Other RNA-containing viruses				
Sindbis	Chick fibroblasts	No cut-off	—	Pfefferkorn & Clifford (1964)
Venezuelan equine encephalitis	L	1·0	Unaffected by actinomycin	Lust (1966)
Newcastle disease	HeLa	4·0	50 % inhibition of protein synthesis at time of virus maturation	Wheelock & Tamm (1961)
Newcastle disease	Chick fibroblasts	6·5	Cut-off unaffected by actinomycin; abolished by azauridine* and puromycin*	Bolognesi & Wilson (1966)
Reovirus	L	No cut-off	—	Gomatos & Tamm (1963); Kudo & Graham (1965)

* Conditions under which viral RNA synthesis is inhibited.

Table 3. *Early inhibition of host-cell RNA and protein*
synthesis during infection with DNA-containing viruses

Virus	Host cell	Description of effect	References
A. Inhibition of RNA synthesis			
Adeno 2	Monkey kidney	60 % inhibition of RNA synthesis by 20 hr. Rapid inhibition of RNA polymerase activity	Ledinko (1966)
Adeno 5	KB	Host messenger RNA synthesis inhibited some time after cut-off of protein synthesis	Bello & Ginsberg (1966)
Herpes simplex	HEp2	50 % inhibition at 3 hr. p.i. ($m = 10$ p.f.u./cell)	Roizman *et al.* (1965)
	Dog kidney	No inhibition ($m = 10$ p.f.u./cell); 50 % inhibition at high multiplicity ($m = 400$ p.f.u./cell)	Aurelian & Roizman (1965)
Herpes simplex (mutant)	Dog kidney	40 % inhibition at 5 hr. p.i. ($m = 10$–20 p.f.u./cell)	Aurelian & Roizman (1965)
Vaccinia	L	Host messenger RNA synthesis inhibited, starting at 3–4 hr. p.i.; 50 % inhibition at 5 hr. p.i.	Becker & Joklik (1964)
Vaccinia	HeLa	Inhibition of nuclear RNA synthesis starts at 3·5 hr. p.i.; 50 % inhibition at 5·5 hr. p.i.	Salzman *et al.* (1964)
B. Inhibition of protein synthesis			
Adeno 2	Monkey kidney	68 % inhibition at 20 hr. p.i.	Ledinko (1966)
Adeno 2	KB	No cut-off seen up to 24 hr. p.i.	Polasa & Green (1965)
Adeno 5	KB	Cut-off coincident with onset of viral antigen synthesis	Bello & Ginsberg (1966)
Herpes simplex	HEp2	30 % inhibition in first 3 hr. p.i., then stimulation. Host polysome breakdown seen	Sydiskis & Roizman (1966)
Vaccinia	L	Slight inhibition, but masked by viral protein synthesis. 50 % inhibition seen at 1–2 hr. p.i. in interferon-treated cells	Joklik & Merigan (1966)
Vaccinia	HeLa	Complete inhibition at 4 hr. p.i. Cut-off not affected by actinomycin	Shatkin (1963)

some doubt as to whether they represent a host-cell function triggered by the virus or a purely viral function, for which all of the necessary information is carried by the virus genome. In this review we will outline the evidence pertaining to this question and to the related problems of mechanism and function, and will attempt to provide a logical explanation for the phenomena. At the outset we should point out that much of the evidence is equivocal and some of it is frankly contradictory. In some instances the apparently conflicting nature of the results may merely reflect differences in experimental approach. The degree to which the process of infection has been synchronized can accentuate or lessen the extent or rate of development of the response. In cells in which the normal rates of macromolecular synthesis have been reduced to a low level, for example by incubation in suboptimal media, the inhibition may be obscured by the synthesis of viral constituents. A change in the proportion of cells actually infected or capable of the inhibitory response would have obvious consequences on measurements of synthetic rates. Impurities which are of themselves inhibitory, or which could cause changes in the intracellular precursor pools, may be present in virus stocks and so cause fallacious results. Ideally, in order to compare responses under different circumstances, it would be desirable to use purified virus synchronously infecting all cells, preferably from a cloned population, under carefully controlled nutritional and growth conditions. Where possible these criteria have been used in assessing the relative importance of conflicting results. In many instances, however, in deciding between alternatives we have had no option but to be guided by probabilities and intuition, but in doing so we have endeavoured to indicate what we believe to be fact and what is opinion.

It is not possible in the space available to discuss in detail all of the many instances of virus-induced changes in macromolecular metabolism. We have, therefore, largely restricted the scope of this review to mammalian cells infected with members of the picornavirus group—in particular, polio-, mengo-, mouse encephalitis (ME) and EMC viruses—as more data is available for this group than any other. Where appropriate, however, evidence from other virus-cell systems has been included. To what extent it is valid to generalize on this limited basis will be considered in the last section of this article.

DESCRIPTION OF THE INHIBITORY PHENOMENA

On infection of mammalian cells with a member of the picornavirus group a progressively severe inhibition in the synthesis of all cellular and viral constituents is seen late in infection, just before death, when titres of intracellular virus have already reached a maximum. Presumably this inhibition is a consequence of the processes which lead to the death of the cell. However, an additional profound inhibition of cellular RNA and protein synthesis occurs almost *immediately after infection* and it is with this that we will be primarily concerned. Examples of this phenomenon, which we will refer to as a 'cut-off', are shown in Figure 1. The extent of the inhibition observed is somewhat variable. This may in part reflect the fact that it is frequently measured in circumstances where synthesis of cellular constituents is obscured by synthesis of viral components. However, in most cases the rate of host-cell RNA and protein synthesis

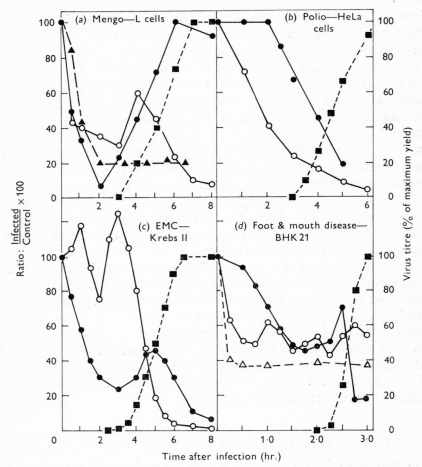

Fig. 1. RNA and protein synthesis in picornavirus-infected cells. (*a*) Mengovirus-infected L cells (Franklin & Baltimore, 1962). (*b*) Poliovirus-infected HeLa cells (Zimmerman, Heeter & Darnell, 1963). (*c*) EMC-infected Krebs II ascites tumour cells (E. M. Martin, unpublished results). (*d*) Foot-and-mouth disease virus-infected BHK 21 cells (Brown, Martin & Underwood, 1966). In each case virus growth is expressed as a percentage of the maximum yield (■ – – ■) and the rate of incorporation of radioactive precursor into RNA (●—●) into protein (○—○) and (*d*) into protein in the presence of guanidine (△ – – △) is expressed as a percentage of the control value. In (*a*) the DNA-dependent RNA polymerase activity of preparations of host-cell nuclei at various times post infection is expressed as a percentage of the initial value (▲ – – ▲). These diagrams are reproduced by permission of the authors and Cold Spring Harbor Laboratory of Quantitative Biology (Fig. 1*a*), Academic Press (Fig. 1*b*), and Elsevier Publishing Company (Fig. 1*d*).

falls below 50 % of the uninfected control well before the appearance of 50 % of the final yield of mature virus (see Fig. 1). In fact this forms a convenient definition of the cut-off phenomenon and, for the purposes of this review, it will be assumed that it has been abolished whenever the 50 % cut-off point is delayed beyond the mid-point of virus maturation. A cut-off in DNA synthesis has also been reported.

MECHANISMS OF INHIBITION

Inhibition of RNA synthesis

An immediate, rapid and extensive cut-off of RNA synthesis is seen in many types of cell infected with picornaviruses. Early work showed no selective inhibition of any one species of RNA—messenger, ribosomal and transfer RNA all appeared to be inhibited to approximately the same extent (Fenwick, 1963; Homma & Graham, 1963; Tobey, 1964). However, more recently, Darnell et al. (1966) have reported that, on infection of HeLa cells with poliovirus, ribosomal precursor RNA synthesis is substantially depressed by 1–1·5 hr. after infection, whereas at this time messenger RNA synthesis is as yet unaffected. Both the synthesis of 45S ribosomal precursor RNA and its conversion in the nucleus to 32S precursor and 16S ribosomal RNA appear to be depressed. Moreover, while formation (from pre-existing 16S RNA) and release from the nucleus to the cytoplasm of the smaller ribosomal subunit appear to be normal, there is a block in the maturation of the larger subunit (containing 28S RNA). This results in an accumulation in the nucleus of the precursor to the larger ribosomal subunit and a decrease in the ratio of newly-formed large (28S) to small (16S) ribosomal RNA in the cytoplasm. The basis for these selective effects is not known, although they do not appear to be secondary to the virus-induced cut-off of protein synthesis.

Early studies showed that the virus-induced cut-off of RNA synthesis is unlikely to be the result of an increased rate of breakdown of RNA and therefore does represent a true inhibition of synthesis. No net changes in the amount of RNA per cell were observed in EMC-infected ascites cells (Martin et al. 1961) or L cells infected with mengovirus (Franklin & Baltimore, 1962). When cells were pre-labelled with RNA precursors, then infected with mengo- or poliovirus in the presence of actinomycin (to prevent further cellular RNA synthesis), there was no abnormal loss of label (Franklin & Baltimore, 1962; Fenwick, 1963, 1964). The cut-off cannot be explained by changes in the rate of entry of labelled precursors into the cellular pools (Martin et al. 1961; Martin

& Work, 1962); nor does it seem likely that an inhibition of the synthesis of the DNA-dependent RNA polymerase of the host cell is the explanation, as this is a relatively stable enzyme (Holland, 1963). However, infection does result in a rapid inhibition of the activity of the host RNA polymerase, at least in L cells infected with mengovirus (Baltimore & Franklin, 1962), and in polio-infected HeLa cells (Holland, 1962; Holland & Peterson, 1964). The decrease in the host-cell polymerase activity parallels the inhibition of incorporation of precursors into cellular RNA (Fig. 1a) and seems the most probable cause of this inhibition. The DNA-dependent polymerase can only be isolated from the nucleus as an impure 'aggregate' containing both the enzyme and its DNA template, either of which may be affected. Franklin & Baltimore (1962) found no evidence for any breakdown of DNA to soluble products during mengovirus infection, while Holland (1962) found that DNA extracted from control and polio-infected HeLa cells was equally able to prime the synthesis of RNA by a bacterial polymerase. However, the priming ability of DNA after deproteinization probably bears little relation to its activity *in situ*, where much of the potential priming capacity of the DNA is masked by chromosomal proteins. In this connection, Holland (1962) could find no difference in the priming ability of a DNA-protein complex prepared from the nuclei of control and polio-infected HeLa cells. In contrast, studies in our own laboratory by R. J. Naftalin (personal communication) showed that chromatin isolated from Krebs II ascites cells 2 hr. after infection with EMC virus had a greatly reduced capacity to prime the synthesis of RNA by a soluble RNA polymerase prepared from *Micrococcus lysodeikticus*. One must have reservations about the validity of measurements of the template activity of isolated chromatin, particularly in the light of the technical difficulties in preparing material of reproducible activity (e.g. see Hotta & Stern, 1966). None the less, a decrease in the proportion of DNA accessible to RNA polymerase remains perhaps the most likely mechanism for the virus-induced cut-off of RNA synthesis. Moreover, the observation by Holland & Peterson (1964) that the small proportion of nuclear RNA polymerase which could be solubilized was unaffected by poliovirus infection would tend to support such a mechanism not involving a direct effect on the enzyme itself.

It has been suggested that a newly synthesized protein is the active agent in bringing about the inhibition of cellular RNA synthesis (Franklin & Baltimore, 1962). Evidence for this hypothesis was provided by Balandin & Franklin (1964), who found that cytoplasm of L cells 3 hr. after infection with mengovirus was able to inhibit the incorporation of ATP into RNA catalysed by the 'aggregate' RNA polymerase from uninfected cell nuclei. The inhibitory factor was destroyed by prior treatment with trypsin, but not with ribonuclease. In similar experiments on polio-infected HeLa cells, Holland (1962) could find no evidence for such a cytoplasmic inhibitor of nuclear DNA-dependent RNA polymerase in infected cells; nor has there been any further report of such an inhibitor. A possible alternative explanation for the results of Balandin & Franklin has been provided by the studies of R. J. Naftalin (personal communication). He found that the addition of cytoplasm from EMC virus-infected BHK 21 cells did not inhibit the incorporation of ATP into RNA during the first 15 min. of incubation with an 'aggregate' polymerase from uninfected cells. However, when the incubation was continued for 1 hr. (the period used by Balandin & Franklin), the rate of ATP incorporation in the presence of infected cytoplasm declined markedly, while in the control it continued at a constant rate. It is possible that ribonuclease could have caused this apparent inhibition of polymerase activity. When the control and infected cytoplasmic extracts were assayed for ribonuclease, the infected extract did in fact show a markedly higher activity. A similar increase in the ribonuclease activity of EMC-infected ascites cells has been reported (Martin, 1961), while Hotham-Iglewski & Ludwig (1966) have noted that infection of L cells with mengovirus caused a release of hydrolytic enzymes from the lysosomes. These results, together with the observation by Balandin & Franklin that the inhibitory activity was located in the lysosome-rich, large-particle fraction, would suggest that the trypsin-sensitive inhibitory factor observed by these authors may merely have reflected an increased level of ribonuclease. This does not necessarily mean that the inhibition of RNA synthesis *in vivo* is caused by ribonuclease. On the contrary, all the evidence would suggest that there is no accelerated rate of RNA breakdown in infected cells (see above).

In summary, we can probably exclude changes in precursor pool sizes, DNA, the polymerase itself or rates of RNA breakdown as being responsible for the cut-off. There is some evidence, however, that the cut-off results from the action of an inhibitor of the host's RNA polymerase, possibly by affecting the capacity of the DNA to serve as a template. As yet the nature of this inhibitor is not known.

Inhibition of protein synthesis

Protein synthesis in uninfected mammalian cells is carried out by aggregates of ribosomal particles attached to strands of messenger RNA to form polysomes (Warner, Knopf & Rich, 1963; Wettstein, Staehelin

& Noll, 1963). After infection of HeLa cells with poliovirus, there is a rapid and complete disaggregation of these host-cell polysomes (Penman, Scherrer, Becker & Darnell, 1963; Penman, Becker & Darnell, 1964; Summers, Maizel & Darnell, 1965). When viral protein synthesis begins, a new population of larger polysomes appears with an increased number of ribosomes per messenger RNA, in this case viral RNA (Penman *et al.* 1963; Scharff, Shatkin & Levintow, 1963). A similar breakdown of host-cell polysomes occurs in mengovirus-infected L cells (Joklik & Merigan, 1966) and in ascites cells infected with EMC virus (Dalgarno, Cox & Martin, 1967). The breakdown in polysomes appears to be coincident with the diminished rate of amino acid incorporation into cellular proteins, and is presumably responsible for it.

The inhibition of host-cell messenger RNA synthesis which accompanies infection (see above) cannot account for the disaggregation of polysomes, since mammalian messenger RNA is relatively stable with a half-life greater than 3–5 hr. Moreover, the rate of polysome breakdown in picornavirus-infected cells is much more rapid than that observed when RNA synthesis is inhibited with actinomycin (Penman *et al.* 1963; Dalgarno *et al.* 1967). In any case, inhibition of protein synthesis occasionally precedes the RNA cut-off (Table 1; Fig. 1*b*). The early observations of Kerr, Martin, Hamilton & Work (1962) and Kerr (1963), on the activity *in vitro* of ribosomes from EMC-infected and control Krebs cells in the presence and absence of added messenger RNA (EMC RNA in this case), indicated no gross change on infection in the basic ability of ribosomes to support protein synthesis. Similar observations on the amino-acid-incorporating activity of ribosomes from polio-infected HeLa cells have been reported by Baltimore, Eggers & Tamm (1963), and Summers, McElvain, Thorén & Levintow (1964). Summers & Maizel (1967) measured the ratio of nascent to completed protein chains in the poliovirus-infected HeLa cell, and found that this was unaltered, thus demonstrating that, throughout infection, those ribosomes which are active function at a normal rate. The recent studies of Willems & Penman (1966) attempt to carry the analysis even further. They used cycloheximide to 'freeze' the ribosomes on messenger RNA during incubation under various conditions. After removal of the inhibitor and further incubation, the polysome pattern of control cells remained unchanged; in infected cells there was a rapid disappearance of polysomes at a rate proportional to the length of pre-incubation with cycloheximide and to the time after infection. They concluded that the inhibition was caused by the inability of preformed messenger RNA in infected cells to re-associate with ribosomes. From this it would appear that it is the host-cell messenger, rather than the ribosomes, which is inactivated on infection. As yet there is no indication of how this inactivation is brought about, although it does not seem to be caused by an endonuclease (Willems & Penman, 1966). On the basis of Willems & Penman's results, perhaps the most likely possibility is that the factor(s) governing the binding of messenger RNA to ribosomes and/or initiation of protein synthesis is blocked by an inhibitory protein. Alternatively, it is possible that a particular species of transfer RNA (not required for viral protein synthesis) may be altered in such a way as to disrupt translation (Summers & Maizel, 1967).

Inhibition of DNA synthesis

Monolayers of HeLa cells infected with type 1 poliovirus show a decreased ability to incorporate labelled precursors into their DNA; inhibition begins at about 3 hr. post-infection (p.i.) and DNA synthesis ceases altogether by 6 hr. (Salzmann et al. 1959; Holland & Peterson, 1964). In mengovirus-infected L cells, inhibition of DNA synthesis begins quite early (1·5 hr.) and is complete by 5 hr. p.i. (Franklin & Baltimore, 1962). A cut-off of DNA synthesis is also seen in ascites cells infected with EMC virus, beginning at about 2·5 hr. p.i. (Martin & Work, 1961), and in cells infected with Newcastle disease virus (Wheelock & Tamm, 1961), reovirus (Gomatos & Tamm, 1963) and some DNA-containing viruses (e.g. Kaplan & Ben-Porat, 1963).

The mechanism of this inhibition is not yet clearly understood. It appears that DNA is not degraded: for example, there was no loss of label from DNA in cells pre-labelled with thymidine, then infected with mengovirus (Franklin & Baltimore, 1962). Infection with polio- or mengoviruses did not affect the ability of DNA to act as a template for a bacterial RNA polymerase (Franklin & Baltimore, 1962; Holland & Peterson, 1964). It is possible that the inhibition reflects a disturbance in the physiology of cell division. DNA synthesis during the logarithmic growth of randomly or synchronously dividing mammalian cells is discontinuous, and occupies only about 45 % of the division cycle (see, for example, Terasima & Tolmach, 1963). Inhibition of RNA or protein synthesis with actinomycin or puromycin prevents the cell from entering the phase of DNA synthesis (Tobey, Petersen, Anderson & Puck, 1966). It seems probable, therefore, that the DNA cut-off results from the virus-induced inhibition of RNA or protein synthesis. The protein cut-off seems the more likely immediate cause, since the DNA cut-off parallels the inhibition of protein synthesis, at least in EMC-infected ascites cells (Martin & Work, 1961) and Newcastle disease virus-infected

HeLa cells (Wheelock & Tamm, 1961), while in the latter instance there is relatively little inhibition of RNA synthesis. Another possibility has been suggested by Ackermann & Wahl (1966), who found that the omission of arginine from the medium, although reducing the rate of DNA synthesis in HeLa cells, prevented the polio-induced DNA cut-off without affecting virus growth. They suggested that the DNA cut-off may be mediated by a histone-like protein, the synthesis of which could be accounted for by a mechanism similar to that suggested below for the RNA cut-off. It would be of interest to know if RNA and protein synthesis was also examined in these arginine-depleted infected cells.

ROLE OF THE VIRUS IN INHIBITING MACROMOLECULAR SYNTHESIS

Virus RNA synthesis and maturation

While the studies mentioned in the previous section suggest explanations of how the inhibition of RNA, protein and DNA synthesis might operate, they give no information about the role of the invading virus in these processes. The first question which arises is whether the replication of the virus is necessary. The kinetics of the appearance of the cut-off would suggest that this is not so; for example, in cells infected with mengo-, ME or EMC viruses (though not with poliovirus) the cut-off of RNA synthesis occurs prior to any detectable synthesis of viral RNA (Fig. 1; Table 1). This lack of requirement has been tested directly in a number of ways. Guanidine blocks the replication of some strains of poliovirus without affecting cellular RNA or protein synthesis. It is thought to do this by specifically inhibiting the production of a functional virus-specific RNA polymerase (Baltimore, Eggers, Franklin & Tamm, 1963; Lwoff, 1965). The initial reports by Holland (1963, 1964) indicated that guanidine interfered with the cut-off of RNA and protein synthesis in polio-infected HeLa cells. However, in further studies with this system, Bablanian, Eggers & Tamm (1965a), Summers et al. (1965) and Penman & Summers (1965) found that the RNA cut-off was only delayed by this drug and that the protein cut-off was unaffected. Guanidine had no effect on the development of the protein cut-off in BHK 21 cells infected with foot-and-mouth disease virus (Brown, Martin & Underwood, 1966; Fig. 1d). Using a different approach, Fenwick (1963) showed that the normal cut-off of RNA synthesis occurred when ERK cells infected with poliovirus were incubated at 40°, conditions under which no detectable viral RNA synthesis took place. In addition, studies on the effects of moderate levels of protein synthetic inhibitors and interferon (see below) give ample evidence that replication of viral RNA and virus maturation can be blocked without affecting either cut-off. Hence, it can be concluded that neither maturation nor viral RNA replication is necessary for the cut-off phenomena to occur.

Initiation of the cut-off—theoretical possibilities

There are a number of possible roles the virus could play in initiating the inhibitory response of the host cell and the studies we are about to discuss are mainly concerned with attempts to select between them. In order to clarify the discussion some definition of the alternatives seems desirable. It will be tacitly assumed throughout that, although it is conceivable that RNA other than the parental viral genome could function as an inhibitor in the cut-off phenomena, this is not in fact the case. If synthesis of an inhibitor is required, it will therefore be assumed that either it is a protein or that protein is required for its production.

(a) Possible roles for the virus not involving protein synthesis

Either the capsid proteins or the RNA of the invading virus particle could inhibit RNA or protein synthesis, either directly or by functioning, for example, as an enzyme in the production of an inhibitor. Alternatively, the mere physical presence of the virus particle might be sufficient to trigger the activation or release of pre-existing inhibitors of cellular RNA and protein synthesis. These mechanisms can be excluded if protein synthesis is required for the development of the cut-off.

(b) Possible role of the virus involving virus protein synthesis

The parental viral genome codes for a number of proteins some of which might directly or indirectly inhibit host RNA and protein synthesis.

(c) Possible role of the virus involving cellular protein synthesis

The cell might respond to infection by synthesizing a protein or proteins causing the inhibition of RNA and protein synthesis. This hypothesis must necessarily provide a feasible explanation as to how infection results in the synthesis of these inhibitory proteins.

Assuming that protein synthesis is required, a distinction between these latter two hypotheses should be possible by examining the effects of specific inhibitors of viral protein synthesis (such as interferon—see below) or the effects of damaging or altering the viral genome. For convenience in considering the data relevant to these different hypotheses, that relating to the RNA cut-off is discussed first.

Factors affecting the RNA cut-off

Effects of non-specific inhibitors of protein synthesis

Both puromycin and cycloheximide inhibit protein synthesis, the former by causing the premature release of the growing peptide chain from the ribosomes, the latter by blocking translation of the messenger. p-Fluorophenylalanine (p-FPA) does not prevent protein synthesis, but is incorporated into protein in place of phenylalanine; its presence thus results in the synthesis of proteins which may or may not be functional. Baltimore, Franklin & Callender (1963) reported that the RNA cut-off in mengo-infected L cells was completely abolished by 200 μg./ml. of puromycin, while 500 μg./ml. of p-FPA caused a partial abolition of the cut-off. In ME-infected L cells as little as 20 μg./ml. of p-FPA was sufficient to block the RNA cut-off. The effect of the drug could be overcome by the addition of excess phenylalanine (Verwoerd & Hausen, 1963). Puromycin (100 μg./ml.) blocked the poliovirus-induced inhibition of HeLa cell RNA polymerase (Holland & Peterson, 1964). Finally, cycloheximide (10 μg./ml.) prevented the development of the cut-off in EMC-infected ascites cells (R. J. Naftalin, personal communication), and also of the depression in ribosomal precursor RNA synthesis in the nucleus of HeLa cells infected with poliovirus (Darnell *et al.* 1966).

It therefore appears that protein synthesis is essential for the development of the cut-off. Even here, however, the evidence is not entirely satisfactory as the RNA cut-off does not appear to be affected by lower levels of puromycin, which none the less cause an appreciable inhibition of total protein synthesis (see below). Although the evidence on this point badly needs strengthening, for the purpose of further discussion we will accept that protein synthesis is essential. It follows that any hypothesis involving the direct inhibitory action of the viral components must be eliminated. This does not mean that such mechanisms are impossible under different circumstances: for instance, Holland (1964) found that very high multiplicities of poliovirus (up to 10^4 p.f.u./cell) caused an early inhibition of RNA synthesis in HeLa cells, and that this inhibition was not affected by puromycin, p-FPA or guanidine. In this case it seems likely that the viral RNA was directly inhibiting the host RNA polymerase, as photo-inactivation of the viral RNA blocked the inhibitory response. However, the response of cells to these very high multiplicities is probably unrelated to the inhibitory phenomena under discussion.

Accepting that protein synthesis is required, the above results obtained with high levels of protein synthetic inhibitors give no indication as to whether it is viral or host-cell protein synthesis which is involved. On the other hand, studies with much lower levels of protein synthetic inhibitors do provide us with a clue to the answer to this problem. 5 μg./ml. of p-FPA inhibits the maturation of mengovirus and delays the synthesis of viral RNA and proteins, while 25 μg./ml. of puromycin, which results in a 95% inhibition of host protein synthesis, causes complete inhibition of virus RNA synthesis and virtually complete inhibition of virus-specific protein synthesis (including that of the viral RNA polymerase) (Baltimore & Franklin, 1963). Despite this, neither of these inhibitors at these lower concentrations affects the cut-off of host RNA synthesis. Furthermore, Farnham (1965) reported that 22 μg./ml. of puromycin had no effect on the RNA cut-off in L cells infected with EMC virus. It would seem, therefore, that the protein synthesis which is required for the RNA cut-off is particularly resistant to the inhibitory effects of puromycin. As this is in marked contrast to the sensitivity of *viral* protein synthesis to this drug (Baltimore & Franklin, 1963; Eggers, Baltimore & Tamm, 1963), it would suggest that it is the synthesis of some cellular protein, rather than a virus-specific one, which is essential.

Effect of interferon

Purified interferon is without effect on the synthesis of cell RNA or proteins (Baron, Merigan & McKerlie, 1966; Levy & Merigan, 1966), yet it inhibits virus replication by a mechanism which probably involves blocking of the translation of viral messenger RNA into virus proteins (Marcus & Salb, 1966; Joklik & Merigan, 1966; Sonnabend, Martin, Mécs & Fantes, 1967; Carter & Levy, 1967). If synthesis of a viral protein is essential for the cut-off, interferon should prevent its formation. When L cells were pre-treated with crude mouse interferon before infection with mengovirus, there was a 1 hr. delay in the appearance of the RNA cut-off; nevertheless, the inhibition of RNA synthesis was as severe as that occurring in untreated infected cells and reached a maximum before the time of synthesis of viral RNA (Levy, 1964). Levy and co-workers also examined the effect of interferon on the inhibition of chick-cell RNA synthesis by an arbovirus (Sindbis) and found that, in contrast to the situation with mengovirus, the cut-off was abolished (Levy, Snellbaker & Baron, 1966). However, it is not clear whether the cut-off in this latter instance is the same phenomenon which we have been discussing, since it was not observed under the more usual mono-layer conditions of cell culture. It is a pity that studies on the effects of interferon on other picornavirus-cell systems have not yet been reported, as interferon is potentially capable of showing whether viral protein synthesis is essential for the cut-off. None the less, Levy's result with mengo-infected L cells does suggest that, for picornaviruses at least, viral protein synthesis is not required for the RNA cut-off.

Effect of agents which damage the viral genome

If viral protein synthesis is not essential, then damage to the viral genome by extensive ultraviolet irradiation should not affect the cut-off. For picornaviruses only one study of this type has been reported: Franklin & Baltimore (1962) found that, when L cells were infected with mengovirus which had been 99·9 % inactivated by ultraviolet light, the usual inhibition of the activity of the polymerase responsible for host RNA synthesis (Fig. 1 a) did not occur. Thus, in contrast to the interferon results, it would appear that an intact viral genome is essential for the cut-off.

Effect of alterations in the host cell

There is ample evidence that the pattern of RNA cut-off caused by a particular virus is dependent on the host cell it infects. Bablanian, Eggers & Tamm (1965 a) found that the RNA cut-off initiated by type 2 poliovirus in HeLa cells differed in the progress of its development from that seen in human embryo lung or ERK cells (although the latter may, in fact, be a strain of human cells; Coombs, Daniel, Gurner & Kelus, 1961). Mengovirus caused a rapid, profound inhibition of RNA synthesis in L cells, but a delayed and less severe inhibition in HeLa cells, where the cut-off pattern was similar to that seen with poliovirus (McCormick & Penman, 1967). The clearest example of cellular modification of the cut-off response is provided by the work of Plagemann & Swim (1966), who studied the growth of mengovirus in two strains of Novikoff rat hepatoma cells. The RNA cut-off in strain 63 was typical of that seen in mengo-infected L cells, while in a second strain (67) the cut-off was entirely absent, despite similar rates of virus growth in both strains. Not only do various cell strains respond differently to the same virus, but even the same host cells can show a different response under altered growth conditions. Levy (1964) found that the RNA cut-off in mengo-infected L cells was delayed by 3 hr., if the cells were fully grown out in monolayers and cell growth had ceased. In EMC-infected L cells the cut-off, which occurs at 1·5 hr. p.i. in spinner cultures, is completely absent in fully-grown-out monolayers, while cortisone, when added to L cells infected in suspension, both delayed the RNA cut-off and considerably reduced its severity (R. J. Naftalin, personal communication).

Virus growth showed a similar pattern in all of these studies. The fact that conditions of growth can modify the inhibitory response may explain the early conflicting results of Ackermann and his colleagues, who observed a stimulation of cellular RNA synthesis on poliovirus infection of HeLa cells (Ackermann, 1958).

Be this as it may, the metabolic state of the host cell clearly plays some role in determining whether or not the cut-off occurs. Although this does not exclude a requirement for viral protein synthesis, it is perhaps conceptionally more in accord with the cut-off being a process initiated by the virus but mediated through the regulatory mechanism of the cell, than by the synthesis of a viral protein directly inhibitory to host RNA synthesis.

Factors affecting the protein cut-off

Effects of inhibitors

In many studies on the response of the RNA cut-off to metabolic inhibitors the effects of these drugs on the protein cut-off were also examined. In general the results were similar. In L cells infected with mengo- or ME-viruses the cut-off was blocked by p-FPA (Baltimore, Franklin & Callender 1963; Verwoerd & Hausen, 1963). These are the only two studies which directly demonstrate a requirement for protein synthesis. However, observations which can best be explained in terms of such a requirement were reported by Penman & Summers (1965), who found that treatment of HeLa cells with puromycin or cycloheximide, followed by infection and removal of the drugs, resulted in a 30 min. delay in the commencement of the polio-induced protein cut-off. However, pretreatment of L cells with crude or purified interferon sufficient to inhibit the synthesis of at least one virus-specific protein—the RNA polymerase (Miner, Ray & Simon, 1966)—had no effect on the development of the protein cut-off caused by mengovirus (Levy, 1964; Joklik & Merigan, 1966). As in the case of the RNA cut-off, these results would imply firstly that protein synthesis is necessary for the protein cut-off, and secondly, that the protein whose synthesis is required is not coded by the viral genome.

With a variety of cells infected with picornaviruses, actinomycin either did not affect or slightly stimulated the development of the protein cut-off (see Tables 2 and 3). Synthesis of host messenger RNA is therefore unnecessary for the cut-off. This means that if synthesis of a host protein is essential, its messenger must be preformed.

Effects of damage to the viral genome

Penman & Summers (1965) reported that inactivation of poliovirus by ultraviolet light before infection completely abolished the cut-off of protein synthesis in HeLa cells. This parallels the observations of Franklin & Baltimore (1962) on the RNA cut-off in suggesting that an intact viral genome is required for the development of the protein cut-off.

Host-controlled factors

Again, the host cell can modify the inhibitory response. Differences in the timing and extent of the cut-off were seen on infection of a variety of cells with type 2 poliovirus (Bablanian *et al.* 1965*a*). Mengovirus produced a sharp, early cut-off in L cells, but it was much delayed in HeLa cells (McCormick & Penman, 1967), and was totally absent in Novikoff-67 cells (Plagemann & Swim, 1966). These limited observations would therefore tend to favour a mechanism which involves the active participation of the host cell.

Discussion

There is a remarkable similarity in the effects of picornavirus infection on host RNA and protein synthesis. Where the two responses have been measured simultaneously, the effects of various factors (inhibitors, cell type, etc.) on them are qualitatively similar. The two responses seem unlikely to be basically identical, however, as for example the RNA cut-off may precede the protein cut-off in one cell but follow it in another (cf. Bablanian *et al.* 1965*a*; Table 1). Nor does the one phenomenon seem to be responsible for the other. We have already pointed out that it is highly improbable that the protein cut-off is caused by the inhibition of RNA synthesis—mammalian messenger RNA is too stable for this. However, since cellular RNA synthesis is known to be affected by protein synthetic inhibitors (Latham & Darnell, 1965; Tamaoki & Mueller, 1965), it is theoretically possible that the RNA cut-off results from a prior inhibition of protein synthesis; but this seems unlikely. The RNA cut-off can occur well before the onset of the protein cut-off (Fig. 1*c*; Levy, 1964; Tobey & Campbell, 1965; Plagemann & Swim, 1966; McCormick & Penman, 1967); also McCormick & Penman (1967) have observed that the pattern of inhibition of RNA synthesis in L and HeLa cells caused by cycloheximide is quite different from that induced by infection of these cells with mengovirus. None the less no instance of one cut-off occurring in the absence of the other has yet been reported for picornaviruses. In view of this and the fact that both responses have many features in common, the mechanisms governing them seem likely to be similar and will be discussed in parallel.

Both processes appear to require the synthesis of protein. It is, of course, not necessary that the protein whose synthesis is essential for the cut-off is the actual inhibitory agent. For example, it could act by modifying a nucleic acid (say a transfer RNA), which in the case of the protein cut-off could then block translation by the ribosomes; or it could be required for the active transport of the inhibitory agent through membranes. However, in the absence of any evidence to the contrary, we will assume that the inhibitors are proteins, and that their synthesis is necessary for the cut-off of both RNA and protein synthesis to occur. This leaves us with the problem of whether the inhibitor is viral or cellular in origin.

The viral-origin theory

The main evidence in favour of a viral origin is the fact that both inhibitory responses are abolished if the virus is first inactivated with ultraviolet light, which would seem to indicate that both require an intact viral genome. Furthermore, other theories face difficulties in accounting for the role of the virus in initiating the cut-off and for the lack of effect of actinomycin on the protein cut-off. Unfortunately, there are no equivalent data for the RNA cut-off, although the use of analogues such as 5-fluorouracil or 2,6-diaminopurine, which result in the synthesis of non-functional RNAs, might make it possible to distinguish proteins coded by pre-existing messenger RNAs, whether viral or cellular.

On the other hand, if the inhibitory proteins are products of the viral genome, why is it that interferon does not affect either cut-off? We may be incorrect in our assumptions about the mode of action of interferon: it may block the translation of some viral cistrons, such as those for the RNA polymerase, without affecting the read-out of others—but this seems improbable, since the evidence favours a total inhibition of messenger function (Marcus & Salb, 1966; Joklik & Merigan, 1966). If the viral RNA were the messenger coding for the synthesis of an inhibitory protein, it is not unreasonable to suppose that the amount of inhibitor, and hence the rate of development and extent of inhibition, should bear some relation to the number of messenger molecules, i.e. to the multiplicity of infection. However, the results of Kaverin & Kukhanova (1967) with EMC-infected L cells suggest that high multiplicities of infection have no such influence on the rate of development or extent of the cut-off, although it is not known how many virus particles actually entered the cell.

The cellular-origin theory

The evidence from studies with interferon and low levels of puromycin, together with the different response of host cells to the same virus, strongly favours the idea that the inhibitory protein is a product of the cell's own genome. Direct proof of this may perhaps come from the isolation and characterization of the inhibitory proteins. It is possible, at least in the case of the inhibitor of RNA synthesis, that it may be a histone, as these are known to be able to inhibit cellular RNA polymerase, at least *in vitro* (Huang & Bonner, 1962; Allfrey, Littau & Mirsky, 1963). Whether they do so *in vivo* is less certain. Holoubek & Rueckert (1964) observed a stimulation in the synthesis of nuclear histones in Ehrlich ascites cells infected with ME virus, and suggested that this might reflect the formation of the RNA synthetic inhibitor. A similar stimulation in the synthesis of histones by HeLa cells infected with poliovirus has also been reported (Sokol, Cox, Dinka & Ackermann, 1965). A careful examination of the changes in patterns of proteins synthesized in infected cells, such as that begun by Summers *et al.* (1965), might lead to the ultimate solution of this problem.

If we postulate that the inhibitory protein is coded by the cell, then we must further explain the failure of actinomycin to inhibit its synthesis, the role of the virus in initiating its synthesis and finally the failure of ultraviolet-inactivated virus to carry out this initiating role. The lack of effect of actinomycin indicates that the messenger must be preformed, at least for the synthesis of the protein synthetic inhibitor. It is feasible that the requirement for protein synthesis may be non-specific, i.e. it may not involve the synthesis of a protein or proteins qualitatively different from those already being formed. Such a protein could be involved in the uncoating of the virus or in the hypothetical triggering mechanism, in which case its messenger may already exist in the cell perhaps in a masked or inactive form (cf. Spirin & Nemer, 1965). It is even possible that the inhibitor is already present and that it is its transport to its target which requires protein synthesis. As to the role of the virus in initiating these processes, it is at least possible that the mere entry of the virus or its physical presence within the cell might trigger, for example, the release of lysosomal enzymes, which in turn could initiate the processes leading to cut-off. This mechanism might seem especially attractive as lysosomes are known to be involved in the penetration and uncoating of at least some viruses (Dales, 1962; Epstein, Hummeler & Berkaloff, 1964). Moreover, infection with mouse hepatitis, vaccinia and certain myxoviruses (Allison & Sandelin, 1963; Mallucci & Allison, 1965) and of

L cells with mengovirus (Hotham-Iglewski & Ludwig, 1966) can cause release of lysosomal hydrolytic enzymes. However, this 'activation' of lysosomal enzymes (which does not seem to require protein synthesis) occurs late in infection (Mallucci & Allison, 1965) and seems more likely to be involved in cell lysis and the release of mature virus than in the early RNA or protein cut-off. Nevertheless, in connection with the requirement for protein synthesis, it is still possible that the enzymes released from lysosomes either during the penetration and uncoating process or during subsequent activation could stimulate the formation of an inhibitory protein by, for example, exposing a pre-existing but inactive messenger. Such a mechanism would explain why cortisone, which is known to stabilize lysosomes (De Duve, 1963) and which prevents the 'activation' of lysosomes by mengovirus (Hotham-Iglewski & Ludwig, 1966), diminishes and delays the RNA cut-off in EMC-infected L cells (R. J. Naftalin, personal communication).

It is not so easy to explain the results with inactivated virus. It is reasonable to suppose that the inhibitory response is not triggered until the viral RNA has become uncoated. Non-infectious ME virus particles do not enter the cell or become uncoated. Irradiation with ultraviolet light may prevent this process, possibly by inducing the formation of covalent bonds between the RNA and protein. Alternatively, the cell may not recognize the RNA damaged by inactivation as a 'threat'. These points could be examined experimentally, particularly the uncoating of irradiated virus. In this connection it would be of great interest to examine the effects of virus neutralized with antibody, as this treatment has been shown to permit penetration of the virus while preventing the release of its complement of viral RNA (Mandel, 1967). Useful information might also be provided if the virus were inactivated by other means: for example, by infecting in the dark with virus grown in the presence of acridines, then exposing the cells to light after penetration of the virus (e.g. Holland, 1964).

It is clearly difficult to decide between the various hypotheses. In order to reach at least one clear-cut conclusion, it is very tempting to exclude at least those hypotheses not involving a requirement for protein synthesis. But even here the results obtained in the studies using low amounts of puromycin must sound a note of caution. Ignoring this, however, the problem reduces to that of choosing between the apparently conflicting results with interferon and irradiated virus—once this is done other contradictory observations can be reconciled with the appropriate hypothesis. It is, of course, possible to take all the data at face value and conclude that the viral RNA specifically stimulates the synthesis of an inhibitory protein without either acting as a messenger itself or stimulating messenger production. This is unlikely, but not impossible. More realistically, we must conclude that we have been misled in interpreting the data from the interferon or inactivated-virus experiments. The arguments are evenly balanced and, with the information at present available, a decision must be based on intuition rather than logic. Possibly because of our interest in interferon and its mode of action, if forced to choose between these alternatives, we would favour a cellular origin for the inhibitory protein.

THE SIGNIFICANCE OF THE INHIBITORY PHENOMENA

The conclusions we have reached have been based almost entirely on work with picornaviruses. Do these have any relevance to similar inhibitory responses seen in infections with other viruses? The RNA cut-off produced by vesicular stomatitis virus bears a strong similarity to that seen in picornavirus-infected cells. It would be of great interest if this resemblance reflected a basically similar mechanism. However, Wagner & Huang (1966) found that the RNA cut-off observed on vesicular stomatitis virus-infection of Krebs ascites cells was unaffected by prior extensive ultraviolet irradiation of the virus. Furthermore, some preparations of this virus contain a smaller non-infectious particle, and the usual cut-off of RNA synthesis was observed on 'infection' of cells with this defective particle irrespective of whether or not it had been subjected to extensive ultraviolet irradiation (Huang, Greenawalt & Wagner, 1966). It would appear, therefore, that for vesicular stomatitis virus an intact viral genome is not essential for the development of the cut-off. This is in contrast to the conclusions reached by Franklin & Baltimore (1962) in their studies with inactivated mengovirus. Apparently, despite their striking similarity, the cut-off phenomena observed with the two virus systems must have different mechanisms. However, it would certainly be intriguing to have the experiments repeated in parallel with the Krebs ascites cell as the host for both viruses. The arboviruses also probably cause a cut-off of RNA and protein synthesis, although this may be masked by the synthesis of viral components (compare fig. 6 with table 1 in Taylor, 1965); however, too little is known about the inhibitory effects initiated by these viruses to warrant comment. Newcastle disease virus causes a marked inhibition of host protein synthesis (Wheelock & Tamm, 1961), but this occurs rather later in the infection cycle than is usually seen with picornaviruses and may be unrelated to the phenomena we have been discussing.

Infection with this virus does not affect host RNA synthesis. In the presence of interferon, which inhibits the viral protein synthesis which would otherwise mask the effect, a well-defined cut-off of host-cell protein synthesis is observed in vaccinia-infected L cells (Joklik & Merigan, 1966). This cut-off could be explained quite adequately, however, by assuming that the virus itself contains an inhibitory protein which is released on uncoating (W. K. Joklik, personal communication). In the absence of knowledge as to whether or not protein synthesis is required for the development of the cut-off in this system, its relation to that observed with picornaviruses remains purely speculative. It would seem, therefore, that until further data are available for other virus cell systems it would be unwise to generalize too much on the basis of results obtained only with picornaviruses.

The factors involved in cell death are complex and it is not intended to discuss them in any detail here. However, as it has been suggested that the cut-off of RNA and/or protein synthesis may be responsible for this phenomenon (e.g. Martin & Work, 1961), some mention of it should perhaps be made. Certainly inhibition of either RNA or protein synthesis alone cannot account for the rapid killing of cells infected with picornaviruses; for example, neither puromycin nor actinomycin will cause the death of Krebs ascites or HeLa cells in less than 24 hr. Possibly the combination of these effects, together with the inhibition of DNA synthesis, may be toxic to the cell. If so, then the mere lack of synthesis of these constituents is probably not responsible *per se* for the death of the cell, as macromolecular synthesis can be inhibited by low temperatures without affecting the viability of cells. The resulting disturbance in metabolism may be the immediate cause of death, but this is begging the question. Bablanian, Eggers & Tamm (1965 b) have suggested that it is the synthesis of viral capsid proteins which is responsible for the cytopathic effects of poliovirus; however, this fails to explain how interferon-treated cells are killed by virus infection even though replication of the virus is inhibited (Joklik & Merigan, 1966; E. M. Martin, unpublished observations). It still remains a possibility that a factor specifically concerned with the cell-killing process is elaborated after infection and that this is independent of the inhibitory factors involved in the cut-off of RNA, DNA or protein synthesis.

Finally, one can ask of what advantage these inhibitory effects might be to the virus or host cell especially as, at least in the case of mengovirus (Plagemann & Swim, 1966), replication can occur perfectly well in their absence. By the very nature of the problem, it is impossible to compare the efficiency of virus infection in the presence or absence of cut-off in a given virus-cell system without imposing some alteration on it, and the general data from different systems in which the cut-off is or is not displayed are too meagre as yet to allow any significant correlation to be made between these phenomena and virus growth. However, it is relatively easy to conceive of reasons why the protein cut-off might be of advantage to the virus in natural infections. It could prevent an initially infected cell from synthesizing interferon and so block a defence mechanism which would otherwise curtail the spread of infection. Furthermore, since the cut-off results in the production of large numbers of messenger-free ribosomes, it could give the viral message a maximum chance of expression. On this basis it would seem likely that the virus plays an active (specific) rather than a passive (non-specific) role in the production of the agent responsible for the protein cut-off, since this would offer a selective evolutionary advantage. There remains the problem of explaining how the translation of host-cell messenger RNA is prevented without affecting that of the viral RNA. Unless one invokes compartmentalization, it seems the protein synthetic mechanisms of the cell must be able to distinguish between host and viral messenger RNA. There is in fact some direct experimental evidence in favour of this. The results of Marcus & Salb (1966) would indicate that ribosomes from interferon pretreated cells are capable *in vitro* of translating host cell messenger RNA, but not viral RNA. Although the basis for this distinction is not yet known, once it is accepted that such a distinction is possible, it is easy to postulate that it could, in response to suitable changes in the host cell, operate equally well in favour of the virus, resulting in the synthesis of viral protein to the exclusion of host protein.

It is difficult to see any advantage, either to the host or the virus, in the RNA cut-off. It is too slow to affect materially the concentration or variety of host messenger or transfer RNAs, since these have a relatively long half-life and there is no evidence for increased breakdown. It is possible that it enhances the protein cut-off; for example, in polio-infected cells the addition of actinomycin accelerates the disaggregation of polysomes (Willems & Penman, 1966) (but this may not be the case in mengovirus infections). The RNA cut-off could be an initial response of the cell to any invasion by a foreign body and the failure to reverse it a result of the protein cut-off. The fact that the response is only manifest under certain growth conditions of the cell might favour this point of view. Certainly it is intriguing to postulate some correlation between the factors involved in the cut-off phenomena and those regulating cell growth and division.

Clearly the interpretation of the present situation is heavily dependent upon the data obtained with protein synthetic inhibitors (puromycin, *p*-FPA, guanidine and interferon), much of which is far from conclusive or even contradictory. However, our object has been to stimulate further work rather than to state dogmatically a point of view. After reviewing the extensive information already available on the causes of the inhibitory responses, it is apparent that further work still needs to be done, particularly on viruses and cells with altered genomes. From such studies it may well transpire that the cell itself is basically responsible for the cut-off phenomena. If so, the final interpretation may have to await our understanding of the factors controlling RNA and protein synthesis in the normal cell. However, it is possible that the further investigation of the virus-induced changes in macromolecular synthesis may yet, in turn, contribute to the achievement of such an understanding.

SUMMARY

Infection of mammalian cells with picornaviruses, or certain other RNA- or DNA-containing viruses, leads to a rapid inhibition of host-cell RNA and protein synthesis, and, somewhat later, to an inhibition of DNA synthesis. Evidence relating to the mechanism by which picornaviruses cause this inhibition of macromolecular synthesis, and the factors controlling its development, are reviewed. The inhibition of host RNA synthesis seems most likely to be by the production of a protein which, probably by combining with the template, interferes with the transcription of DNA by the nuclear RNA polymerase. The inhibition of protein synthesis results from an accelerated breakdown of polysomes, possibly the consequence of an inactivation of host-cell messenger RNA, while the inhibition of DNA synthesis may arise indirectly from either of these other inhibitory responses. It is not possible to come to any definite conclusion about the genetic origin of the information governing these inhibitory responses. The results of the potentially most definitive experiments with u.v.-inactivated virus on the one hand, and with interferon on the other, give conflicting answers to this question. It is finally concluded that the proteins responsible for inhibition are a product of the cell genome, but the arguments in favour of a viral origin are also persuasive. The relevance of the phenomena seen in picornavirus-infected cells to the inhibitory responses observed during infection with other viruses, and the significance of these phenomena, are also discussed.

ACKNOWLEDGEMENTS

We would like to thank Drs J. E. Darnell, R. M. Franklin and S. J. Martin for permission to use their data in the preparation of Figure 1. We are especially grateful to Mr R. J. Naftalin and Dr J. E. Darnell for providing us with data prior to publication and to Mr R. J. Naftalin, Dr J. A. Sonnabend and Dr T. S. Work for much useful discussion in the preparation of this manuscript.

REFERENCES

ACKERMANN, W. W. (1958). Cellular aspects of the cell-virus relationship. *Bact. Rev.* **22**, 223.

ACKERMANN, W. W., LOH, P. C. & PAYNE, F. E. (1959). Studies of the biosynthesis of protein and ribonucleic acid in HeLa cells infected with poliovirus. *Virology*, **7**, 170.

ACKERMANN, W. W. & WAHL, D. (1966). Programming of poliovirus inhibition of deoxyribonucleic acid synthesis in HeLa cells. *J. Bact.* **92**, 1051.

ALLFREY, V. G., LITTAU, V. C. & MIRSKY, A. E. (1963). On the role of histones in regulating ribonucleic acid synthesis in the cell nucleus. *Proc. natn. Acad. Sci. U.S.A.* **49**, 414.

ALLISON, A. C. & SANDELIN, K. (1963). Activation of lysosomal enzymes in virus-infected cells and its possible relationship to cytopathic effects. *J. exp. Med.* **117**, 879.

AURELIAN, L. & ROIZMAN, B. (1965). Abortive infection of canine cells by Herpes simplex virus. II. Alternative suppression of synthesis of interferon and viral constituents. *J. molec. Biol.* **11**, 539.

BABLANIAN, R., EGGERS, H. J. & TAMM, I. (1965a). Studies on the mechanism of poliovirus-induced cell damage. I. The relation between poliovirus-induced metabolic and morphological alterations in cultured cells. *Virology*, **26**, 100.

BABLANIAN, R., EGGERS, H. J. & TAMM, I. (1965b). Studies on the mechanism of poliovirus-induced cell damage. II. The relation between poliovirus growth and virus-induced morphological changes in cells. *Virology*, **26**, 114.

BALANDIN, I. G. & FRANKLIN, R. M. (1964). The effect of mengovirus infection on the activity of the DNA-dependent RNA polymerase of L-cells. II. Preliminary data on the inhibitory factor. *Biochem. biophys. Res. Commun.* **15**, 27.

BALTIMORE, D., EGGERS, H. J., FRANKLIN, R. M. & TAMM, I. (1963). Poliovirus-induced RNA polymerase and the effects of virus-specific inhibitors on its production. *Proc. natn. Acad. Sci. U.S.A.* **49**, 843.

BALTIMORE, D., EGGERS, H. J. & TAMM, I. (1963). Altered location of protein synthesis in the cell after poliovirus infection. *Biochim. biophys. Acta*, **76**, 644.

BALTIMORE, D. & FRANKLIN, R. M. (1962). The effect of mengovirus infection on the activity of the DNA-dependent RNA polymerase of L cells. *Proc. natn. Acad. Sci. U.S.A.* **48**, 1383.

BALTIMORE, D. & FRANKLIN, R. M. (1963). Effects of puromycin and p-fluoro-phenylalanine on mengovirus ribonucleic acid and protein synthesis. *Biochim. biophys. Acta*, **76**, 431.

BALTIMORE, D., FRANKLIN, R. M. & CALLENDER, J. (1963). Mengovirus-induced inhibition of host ribonucleic acid and protein synthesis. *Biochim. biophys. Acta*, **76**, 425.

BARON, S., MERIGAN, T. C. & MCKERLIE, M. L. (1966). Effect of crude and purified interferons on the growth of uninfected cells in culture. *Proc. exp. Biol. Med.* **121**, 50.

BECKER, Y. & JOKLIK, W. K. (1964). Messenger RNA in cells infected with vaccinia virus. *Proc. natn. Acad. Sci. U.S.A.* **51**, 577.

BELLO, L. J. & GINSBERG, H. S. (1966). Inhibition of host m-RNA and protein synthesis in type 5 adenovirus-infected cells. *Fedn. Proc.* **25**, 652.

BOLOGNESI, D. P. & WILSON, D. E. (1966). Inhibitory proteins in the Newcastle disease virus-induced suppression of cell protein synthesis. *J. Bact.* **91**, 1896.

BROWN, F., MARTIN, S. J. & UNDERWOOD, B. (1966). A study of the kinetics of protein and RNA synthesis induced by foot-and-mouth disease virus. *Biochim. biophys. Acta*, **129**, 166.

CARTER, W. A. & LEVY, H. B. (1967). Ribosomes: effect of interferon on their interaction with rapidly labelled cellular and viral RNAs. *Science*, **155**, 1254.

COOMBS, R. R. A., DANIEL, M. R., GURNER, B. W. & KELUS, A. (1961). Species-characterizing antigens of 'L' and 'ERK' cells. *Nature, Lond.* **189**, 503.

DALES, S. (1962). An electron microscope study of the early association between two mammalian viruses and their hosts. *J. cell Biol.* **13**, 303.

DALGARNO, L., COX, R. A. & MARTIN, E. M. (1967). Polyribosomes in normal Krebs 2 ascites tumour cells and in cells infected with encephalomyocarditis virus. *Biochim. biophys. Acta*, **138**, 316.

DARNELL, J. E., GIRARD, M., BALTIMORE, D., SUMMERS, D. F. & MAIZEL, J. V. (1966). The synthesis and translation of poliovirus RNA. *Proc. Symp. molec. Biol. Vir.*, *University of Edmonton, Alberta, Canada*, June 1966 (in the Press).

DE DUVE, C. (1963). The lysosome concept. *Ciba Foundation Symp. on 'Lysosomes'*, p. 1. Ed. A. V. S. de Reuck and M. P. Cameron. London: J. and A. Churchill.

EGGERS, H. J., BALTIMORE, D. & TAMM, I. (1963). The relation of protein synthesis to formation of poliovirus RNA polymerase. *Virology*, **21**, 281.

EPSTEIN, M. A., HUMMELER, K. & BERKALOFF, A. (1964). The entry and distribution of herpes virus and colloidal gold in HeLa cells after contact in suspension. *J. exp. Med.* **119**, 291.

FARNHAM, A. E. (1965). Effect of aminonucleoside (of puromycin) on normal and encephalomyocarditis virus-infected L cells. *Virology*, **27**, 73.

FENWICK, M. L. (1963). The influence of poliovirus infection on RNA synthesis in mammalian cells. *Virology*, **19**, 241.

FENWICK, M. L. (1964). The fate of rapidly labelled ribonucleic acid in the presence of actinomycin in normal and virus-infected animal cells. *Biochim. biophys. Acta*, **87**, 388.

FRANKLIN, R. M. & BALTIMORE, D. (1962). Patterns of macromolecular synthesis in normal and virus-infected mammalian cells. *Cold Spring Harb. Symp. quant. Biol.* **27**, 175.

GOMATOS, P. J. & TAMM, I. (1963). Macromolecular synthesis in reovirus-infected L cells. *Biochim. biophys. Acta*, **72**, 651.

HAUSEN, P. & VERWOERD, D. W. (1963). Studies on the multiplication of a member of the Columbia SK group (ME virus) in L cells. III. Alteration of RNA and protein synthetic patterns in virus-infected cells. *Virology*, **21**, 617.

HOLLAND, J. J. (1962). Inhibition of DNA-primed RNA synthesis during poliovirus infection of human cells. *Biochem. biophys. Res. Commun.* **9**, 556.

HOLLAND, J. J. (1963). Depression of host controlled RNA synthesis in human cells during poliovirus infection. *Proc. natn. Acad. Sci. U.S.A.* **49**, 23.

HOLLAND, J. J. (1964). Inhibition of host cell macromolecular synthesis by high multiplicities of poliovirus under conditions preventing virus synthesis. *J. molec. Biol.* **8**, 574.

HOLLAND, J. J. & PETERSON, J. A. (1964). Nucleic acid and protein synthesis during poliovirus infection of human cells. *J. molec. Biol.* **8**, 556.

HOLOUBEK, V. & RUECKERT, R. R. (1964). Studies on nuclear protein metabolism after infection of Ehrlich ascites cells with Maus-Elberfeld (ME) virus. *Biochem. biophys. Res. Commun.* **15**, 166.

HOMMA, M. & GRAHAM, A. F. (1963). Synthesis of RNA in L cells infected with Mengo Virus. *J. cell comp. Physiol.* **62**, 179.

HOTHAM-IGLEWSKI, B. & LUDWIG, E. H. (1966). Effect of cortisone on activation of lysosomal enzymes resulting from mengovirus infection of L-929 cells. *Biochem. biophys. Res. Commun.* **22**, 181.

HOTTA, Y. & STERN, H. (1966). Ribonucleic acid polymerase activity in extended and contracted chromosomes. *Nature, Lond.* **210**, 1043.

HUANG, R. C. & BONNER, J. (1962). Histone, a suppressor of chromosomal RNA synthesis. *Proc. natn. Acad. Sci. U.S.A.* **48**, 1216.

HUANG, A. S., GREENAWALT, J. W. & WAGNER, R. R. (1966). Defective T particles of vesicular stomatitis virus. I. Preparation, morphology and some biologic properties. *Virology*, **30**, 161.

JOKLIK, W. K. & MERIGAN, T. C. (1966). Concerning the mechanism of action of interferon. *Proc. natn. Acad. Sci. U.S.A.* **56**, 558.

KAPLAN, A. S. & BEN-PORAT, T. (1963). The pattern of viral and cellular DNA synthesis in pseudorabies virus-infected cells in the logarithmic phase of growth. *Virology*, **19**, 205.

KAVERIN, N. V. & KUKHANOVA, M. K. (1967). Effect of multiplicity of infection on the rate of protein synthesis in Encephalomyocarditis virus-infected L cells. *Acta virol.* **11**, 195.

KERR, I. M. (1963). The relation between RNA and protein metabolism in EMC virus-infected mouse ascites tumour cells. Ph.D. Thesis, University of London.

KERR, I. M., MARTIN, E. M., HAMILTON, M. G. & WORK, T. S. (1962). The initiation of virus protein synthesis in Krebs ascites tumor cells infected with EMC virus. *Cold Spring Harb. Symp. quant. Biol.* **27**, 259.

KUDO, H. & GRAHAM, A. F. (1965). Synthesis of reovirus ribonucleic acid in L cells. *J. Bact.* **90**, 936.

LATHAM, H. & DARNELL, J. E. (1965). Distribution of mRNA in the cytoplasmic polyribosomes of the HeLa cell. *J. molec. Biol.* **14**, 1.

LEDINKO, N. (1966). Changes in metabolic and enzymatic activities of monkey kidney cells after infection with adenovirus 2. *Virology*, **28**, 679.

LEVY, H. B. (1961). Intracellular sites of poliovirus reproduction. *Virology*, **15**, 173.

LEVY, H. B. (1964). Studies on the mechanism of interferon action. II. The effect of interferon on some early events in mengo virus infection in L cells. *Virology*, **22**, 575.

LEVY, H. B. & MERIGAN, T. C. (1966). Interferon and uninfected cells. *Proc. Soc. exp. Biol. Med.* **121**, 53.

LEVY, H. B., SNELLBAKER, L. F. & BARON, S. (1966). Effect of interferon on RNA synthesis in Sindbis virus-infected cells. *Proc. Soc. exp. Biol. Med.* **121**, 630.

LUST, G. (1966). Alterations of protein synthesis in arbovirus-infected L cells. *J. Bact.* **91**, 1612.

LWOFF, A. (1965). The specific effectors of viral development. *Biochem. J.* **96**, 289.

MALLUCCI, L. & ALLISON, A. C. (1965). Lysosomal enzymes in cells infected with cytopathic and non-cytopathic viruses. *J. exp. Med.* **121**, 477.

MANDEL, B. (1967). The interaction of neutralized poliovirus with HeLa cells. II. Elution, penetration and uncoating. *Virology*, **31**, 248.

MARCUS, P. I. & SALB, J. M. (1966). Molecular basis of interferon action: inhibition of viral RNA translation. *Virology*, **30**, 502.

MARTIN, E. M. (1961). Biochemical studies on Krebs II mouse ascites tumour cells infected with encephalomyocarditis virus Ph.D. Thesis, University of London.

MARTIN, E. M. (1967). Replication of small RNA viruses. *Br. med. Bull.* **23**, 192.

MARTIN, E. M., MALEC, J., SVED, S. & WORK, T. S. (1961). Studies on protein and nucleic acid metabolism in virus-infected mammalian cells. I. Encephalomyocarditis virus in Krebs II mouse-ascites-tumour cells. *Biochem. J.* **80**, 585.

MARTIN, E. M. & WORK, T. S. (1961). Studies on protein and nucleic acid metabolism in virus-infected mammalian cells. 4. The localization of metabolic changes within subcellular fractions of Krebs II mouse-ascites-tumour cells infected with encephalomyocarditis virus. *Biochem. J.* **81**, 514.

MARTIN, E. M. & WORK, T. S. (1962). Studies on protein and nucleic acid metabolism in virus-infected mammalian cells. 5. The kinetics of synthesis of virus protein and of virus ribonucleic acid in Krebs II mouse-ascites-tumour cells infected with encephalomyocarditis virus. *Biochem. J.* **83**, 574.

McCORMICK, W. & PENMAN, S. (1967). Inhibition of RNA synthesis in HeLa and L cells by mengovirus. *Virology*, **31**, 135.

MINER, N., RAY, W. J., JUN. & SIMON, E. H. (1966). Effect of interferon on the production and action of viral RNA polymerase. *Biochem. biophys. Res. Commun.* **24**, 264.

PENMAN, S., BECKER, Y. & DARNELL, J. E. (1964). A cytoplasmic structure involved in the synthesis and assembly of poliovirus components. *J. molec. Biol.* **8**, 541.

PENMAN, S., SCHERRER, K., BECKER, Y. & DARNELL, J. E. (1963). Polyribosomes in normal and poliovirus infected HeLa cells and their relationship to messenger RNA. *Proc. natn. Acad. Sci. U.S.A.* **49**, 654.

PENMAN, S. & SUMMERS, D. F. (1965). Effects on host cell metabolism following synchronous infection with poliovirus. *Virology*, **27**, 614.

PFEFFERKORN, E. R. & CLIFFORD, R. L. (1964). The origin of the protein of Sindbis virus. *Virology*, **23**, 217.

PLAGEMANN, P. G. W. & SWIM, H. E. (1966). Replication of mengovirus. I. Effect on synthesis of macromolecules by host cells. *J. Bact.* **91**, 2317.

POLASA, H. & GREEN, M. (1965). Biochemical studies on adenovirus multiplication. VIII. Analysis of protein synthesis. *Virology*, **25**, 68.

ROIZMAN, B., BORMAN, G. S. & ROUSTA, M. K. (1965). Macromolecular synthesis in cells infected with herpes simplex virus. *Nature, Lond.* **206**, 1374.

SALZMAN, N. P., LOCKART, R. Z., JUN. & SEBRING, E. D. (1959). Alterations in HeLa cell metabolism resulting from poliovirus infection. *Virology*, **9**, 244.

SALZMAN, N. P., SHATKIN, A. J. & SEBRING, E. D. (1964). The synthesis of DNA-like RNA in the cytoplasm of HeLa cells infected with vaccinia virus. *J. molec. Biol.* **8**, 405.

SCHARFF, M. D., SHATKIN, A. J. & LEVINTOW, L. (1963). Association of newly formed viral protein with specific polyribosomes. *Proc. natn. Acad. Sci. U.S.A.* **50**, 686.

SHATKIN, A. J. (1963). Actinomycin D and vaccinia virus infection of HeLa cells. *Nature, Lond.* **199**, 357.

SOKOL, F., COX, D. C., DINKA, S. & ACKERMANN, W. W. (1965). Effect of poliovirus infection on histone synthesis in the HeLa cell. *Proc. Soc. exp. Biol. Med.* **119**, 1015.

SONNABEND, J. A., MARTIN, E. M., MÉCS, E. & FANTES, K. H. (1967). The effect of interferon on the synthesis and activity of an RNA polymerase isolated from chick cells infected with Semliki forest virus. *J. gen. Virol.* **1**, 41.

SPIRIN, A. S. & NEMER, M. (1965). Messenger RNA in early sea urchin embryos: cytoplasmic particles. *Science*, **150**, 214.

SUMMERS, D. F. & MAIZEL, J. V., JUN. (1967). Disaggregation of HeLa cell polysomes after infection with poliovirus. *Virology*, **31**, 550.

SUMMERS, D. F., MAIZEL, J. V., JUN. & DARNELL, J. E., JUN. (1965). Evidence for virus-specific non-capsid proteins in poliovirus-infected HeLa cells. *Proc. natn. Acad. Sci. U.S.A.* **54**, 505.

SUMMERS, D. F., MCELVAIN, N. F., THORÉN, M. M. & LEVINTOW, L. (1964). Incorporation of amino acids into polyribosome-associated protein in cytoplasmic extracts of poliovirus-infected HeLa cells. *Biochem. biophys. Res. Commun.* **15**, 290.

SYDISKIS, R. J. & ROIZMAN, B. (1966). Polysomes and protein synthesis in cells infected with a DNA virus. *Science*, **153**, 76.

TAMAOKI, T. & MUELLER, G. C. (1965). The effects of actinomycin D and puromycin on the formation of ribosomes in HeLa cells. *Biochim. biophys. Acta*, **108**, 73.

TAYLOR, J. (1965). Studies on the mechanism of action of interferon. I. Interferon action and RNA synthesis in chick embryo fibroblasts infected with Semliki forest virus. *Virology*, **25**, 340.

TERASIMA, T. & TOLMACH, L. J. (1963). Growth and nucleic acid synthesis in synchronously dividing populations of HeLa cells. *Exp. Cell Res.* **30**, 344.

TOBEY, R. A. (1964). Mengovirus replication. I. Conservation of virus RNA. *Virology*, **23**, 10.

TOBEY, R. A. & CAMPBELL, E. W. (1965). Mengovirus replication. III. Virus reproduction in Chinese hamster ovary cells. *Virology*, **27**, 11.

TOBEY, R. A., PETERSEN, D. F., ANDERSON, E. C. & PUCK, T. T. (1966). Life cycle analysis of mammalian cells. III. The inhibition of division of Chinese hamster cells by puromycin and actinomycin. *Biophys. J.* **6**, 567.

VERWOERD, D. W. & HAUSEN, P. (1963). Studies on the multiplication of a member of the Columbia SK group (Me virus) in L cells. IV. Role of 'early proteins' in virus induced metabolic changes. *Virology*, **21**, 628.

WAGNER, R. R. & HUANG, A. S. (1966). Inhibition of RNA and interferon synthesis in Krebs-2 cells infected with vesicular stomatitis virus. *Virology*, **28**, 1.

WARNER, J. R., KNOPF, P. M. & RICH, A. (1963). A multiple ribosomal structure in protein synthesis. *Proc. natn. Acad. Sci. U.S.A.* **49**, 122.

WETTSTEIN, F. O., STAEHELIN, T. & NOLL, H. (1963). Ribosomal aggregate engaged in protein synthesis: characterization of the ergosome. *Nature, Lond.* **197**, 430.

WHEELOCK, E. F. & TAMM, I. (1961). Biochemical basis for alterations in structure and function of HeLa cells infected with Newcastle disease virus. *J. exp. Med.* **114**, 617.

WILLEMS, M. & PENMAN, S. (1966). The mechanism of host cell protein synthesis inhibition by poliovirus. *Virology*, **30**, 355.

ZIMMERMAN, E. F., HEETER, M. & DARNELL, J. E. (1963). RNA synthesis in poliovirus-infected cells. *Virology*, **19**, 400.

VIRUS-INDUCED CHANGES IN
TRANSLATION MECHANISMS

H. SUBAK-SHARPE

Medical Research Council Experimental Virus Research Unit,
Institute of Virology, Glasgow

It may facilitate the discussion of virus-induced changes in the translation mechanism of infected cells to describe first the essentials of the transcription and translation mechanisms of non-infected cells.

THE ESSENTIALS OF INFORMATION TRANSFER

Transcription

The genetic information for the synthesis of proteins is now known to be stored in the base sequence of nucleic acid molecules. Such nucleic acid may be double-stranded or single-stranded DNA or RNA, but only double-stranded DNA will be considered here. DNA molecules are not direct templates for protein synthesis, but the genetic information must first be *transcribed* to molecules of single-stranded RNA by an enzyme (DNA-dependent RNA polymerase). *In vivo* this enzyme copies only one of the two DNA strands of a given gene, producing messenger RNA, whenever the gene codes for a polypeptide. Not all the RNA produced is however messenger RNA. The primary precursors of ribosomal RNAs and of transfer RNAs are similarly transcribed from DNA, but subsequently several bases in these RNAs are modified by different specific enzymes. The resultant ribosomal RNAs and transfer RNAs are not templates for protein synthesis, but have different essential functions in protein synthesis.

Translation

Transfer RNA

Polypeptides are not formed simply from amino acids which directly attach to messenger RNA templates. Amino acids as such have no known affinity for the nucleotide sequences of these RNA templates. All cells contain specific adaptor molecules called transfer RNA (tRNA, sRNA) to which amino acids can be covalently attached. When an amino acid is so attached to a tRNA this is referred to as aminoacyl tRNA.

A given tRNA is characterized by extreme specificity, both for a particular amino acid and for particular groupings of three nucleotides (a codon) along a messenger RNA template. There are generally more than one molecular species of tRNA for each of the 20 different amino acids found in proteins. The total number of tRNA species in a cell is not yet established but is believed to be about 40. tRNA molecules contain several different unusual bases, which are formed through enzymic modification after the primary ($\sim$ 80 nucleotide long) polynucleotide chain of tRNA has been made. Of the different modifying enzymes in *Escherichia coli* six different methylases and a thiolase have been identified.

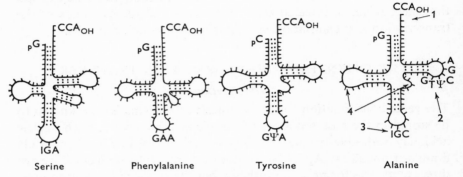

Fig. 1. Clover-leaf models of four yeast tRNAs whose complete structure is known. 1, –CCA terminal present in all tRNAs; 2, probable ribosomal surface orientating site containing -G-T-Ψ-C- sequence; 3, the anticodon; 4, probable aminoacyl-tRNA synthetase recognition site is composed of one or both of these loops.

In its native state tRNA seems to have a specially ordered three-dimensional structure which is currently thought to be of the clover-leaf type. Each tRNA has at least three, topographically separable, specific recognition sites: the codon recognition site (the anticodon); the aminoacyl-tRNA synthetase recognition site; and the ribosomal surface orientating site (Fig. 1). Furthermore all tRNAs terminate at the 3′-hydroxyl end with the trinucleotide –CCA. Aminoacyl-tRNA has the carboxylic acid group of the amino acid covalently attached via a high energy bond to the terminal A of the tRNA molecule, although the aminoacyl-tRNA synthetases specifically recognize the tRNA's characteristic recognition site which is many nucleotides distant from the –CCA terminal. The anticodon of aminoacyl-tRNA recognizes the corresponding codon in messenger RNA and thus controls the position where the carried amino acid is added to the growing polypeptide chain. If, say, a cysteinyl-tRNA for some reason is charged with the wrong amino acid, say alanine, this will cause alanine to be inserted in the growing polypeptide where cysteine should be. The maximum precision of polypeptide synthesis clearly cannot exceed the accuracy with which the specific aminoacyl-tRNA synthetases acylate their particular tRNA with the correct amino acid.

Protein synthesis

Polypeptide bonds are formed, that is protein synthesis takes place, on the ribosomes, which are poorly understood complex particles consisting of about one half protein and one half RNA.

Messenger RNA molecules act as the primary templates which order the amino acid sequences in polypeptides. Successive codons along the messenger RNA chain programme the succession of individual amino acids linked by peptide bonds in the specified polypeptide chain. The messenger RNA attaches at a specific location to the ribosomes and the message is read in the 5' → 3' direction one codon at a time, with the result that polypeptide chains grow stepwise, amino acid by amino acid, beginning at the amino terminal end. The growing polypeptide is, at any time, covalently attached at the carboxyl group of the last amino acid to the –CCA end of the aminoacyl-(in this case called polypeptidyl-) tRNA that brought this amino acid into proper alignment on the ribosome. Once the amino acid is detached from the –CCA end, tRNA is released from the ribosome and cannot participate in further polypeptide synthesis until it has again been aminoacylated by synthetase. Full details of peptide bond formation are at present still lacking. Figure 2 illustrates these operations in diagrammatic form.

These, in very brief outline, are the essentials of the translation mechanism of cells which concern us here.

THE TRANSLATION OF VIRAL GENETIC INFORMATION

Until very recently it was generally accepted that a virus which multiplies in a cell does so by utilizing the unaltered transcription and translation mechanism of the cell. In broad outline the genetic information of the virus was assumed to function through viral messenger RNAs becoming attached to the ribosomes present in the host cell, which then synthesized virus specified polypeptides by utilizing the pre-existing host-specified population of aminoacyl-tRNAs. The supply of aminoacyl-tRNAs was assumed to be constantly replenished through the continued normal functioning of the unaltered population of aminoacyl-tRNA synthetases typically found in the uninfected cell. Thus the genetic information of the virus was believed to succeed in reaching expression merely by using

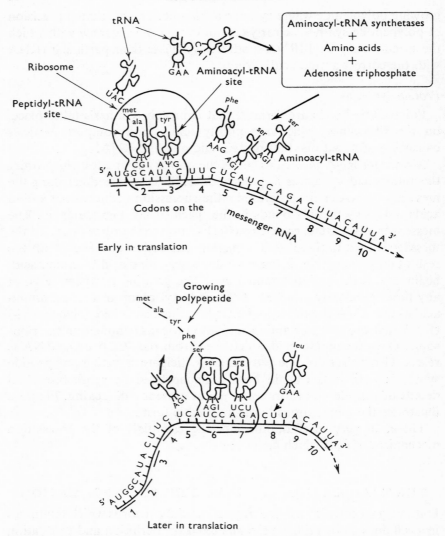

Fig. 2. Diagrammatic representation of the translation mechanism in operation.

the host's protein synthesizing apparatus, without altering its nature. The only virus-directed interference with the protein synthesizing machinery which was envisaged was that host messenger RNA became, in some unspecified way, displaced from the ribosomes where it had been directing protein synthesis and that, in the case of some viruses, host-specified RNA synthesis was also shut off.

This uncomplicated view is unfortunately no longer tenable. Recent experiments carried out in several laboratories show that some mammalian and bacterial viruses induce changes in the translation mechanism of their host cells. Moreover, they do so at several different points: at the level of the tRNAs themselves, at the level of aminoacyl-tRNA synthetases, at the level of the tRNA methylases, and conceivably also at the level of ribosomal '5S' RNA.

It is our purpose to present in outline and then to discuss collectively the recent evidence for changes in the translation mechanism of cells following virus infection.

Changes at the level of tRNA

Evidence that the population of tRNAs becomes modified in virus-infected cells comes from several different laboratories and involves both bacterial and mammalian viruses.

Sueoka & Kano-Sueoka (1964) first reported a specific modification of leucyl-tRNA of *E. coli* after phage T2 infection and have recently published further extensive studies of this system (Kano-Sueoka & Sueoka, 1966; Sueoka, Kano-Sueoka & Gartland, 1966). The elution profiles of 17 aminoacyl-tRNAs from T2 infected and non-infected *E. coli* B were investigated by chromatography of doubly-labelled preparations on columns of methylated albumin kieselguhr (MAK). Only in the case of leucyl-tRNA was a clear difference found between preparations before and after infection, but there remains some ambiguity in the cases of isoleucyl-tRNA and seryl-tRNA.

The leucyl-tRNA present in preparations made from non-infected cells elutes from MAK as two major overlapping peaks. The main alteration observed at 8 min. after infection with T2 was a decrease of peak I, with apparent concomitant increase in peak II. Furthermore, a transient new small peak eluting in front of peak I was observed only at 1–2 min. post-infection. This change in leucyl-tRNA is one of the earliest events which characteristically follow infection with T-even phage. It is not found after infection with T-odd phages or after induction of phage λ. Leucyl-tRNA modification following T-even infection was observed in various strains of *E. coli* and also in a strain of *Shigella*.

The coding properties of the altered elution peaks at 1 and 8 min. post-infection have been examined by binding of the different fractions to ribosomes (Nirenberg *et al.* 1966). Marked changes were observed, but they are rather complex and attempts at interpretation appear to be somewhat premature. Mainly on the strength of the finding that some of the fractions can be aminoacylated by both yeast and *E. coli* synthetase, while other fractions can be aminoacylated by *E. coli* synthetase only, it has been suggested that there may be two different leucyl-tRNA cistrons and that the different fractions observed result from virus induced modification of the 'precursor' products of the two cistrons. This is in line with the view originated by Sueoka that the tRNA is of host origin and modified as a consequence of T-even phage infection, but not coded by the T-even phage genome. Although this view may well prove to be correct no critical experiments appear to have been made to settle whether or not the leucyl-tRNA is virus coded. No results of sRNA-DNA hybridization studies or nucleotide sequence studies have as yet been published. The 'adaptor modification' interpretation, which has been advanced by Sueoka, appears to rest mainly on the following observations:

(1) Protein synthesis is required, as the changes do not occur if chloramphenicol is present at the time of phage infection.

(2) The amount of leucyl-tRNA relative to other amino acid acceptor RNAs does not change appreciably by 8 min. after phage infection. (Kano-Sueoka & Sueoka (1966) in their table 1 show slight differences which favour leucyl-tRNA increase in 12 cases out of 15.)

(3) The new front and rear peaks can be charged by yeast enzyme which normally only charges one of the two peaks of the non-infected *E. coli* leucyl-tRNA. This evidence seems insufficient to establish conclusively that the observed changes are due to modification of pre-existing molecules, rather than due to production of new virus-coded tRNA. Of course, there can remain no doubt that changes take place in the population of leucyl-tRNA of *E. coli* following infection by T-even phages.

Waters & Novelli (1967) have studied the same host-virus system, but with a chromatographic technique of greater resolving power than MAK column chromatography. Their 'reverse phase column' can actually resolve leucyl-tRNA into five peaks (Kelmers, Novelli & Stulberg, 1965), although at the actual pH chosen for their study it resolves leucyl-tRNA into only three peaks. In their experiments the period for observation of changes was also extended well beyond the 8 min. limit mainly employed by Sueoka *et al.* (1966) up to 2 hr. after infection. Quantitative changes between peaks I, II and III, were observed 6 min. after infection and these probably correspond to the changes observed on MAK by Sueoka & Kano-Sueoka at 8 min. after infection. At 32–120 min. after infection, Waters & Novelli found at least one, possibly two, additional peaks eluting after peak III which, by then, predominates. The changes were shown to be phage-specific and not due to undermethylation or changed growth conditions. The late-appearing leucyl-tRNAs in the new peaks IV and V may mean that they are involved with some late phage function. At present it is unknown whether the material in these peaks is synthesized *de novo* or due to modification of pre-existing leucyl-tRNA. This work demonstrates, first, quantitative changes in peaks I, II and III and, secondly, the late appearance of two leucyl-tRNA species which are undetectable in normal cells.

Smith *et al.* (1966) have conclusively demonstrated that bacterio-phages can carry genetic information for tRNA and introduce it success-fully into host cells, thus altering the population of tRNAs by *de novo* addition of virus-specified tRNA molecules. By ingenious experiments they obtained a defective transducing phage 80 which carried the Su-3 gene. The Su-3 gene is the structural gene for an 'amber' suppressor tRNA which permits the translation of the 'amber' UAG triplet (which normally results in chain termination) as the amino acid tyrosine. Cells were infected under conditions which allowed phage DNA and RNA synthesis to continue for 2 hr. This was done by blocking protein synthesis with chloramphenicol at 60 or 80 min. after induction. By this enrichment technique they showed that the amount of tyrosinyl-tRNA increased about sevenfold relative to the amounts of the other amino-acyl-tRNAs. The increased amount of tyrosinyl-tRNA was found when either the wild-type or the amber-suppressor allele of the Su-3 gene were used in this system. The only difference between the two alleles was the codon recognition of the phage-carried tyrosinyl-tRNA. The tyrosinyl-tRNA of the wild-type allele Su-3⁻ does not recognize UAG and binds only with UAU and UAC (the normal tyrosine codons) while the tyrosinyl-tRNA of the Su-3⁺ allele binds only with UAG and does not recognize UAU or UAC. Subsequently Smith *et al.* (1967) have shown that the only difference between the tRNA products of the Su-3⁺ and Su-3⁻ genes is a single nucleotide change in the anticodon. Normally the Su-3 gene specified tRNA appears to constitute a minor fraction of the tyrosinyl-tRNA of *E. coli*.

In these experiments there can be no doubt that the tRNA population of the cell is being altered in a qualitative way following the introduction of a virus genome with information for a new tRNA. Yet it must be stressed that no significant changes in the relative proportion of tyro-sinyl-tRNA were observed in samples taken right up to cell lysis (that is without artificially extended RNA synthesis) under their original conditions. In experiments where chloramphenicol was used to inhibit protein synthesis tyrosinyl-tRNA was transcribed from the transducing phage DNA only when a 40 min. period was allowed to elapse between induction and chloramphenicol administration. If cells were then further incubated for 2 hr. the significant relative increase in tyrosinyl-tRNA was found. These findings compel one to re-evaluate the argument, that qualitative changes in the population of leucyl-tRNA are unlikely because protein synthesis is necessary after infection with T-even phages for the change to occur. The same is true for arguments which depend on failure to observe relative changes in tRNAs following infection.

A different change in the population of tRNA of *E. coli* infected with bacteriophage T4 has been reported by Brenner, Kaplan & Stretton (1966). They observed that suppression of the chain-terminating codon UAA (ochre) by Su-4+ is modified after infection of the cells by T4. (Evidence for the presence of a suppressing tRNA in the case of Su-4 has been obtained by Wilhelm (1966).) Similar modification of suppressor tRNAs may also affect suppressors of the chain-terminating codon UGA (J. F. Sambrook, personal communication). Thus it seems that some minor tRNAs (with coding properties apparently similar to those of the major tRNAs) whose normal function in the cell is uncertain, may be modified or destroyed as a consequence of T-even phage infection.

Evidence that some viruses may code for new transfer-RNAs has come from studies on herpes virus infection of mammalian cells (Subak-Sharpe & Hay, 1965; Subak-Sharpe, Shepherd & Hay, 1966). Using chromatographically purified sRNA from herpes-infected and non-infected cells it was first shown that herpes DNA hybridized only with sRNA from infected and not from non-infected cells. Next it was shown by MAK column co-chromatography of arginyl-tRNA (differentially labelled with ^{3}H and ^{14}C in the amino acid) that some arginyl-tRNA from the infected preparations eluted at higher ionic strength than did that from the control. Thus the elution profile of arginyl-tRNA alters following infection. Finally it was demonstrated that preparations obtained from infected cells contained some arginyl-tRNA which differed in nucleotide sequence from any found in non-infected cells. This was done by co-chromatography on DEAE-cellulose columns of doubly labelled mixtures of arginyl-tRNA which had been digested with T1 RNase (this enzyme specifically cuts guanosine phosphate bonds). The observations of earlier eluting labelled fragments suggested that a proportion of the arginyl-tRNA in herpes-infected cells had its first guanylic acid residue nearer to the –CCA terminal than has any arginyl-tRNA normally found in mammalian non-infected cells. It is not easy to maintain that this could result from modification of pre-existing, host-specified material. The most reasonable interpretation is that this arginyl-tRNA is coded by the herpesvirus genome. Analogous experiments (Hay, Shepherd & Subak-Sharpe, 1967, in preparation) have led to a similar finding with seryl-tRNA except that the new fragments elute later, indicating that the first guanylic acid residue is more distal from the –CCA terminal than in the non-infected cell preparations. No differences were observed when glycyl-tRNA was similarly co-chromatographed.

Specific modification of the biological function of tRNA from *E. coli* after infection with the RNA bacteriophage Qβ has been briefly reported (Hung, 1966). As full experimental details have not been published this cannot yet be evaluated.

Changes at the level of the aminoacyl-tRNA synthetases

Several investigators have concluded that *E. coli* possesses only one aminoacyl-tRNA synthetase for each amino acid, although there may be several different molecular species of tRNA able to accept that amino acid (Yamane & Sueoka, 1964; Baldwin & Berg, 1966). Nevertheless, recent investigations of Yu (1966) have demonstrated that multiple forms of leucyl-tRNA synthetase (which can be fractionated on hydroxyapatite columns) are present in this organism. Furthermore, Barnett & Epler (1966) have identified two phenylalanyl-tRNA synthetases and two aspartyl-tRNA synthetases in *Neurospora*. In each case the 'minor' synthetase acylates a tRNA of mitochondrial origin (Barnett & Brown, 1967). Thus evidence is beginning to accumulate of the presence of multiple aminoacyl-tRNA synthetases in some organisms.

It has been shown by use of a temperature-sensitive mutant of valyl-tRNA synthetase that infection with T4 bacteriophages leads to the appearance of a different valyl-tRNA synthetase activity in *E. coli KB* (Neidhardt & Earhart, 1966). Two minutes after infection the new activity begins to appear and then increases to reach a maximum at 9–12 min. The increase in activity, which requires concomitant protein synthesis, is found after T-even phage infection but not after T5 or λ infection. When u.v. irradiated T4 phage is used, enzyme activity continues to increase linearly over 25 min., whereas with non-irradiated phage a plateau is reached in 10 min. Thus the synthetase behaves like an 'early' enzyme induced by T-even in *E. coli*. The new enzyme activity is temperature-resistant both *in vivo* and *in vitro*.

By fractionation on a DEAE column new valyl-tRNA synthetase activity could be demonstrated also in wild-type *E. coli KB*. The new peak in phage-infected *E. coli KB* comprised over half of the total valyl-tRNA synthetase activity, even though there was no net increase of total activity following phage infection. The results therefore establish that a qualitatively different valyl-tRNA synthetase activity appears as a result of protein synthesis in T4 infected cells. This could be due either to a new phage-coded synthetase, or to a phage modified cell-coded synthetase.

Similar experiments using as host cells a strain of *E. coli KB* with a temperature sensitive phenylalanyl-tRNA synthetase indicate that T4 does not induce different activity of this enzyme.

Changes at the level of the tRNA methylases

Methylation appears to be able to influence the acceptor, transfer, and perhaps even the coding functions of tRNA (Littauer, Revel & Stern, 1966). It is relevant to recall: (*a*) that six different tRNA methylases, each with a definite base specificity, have been purified from *E. coli* extracts (Hurwitz, Gold & Anders, 1964*a*, *b*); (*b*) that in the case of methyl-deficient phenylalanyl-tRNA Littauer *et al.* (1966) observed in ribosome-binding experiments some recognition of codons which are not read by normal phenylalanyl-tRNA; and (*c*) that Peterkofsky, Jesensky & Capra (1966) have shown that methylation has a profound effect on the codon recognition of leucyl-tRNA.

Wainfan, Srinivasan & Borek (1965) have studied the effect of T2 phage infection of *E. coli B* and found a 50 % increase in total tRNA methylase capacity within 10 min. This increase is not uniform for the different base specific methylases. No changes were observed with bacteriophage T1.

In *E. coli K*12 (λ) total tRNA methylase activity is reduced to about 20 % of controls 10 min. after phage λ induction by u.v. (enzyme was at saturation level in these experiments). Although the total tRNA methylase capacity subsequently returns to its former level, the relative levels of base specific enzymes are changed. A dialysable inhibitor of enzyme activity has been detected in these u.v. induced cells (Wainfan, Srinivasan & Borek, 1966). Thus evidence has been obtained in two very different bacteriophage systems, that the population of tRNA-methylating enzymes is in some way changed, or modified, as an early consequence of virus infection or induction. Obviously this is likely to affect the tRNA population and therefore translation in these cells.

In *E. coli* tRNA is also known to be modified due to enzymic thiolation—that is transfer of sulphur from cysteine to uridine residues in the polynucleotide to produce 4-thiouridine (Hayward & Weiss, 1966;

Lipsett & Peterkofsky, 1966; Lipsett, 1966). However, so far no experiments have been reported which investigate the effect of bacteriophage infection on disulphide bonds in sRNA or on sRNA sulphur transferase activity, nor for that matter on the activity of other enzymes known to modify nucleotides in tRNA.

Changes at the level of '5S' RNA

An active 70S ribosome in *E. coli* consists of one 30S and one 50S subunit. There are about twenty different kinds of ribosomal polypeptide on the former and thirty different kinds on the latter. The 30S subunit contains one molecule of 16S ribosomal RNA. The 50S subunit contains one molecule of 23S ribosomal RNA and one molecule of firmly attached '5S' ribosomal RNA.

Recently several workers have detected small amounts of '5S' RNA strongly associated with the ribosomes from uninfected mammalian, invertebrate, fungal and bacterial cells (Rosset, Monier & Julien, 1964; Galibert, Larsen, Lelong & Boiron, 1965; Comb, Sarkar, De Vallet & Pinzino, 1965; Comb & Zehavi-Willner, 1967). The function of this RNA which does not seem to contain unusual bases is not fully understood, but it appears to be an integral part of the ribosome complex. The structure of '5S' RNA from *E. coli* has recently been determined (Brownlee & Sanger, 1967). Although it is by no means clear that '5S' RNA is an indispensable part of the translation apparatus, this seems probable.

'5S' RNA is included in this survey because a similar virus-specified 'VA' RNA has been identified in KB (mammalian) cells following infection with adenovirus types 1 and 2 (Reich, Forget, Weissman & Rose, 1966), and a low molecular weight RNA has also been found after T4 phage infection of *E. coli* (Baguley, Berguist & Ralph, 1967). The main difference observed with 'VA' RNA is that it appears predominantly in the supernatant fraction of the cell, whereas '5S' RNA is always firmly bound to the ribosomal fraction. 'VA' RNA is not detectable during the first 10 hr. after infection and its appearance coincides with that of infectious adenovirus. It is more resistant to pancreatic RNase than sRNA, differs in base composition from '5S' RNA, and is produced only if RNA and DNA syntheses are allowed to take place after infection. Once formed it appears to be stable. It is not known whether 'VA' RNA is virus or host specified, or what function it has. The low molecular weight T4 phage specific RNA is found mainly associated with ribosomes, has a base composition unlike tRNA, and is hybridizable with T4 DNA but not with

E. coli DNA. It is not likely to be randomly degraded phage-specified messenger RNA, and its function is unknown.

It is conceivable that either or both RNAs may modify the translation mechanism at the level of the ribosomes.

DISCUSSION

The preceding survey has shown that there is already a considerable amount of evidence that viruses can induce changes at a variety of different points in the translation mechanism of their host cells. This poses two fundamental questions. Firstly, what is the nature of the observed virus-induced changes and, secondly, why should a virus change the translation mechanism of its host?

What is the nature of the observed virus induced changes?

Is the entire linear sequence of these tRNA or synthetase molecules specified by the virus genome? Or are these molecules initially coded by the host genome but subsequently modified (either directly or indirectly) by other virus-specified molecules? The first alternative will be referred to as virus-specification, the second as virus-modification.

How can the alternatives be distinguished? The cases of tRNA and then of protein will be considered in turn.

tRNA: general theoretical considerations

Incontrovertible proof that a tRNA is virus-specified will not be possible until that tRNA's complete nucleotide sequence has been elucidated and molecules of it can be produced *in vitro* using purified viral nucleic acid as template. Less rigorous, but acceptable, evidence could be obtained from DNA/tRNA hybridization studies if the tRNA in the hybrid can be made either to retain its covalently attached amino acid, or to retain sufficient of its integrity to allow its subsequent enzymatic aminoacylation. Unfortunately DNA/tRNA hybridization studies as carried out at present can often be criticized on one or more of three counts: (1) possible contamination of the viral DNA by host DNA; (2) possible sRNA contamination by degraded viral messenger RNA; (3) the possibility that fortuitous short sequence homologies exist between viral DNA and host RNA, which however are sufficiently long to give hybridization. Thus hybridization alone is not a sufficient criterion.

Suggestive, though not conclusive, evidence can come from the analysis of the –CCA terminal oligonucleotides after T1 RNase digestion of doubly-labelled aminoacyl-tRNA. The reasons why the possibility cannot be ruled out, that a change in aminoacylation of a pre-existing tRNA is responsible for any new fragments that are observed, will be given later in the discussion.

Consistent differences found by MAK column chromatography and reverse phase column chromatography demonstrate conclusively only that the population of tRNA has been altered. If differences are not found it would be wrong to conclude that the tRNA population has not been altered. No amount of positive results accumulated by MAK or reverse phase column chromatography could allow one to discriminate between the alternative hypotheses that the changed tRNA peaks are due to virus-specified or virus-modified host-specified molecules.

tRNA: the observed changes

The leucyl-tRNA changes induced by the T-even phages can now be considered. At first sight, it seems as though the failure to find appreciable changes in the total quantity and in the relative amount of aminoacyl-tRNA at 8 min. after infection with T2 provides support for the virus-modification alternative. But the experiments of Smith et al. (Fig. 3, p. 481, 1966, 1967) expose this as a fallacy. Their virus-induced change in the translation mechanism is known to be due to a new tRNA molecular species coded by the introduced virus genome. Nevertheless, they could not detect relative changes in the amount of tyrosinyl-tRNA under their experimental conditions when they induced virus and allowed it to multiply up to the time of normal cell lysis. Nor were relative changes detected up to the time of normal cell lysis in experiments where cells which had been starved for 30 min. were then infected with Su-3-carrying phage at a multiplicity of five. Tests for changes in total quantity or relative amounts are therefore not very sensitive, for some new tyrosinyl-tRNA must have been present. Only by extensively prolonging DNA and RNA synthesis were the relative quantities of the tRNAs substantially altered in these cells. Moreover, one should recall that protein synthesis over the first 40 min. was also necessary for production of detectable amounts of new tyrosinyl-tRNA under their conditions.

Recently it has been found that new tyrosinyl-tRNA does increase 1·5–2·0-fold by 15 min. after infection with a multiplicity of 20 phages per cell, provided that the cells are *not* pre-starved. Some increase can then even be detected in the presence of chloramphenicol (J. D. Smith & J. Abelson, personal communications). These new findings demonstrate how easily variation in the experimental conditions can affect the criteria of relative increase and of need for protein synthesis.

Neither the experiments with yeast and *E. coli* leucyl-tRNA synthetase nor the ribosome binding experiments that have been done can discriminate between the alternative hypotheses. Thus the results of Kano-Sueoka & Sueoka (1966), Sueoka *et al.* (1966), Nirenberg *et al.* (1966) and those of Waters & Novelli (1967) are, on the present evidence, just as likely to be due to virus-specified as to virus-modified, host-specified tRNA. All these investigators appear to favour Sueoka's original interpretation that the leucyl-tRNA is virus-modified but host-specified in origin. This could be right, but it has yet to receive incontrovertible experimental support.

Next to be considered are the observations of Brenner *et al.* (1966). Although there is little detailed evidence published about the disappearance of ochre-suppressing activity of Su-4[+] following T4 infection, the nature of the observations seems to be compatible only with the hypothesis that the change is due to virus-directed modification of host-specified tRNA. The means by which this is achieved are entirely unknown.

In the case of herpesvirus infection of mammalian cells the evidence from MAK chromatography demonstrates that the arginyl-tRNA population has been altered. But here the investigation was extended by use of independent techniques, and the results of the T1 RNase experiments give support to the interpretation that the virus specifies both new arginyl-tRNA and seryl-tRNA. However, before this is accepted it needs further substantiation, because other interpretations are possible.

For example it could be argued that the virus modifies a pre-existing guanine in host-specified tRNA. In the case of arginyl-tRNA this would mean alteration of an unusual base into guanine, for the new fragments elute earlier. Conversely in the case of seryl-tRNA the guanine nearest the –CCA terminal would have to be modified as a consequence of virus infection—for here the new fragments elute later. When one also takes into account that in the five known tRNA sequences from yeast and in the one from *E. coli* no unusual base occurs among the first 19 bases from the –CCA terminal, then this type of explanation seems highly improbable (Holley *et al.* 1965; Madison, Everett & Kung, 1966; Zachau *et al.* 1966; Raj Bhandary *et al.* 1966; Venkstern *et al.* 1967; Smith *et al.* 1967).

Another possibility is that the synthetase recognition site of some other host-specified tRNA is modified as a consequence of virus infection, resulting in a change of amino acid acceptance. This hypothesis would not explain the hybridization results, but, as has already been pointed out, these could be questioned on technical grounds. There is no other direct evidence against this hypothesis but indirect evidence comes from comparisons of the DNAs of herpesvirus and its mammalian hosts (Subak-Sharpe, Bürk, Crawford, Morrison, Hay & Keir, 1966; Subak-Sharpe, 1967). Nearest-neighbour analysis demonstrated that herpesvirus DNA contained comparatively more than ten times the proportion of the sequence CG. It can be reasoned that the population of tRNAs of the host cell, being the result of natural selection, will be optimally adapted to the normal translation requirements of the host cell's polypeptide-specifying DNA. This suggests difficulties at translation for the foreign DNA of herpesvirus, unless this DNA brings in genetic information which redresses the balance, that is increases the proportion of tRNAs which recognize codons that contain CG. There are eight such codons, four of them code for arginine (CGU, CGC, CGA, CGG) and one each for serine (UCG), proline (CCG), threonine (ACG) and alanine (GCG). If these considerations are accepted as reasonable, then the observations of new arginyl-tRNA and seryl-tRNA make sense if they are virus-specified, but not if they are host-specified tRNAs (recognizing other than CG-containing codons) with a virus-modified synthetase recognition site.

Thus the most satisfactory interpretation of the herpesvirus data is that the observed arginyl-tRNA and seryl-tRNA are both virus-coded.

Protein: general theoretical considerations

Proteins, newly observed following virus infection, could also be either virus-specified or virus-induced but host-specified.

Incontrovertible proof that a particular protein is virus-specified will not be possible until the active protein can be produced in an *in vitro* system on supplied messenger RNA templates, themselves made *in vitro* on purified viral DNA.

Less rigorous but acceptable evidence could come from combined genetical, immunological and biochemical approaches. Such evidence would be: (*a*) that an immunologically recognizable protein with clearly defined biochemical characteristics can be produced in several different host species but only following virus infection; (*b*) that mutant forms of the virus (e.g. temperature sensitive) are available, which result, under appropriate conditions, in the production of an altered protein; and (*c*) that infection with unrelated viruses does not result in the production of the identical immunologically recognizable protein.

Suggestive but not conclusive evidence could be provided by investigations which meet only part of the requirements outlined above.

Protein: the observed changes

Thus the case of the temperature-resistant valyl-tRNA synthetase found after T-even phage infection of *E. coli* (Neidhardt & Earhart, 1966), though it strongly suggests that the synthetase is virus-specified, is not quite conclusive. The only observation which seems to militate against this interpretation is that no net increase of total valyl-tRNA synthetase activity followed phage infection in *E. coli KB*, even though over half the total enzymic activity was found in a new peak of the chromatogram. The only reasonable alternative hypothesis to virus specification is that the host's own temperature-sensitive valyl-tRNA synthetase, but not its phenylalanyl-tRNA synthetase, becomes some-how modified as a consequence of virus infection. This type of explana-tion is not very attractive for it is both elaborate and imprecise.

What can be concluded about the observed changes in methylase activity? It seems that changes in the methylation pattern of tRNA can profoundly alter the codon recognition, presumably by changing the three-dimensional configuration of the tRNA molecule. So far the changes which have been experimentally observed appear to be of a type which suggest some loss of fidelity in codon recognition, but it is still possible that more profound changes are also produced. Thus it is quite conceivable that some viruses might use these means to regulate, or change the regulation of, the host-specified translation machinery. In this context the observations by Wainfan *et al.* (1966) of changes in the total amount and in the relative amount of methylase activity following infection with T2 phage and also induction of phage λ are highly suggestive. But at present it is not known by what mechanism phage infection produces the observed changes, or whether the T2 and λ cases are more than superficially similar. The recent discovery of a dialysable inhibitor (Wainfan *et al.* 1966) may shed considerable light on this problem.

Why should a virus change the translation mechanism of its host?

Virus-imposed changes of the translation mechanism could resolve two basic problems.

Obligatory change of the translation mechanism

It may be necessary to remove restraints on translation which are inherent in the host's unaltered mechanism. For example, a particular codon may occur infrequently or never in host-specified messenger RNA and, consequently, tRNA capable of recognizing this codon may be rare or absent in the host's pre-existing population of tRNAs. A virus that makes frequent use of this codon could not replicate in this host unless it specified such tRNA, which would remove the inherent restraint. At first sight it seems that this end could be achieved as readily by virus-directed modification or by induction of host-specified tRNA. But there would be the problem that the postulated modifying or inducing protein must somehow have been synthesized on virus-specified template RNA by the still unaltered translation mechanism of the host.

Change to facilitate the regulation of protein synthesis

A virus might alter the translation mechanism of the host and so regulate protein synthesis. Two types of regulatory change are easy to imagine: one directed against continued host-programmed protein synthesis and favouring virus-programmed protein synthesis: the other switching virus-directed synthesis from the manufacture of 'early' proteins to that of 'late' proteins.

Any theory which postulates regulatory change operating efficiently at the level of translation must inherently account for release of undesirable messenger RNA from the ribosome, to allow it to be used by desirable messenger RNA. If there is not concomitant release of undesirable messenger RNA, then the virus-induced change will merely block the ribosomes with unreadable messenger RNA or reduce the amount of active, but not of total, unwanted protein produced by insertion of mis-sense into the unwanted protein. Only some of the theories of regulatory change seem to take care of this aspect of the problem.

For example the 'adaptor modification hypothesis' (Sueoka & Kano-Sueoka, 1964) starts from the truism 'The translation of messenger RNA should be affected if a change in the tRNA is introduced' and then postulates: (1) changed codon recognition of a particular adaptor out of a set of degenerate adaptors due to structural modification; (2) consequent mistranslation of messenger RNA which accommodates the corresponding codon while other messenger RNA is normally translated; (3) resulting in selective shut-off of some gene functions (but not of their transcription).

Previously, Ames & Hartman (1963) had proposed their 'modulation hypothesis' to explain polarity in the histidine operon of *Salmonella*. This proposed control of translation by use of rare codons (modulating triplets) at strategic positions along a polycistronic message. It was suggested that the tRNA species appropriate to modulating triplets affected the reading of the message either by increasing the probability that ribosomes fell off at that point, or because they were rate limiting. This 'modulation hypothesis' was further extended by Stent (1964) to couple transcription to translation.

The 'adaptor modification hypothesis' was originally meant to explain not only the transition from 'early' to 'late' phage protein synthesis, but also the shut-off of translation of preformed host messenger RNA following infection. The hypothesis was later extended to cover (a) any modification which may eliminate the adaptor of a particular codon and prevent translation of messenger RNA of genes which contain the codon but not that of others; and (b) mis-sense translation due to modification (Kano-Sueoka & Sueoka, 1966). In the case of leucyl-tRNA modification following T-even phage infection the time-course of events has shown that this cannot be related to the transition from the 'early' to the 'late' phase of phage protein synthesis and, more recently, Sueoka *et al.* (1966) have suggested that the modification may either be involved in switch-off of host protein synthesis, or may be the secondary effect of some other unknown phenomenon.

CONCLUSIONS

In conclusion, there is ample evidence to show that virus-induced changes in translation mechanisms occur both at the level of transfer RNA and at the level of proteins involved in the translation mechanism. However, in the majority of cases it is quite uncertain whether the observed change is due to virus-specified molecules or to virus-modified but host-specified molecules, and in no case has this question been answered entirely satisfactorily.

REFERENCES

AMES, B. N. & HARTMAN, P. E. (1963). The histidine operon. *Cold Spring Harb. Symp. quant. Biol.* **28**, 349.

BAGULEY, B. C., BERGQUIST, P. L. & RALPH, R. K. (1967). Low molecular-weight T4 phage-specific RNA. *Biochim. biophys. Acta*, **138**, 51.

BALDWIN, A. N. & BERG, P. (1966). Purification and properties of isoleucyl ribonucleic acid synthetase from *Escherichia coli*. *J. biol. Chem.* **241**, 831.

BARNETT, W. E. & BROWN, D. H. (1967). Mitochondrial transfer ribonucleic acids. *Proc. natn. Acad. Sci. U.S.A.* **57**, 452.

BARNETT, W. E. & EPLER, J. L. (1966). Fractionation and specificities of two aspartyl-ribonucleic acid and two phenylalanyl-ribonucleic acid synthetases. *Proc. natn. Acad. Sci. U.S.A.* **55**, 184.

BRENNER, S., KAPLAN, S. & STRETTON, A. O. W. (1966). Identity of N2 and *Ochre* nonsense mutants. *J. molec. Biol.* **19**, 574.

BROWNLEE, G. G. & SANGER, F. (1967). Nucleotide sequences from the low molecular weight ribosomal RNA of *Escherichia coli*. *J. molec. Biol.* **23**, 337.

COMB, D. G., SARKAR, N., DE VALLET, J. & PINZINO, C. J. (1965). Properties of transfer-like RNA associated with ribosomes. *J. molec. Biol.* **12**, 509.

COMB, D. G. & ZEHAVI-WILLNER, T. (1967). Isolation, purification and properties of 5 S ribosomal RNA: a new species of cellular RNA. *J. molec. Biol.* **23**, 441.

GALIBERT, F., LARSEN, C. J., LELONG, J. C. & BOIRON, M. (1965). RNA of low molecular weight in ribosomes of mammalian cells. *Nature, Lond.* **207**, 1039.

HAYWARD, R. S. & WEISS, S. B. (1966). RNA thiolase: the enzymatic transfer of sulfur from cysteine to sRNA in *Escherichia coli* extracts. *Proc. natn. Acad. Sci. U.S.A.* **55**, 1161.

HOLLEY, R. W., APGAR, J., EVERETT, G. A., MADISON, J. T., MARQUISEE, M., MERRILL, S. H., PENSWICK, J. R. & ZAMIR, A. (1965). Structure of a ribonucleic acid. *Science*, **147**, 1462.

HUNG, P. P. (1966). Specific modification of biological function of sRNA from *E. coli* after infection with an RNA bacteriophage. *Fedn. Proc.* **25**, 404.

HURWITZ, J., GOLD, M. & ANDERS, M. (1964a). The enzymatic methylation of ribonucleic acid and deoxyribonucleic acid. III. Purification of soluble ribonucleic acid-methylating enzymes. *J. biol. Chem.* **239**, 3462.

HURWITZ, J., GOLD, M. & ANDERS, M. (1964b). The enzymatic methylation of ribonucleic acid and deoxyribonucleic acid. IV. The properties of the soluble ribonucleic acid-methylating enzymes. *J. biol. Chem.* **239**, 3474.

KANO-SUEOKA, T. & SUEOKA, N. (1966). Modification of leucyl-sRNA after bacteriophage infection. *J. molec. Biol.* **20**, 183.

KELMERS, A. D., NOVELLI, G. D. & STULBERG, M. P. (1965). Separation of transfer ribonucleic acids by reverse phase chromatography. *J. biol. Chem.* **240**, 3979.

LIPSETT, M. N. (1966). Disulfide bonds in sRNA. *Cold Spring Harb. Symp. quant. Biol.* **31**, 449.

LIPSETT, M. N. & PETERKOFSKY, A. (1966). Enzymatic thiolation of *E. coli* sRNA. *Proc. natn. Acad. Sci. U.S.A.* **55**, 1169.

LITTAUER, U. Z., REVEL, M. & STERN, R. (1966). Coding properties of methyl-deficient phenylalanyl transfer RNA. *Cold Spring Harb. Symp. quant. Biol.* **31**, 501.

MADISON, J. T., EVERETT, G. A. & KUNG, H. K. (1966). On the nucleotide sequence of yeast tyrosine transfer RNA. *Cold Spring Harb. Symp. quant. Biol.* **31**, 409.

NEIDHARDT, F. C. & EARHART, C. F. (1966). Phage-induced appearance of a valyl-sRNA synthetase activity in *Escherichia coli*. *Cold Spring Harb. Symp. quant. Biol.* **31**, 557.

NIRENBERG, M., CASKEY, T., MARSHALL, R., BRIMACOMBE, R., KELLOGG, D., DOCTOR, B., HATFIELD, D., LEVIN, J., ROTTMAN, F., PESTKA, S., WILCOX, M. & ANDERSON, F. (1966). The RNA code and protein synthesis. *Cold Spring Harb. Symp. quant. Biol.* **31**, 11.

PETERKOFSKY, A., JESENSKY, CELIA & CAPRA, J. D. (1966). The role of methylated bases in the biological activity of *E. coli* leucine tRNA. *Cold Spring Harb. Symp. quant. Biol.* **31**, 515.

RAJ BHANDARY, U. L., STUART, A., FAULKNER, R. D., CHANG, S. H. & KHORANA, H. G. (1966). Nucleotide sequence studies on yeast phenylalanine sRNA. *Cold Spring Harb. Symp. quant. Biol.* **31**, 425.

REICH, P. R., FORGET, B. G., WEISSMAN, S. M. & ROSE, J. A. (1966). RNA of low molecular weight in KB cells infected with adenovirus type 2. *J. molec. Biol.* **17**, 428.

ROSSET, R., MONIER, R. & JULIEN, J. (1964). Les ribosomes d'*Escherichia coli*. I. Mise en évidence d'un RNA ribosomique de faible poids moléculaire. *Bull. Soc. chim. Biol.* **46**, 87.

SMITH, J. D., ABELSON, J. N., CLARK, B. F., GOODMAN, H. M. & BRENNER, S. (1966). Studies on *amber* suppressor tRNA. *Cold Spring Harb. Symp. quant. Biol.* **31**, 479.

SMITH, J. D., ABELSON, J. N., LANDY, A., GOODMAN, H. M. & BRENNER, S. (1967). The sequence of *amber* suppressor Su-3 tRNA. *Nature, Lond.* (in the Press).

STENT, G. (1964). The operon on its third anniversary. *Science*, **144**, 816.

SUBAK-SHARPE, H. (1967). Doublet patterns and evolution of viruses. *Br. med. Bull.* **23**, 161.

SUBAK-SHARPE, H., BÜRK, R. R., CRAWFORD, L. V., MORRISON, J. M., HAY, J. & KEIR, H. M. (1966). An approach to evolutionary relationships of mammalian DNA viruses through analysis of the pattern of nearest neighbor base sequences. *Cold Spring Harb. Symp. quant. Biol.* **31**, 737.

SUBAK-SHARPE, H. & HAY, J. (1965). An animal virus with DNA of high guanine + cytosine content which codes for sRNA. *J. molec. Biol.* **12**, 924.

SUBAK-SHARPE, H., SHEPHERD, WILMA M. & HAY, J. (1966). Studies on sRNA, coded by herpes virus. *Cold Spring Harb. Symp. quant. Biol.* **31**, 583.

SUEOKA, N. & KANO-SUEOKA, T. (1964). A specific modification of leucyl-sRNA of *Escherichia coli* after phage T2 infection. *Proc. natn. Acad. Sci. U.S.A.* **52**, 1535.

SUEOKA, N., KANO-SUEOKA, T. & GARTLAND, W. J. (1966). Modification of sRNA and regulation of protein synthesis. *Cold Spring Harb. Symp. quant. Biol.* **31**, 571.

VENKSTERN, T. V., BAJEV, A. A., MIRSABEKOV, A. D., KRUTILINA, A. I., LI, L. & AXELROD, V. D. (1967). The primary structure of $tRNA^{val}$. *Fourth Fed. of Europ. Biochem. Soc. Abstracts*, no. 364, p. 91.

WAINFAN, E., SRINIVASAN, P. R. & BOREK, E. (1965). Alterations in the transfer ribonucleic acid methylases after bacteriophage infection or induction. *Biochemistry*, **4**, 2845.

WAINFAN, E., SRINIVASAN, P. R. & BOREK, E. (1966). Can methylation of tRNA serve a regulatory function? *Cold Spring Harb. Symp. quant. Biol.* **31**, 525.

WATERS, L. C. & NOVELLI, G. D. (1967). A new change in leucine transfer RNA observed in *Escherichia coli* infected with bacteriophage T2. *Proc. natn. Acad. Sci. U.S.A.* **57**, 979.

WILHELM, R. C. (1966). Discussion of paper by Carbon, J., Berg, P. & Yanofsky, C. *Cold Spring Harb. Symp. quant. Biol.* **31**, 496.

YAMANE, T. & SUEOKA, N. (1964). Enzymic exchange of leucine between different components of leucine acceptor RNA in *Escherichia coli*. *Proc. natn. Acad. Sci. U.S.A.* **51**, 1178.

YU, C-T. (1966). Multiple forms of leucyl sRNA synthetase of *E. coli*. *Cold Spring Harb. Symp. quant. Biol.* **31**, 565.

ZACHAU, H. G., DÜTTING, D., FELDMANN, H., MELCHERS, F. & KARAU, W. (1966). Serine specific transfer ribonucleic acids. XIV. Comparison of nucleotide sequences and secondary structure models. *Cold Spring Harb. Symp. quant. Biol.* **31**, 417.

VIRUS-INDUCED ENZYMES IN MAMMALIAN
CELLS INFECTED WITH DNA-VIRUSES

H. M. KEIR

Department of Biochemistry, University of Glasgow,
Glasgow, W.2

INTRODUCTION

During the past ten years a vast pool of information has become available on the structural, physical and chemical properties of mammalian viruses. Much of the recent emphasis has been laid upon the mechanism of replication of viral nucleic acid and of effects exerted upon the metabolism of host-cell nucleic acids throughout the infective process. The data assembled from these studies have provided the basis and presented the problems for a phase of research directed principally to characterization of the coding potential of viral nucleic acid. Since viral nucleic acids individually constitute a limited amount only of genetic material (relative to the genetic potential of their host cells), the important possibility emerges that the entire coding potential of a viral genome can be defined. An experimental approach, which promises to contribute substantially in that respect, consists of identification and characterization of proteins synthesized in the infected cell under the direction of the invading viral genome. These include not only structural proteins destined to become part of the mature progeny virions, but also enzymic and other proteins whose functions are to collaborate in the over-all process of synthesis of viral components and which do not constitute part of the virion.

There is no doubt that these concepts have been strongly influenced by the enzymic alterations which take place in bacterial cells infected by bacteriophage, in particular the bacteriophages T2, T4 and T6. However, progress is likely to be slower with the mammalian systems because of their greater complexity and because these systems are much less favourable in the technical sense than their bacterial counterparts. For example, certain difficulties of production of material impose restrictions on some of the more traditional biochemical approaches to the purification and determination of the properties of proteins.

Nevertheless, mammalian experimental systems have been developed which are susceptible to biochemical analysis. In this paper, I propose to describe some of these systems, dealing almost exclusively with

DNA-viruses and placing greatest emphasis on our own recent work on enzyme induction in cells infected by herpes simplex virus.

The following abbreviations are used: RNase, ribonuclease; DNase, deoxyribonuclease; MP, monophosphate; DP, diphosphate; TP, triphosphate; A, C, G, T and U, adenosine, cytidine, guanosine, thymidine and uridine; the d- prefix indicates deoxy and refers exclusively to the deoxyribose component of nucleosides and nucleotides; tRNA, transfer RNA.

THE RATIONALE FOR EXPERIMENTATION

In considering the coding potential of viral nucleic acid, the DNA-viruses present a greater challenge than do the RNA-viruses, since their genomes (with the possible exception of the polyoma group of viruses) have the capacity theoretically to programme much more in terms of synthesis of polypeptide. It seems reasonable to predict that virus-induced synthesis of protein must be geared to production of progeny virus, and therefore should involve the action of enzymes concerned in synthesis of protein and nucleic acid.

Although the normal mammalian cell is equipped with these enzymes in some measure or another, there is no guarantee that their activity levels and specificities are suitable for efficient and accurate production of new virus particles after initiation of infection. The logical projection of these considerations is therefore that some of the programming potential of the invading viral genome might well be concerned with the production of enzymes involved in the synthesis of these macromolecules. An abbreviated pathway of synthesis of nucleic acids (Fig. 1) shows the enzymic reactions leading to the provision of the proximal purine and pyrimidine precursors (the nucleoside triphosphates) for the enzymic synthesis of DNA and RNA. A minimum of some 40 enzymes are directly implicated in that biosynthetic sequence. Clearly there is a possibility that all of these enzymes might be involved in replication and transcription of viral DNA. In the case of the RNA-viruses, there is no requirement (at least superficially) for induction of the block of enzymes concerned in synthesis of the deoxyribonucleotides. If the soluble nucleotide pools in the host cells were sufficiently large to provide all the monomers for nucleic acid synthesis in the infected cell, then a large number (some 20 to 30) of the enzymes involved in the biosynthetic pathway might be regarded as non-essential. Therefore, any search for virus-induced enzymes might most fruitfully be directed to a survey of those enzymes concerned in the terminal stages of synthesis of nucleic acids. In the case of DNA-viruses, the most likely candidates to meet these requirements would be DNA polymerase, deoxyribonucleoside monophosphate and diphosphate kinases, dTMP synthetase, tetra-hydrofolate dehydrogenase (required in production of 5,10-methylene-tetra-hydrofolate for the dTMP synthetase reaction), dCMP deaminase, and enzymes involved in 'scavenger-type' reactions (e.g. thymidine and

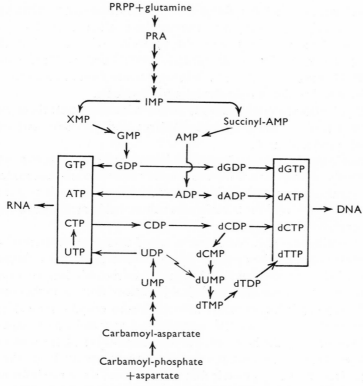

Fig. 1. Abbreviated biosynthetic pathway leading to provision of the proximal precursors for biosynthesis of DNA and RNA. PRPP, 5-phosphoribosylpyrophosphate; PRA, 5-phosphoribosyl-1-amine; XMP, xanthosine 5'-monophosphate; IMP, inosine 5'-monophosphate.

other deoxyribonucleoside kinases, nucleoside monophosphate pyrophosphorylases). If there is a net increase in the rate of DNA synthesis after infection with some DNA-viruses there may be a need to invoke the participation of the ribonucleotide reductases to boost the deoxyribonucleotide pool, and thus, these enzymes could well be included in the list above.

Since replication of DNA-viruses demands transcription from viral DNA templates (Becker & Joklik, 1964; Salzman, Shatkin & Sebring,

1964; Rose, Reich & Weissman, 1965; Hay, Köteles, Keir & Subak-Sharpe, 1966; Benjamin, 1966; Thomas & Green, 1966), the actions of DNA-dependent RNA polymerase and ribonucleoside mono- and diphosphate kinases are additionally implicated.

Certain enzymes concerned with modification of nucleic acid during and after its maturation should be considered. These include methylating enzymes (methyltransferases and methionine adenosyltransferase) and enzymes which phosphorylate, thiolate and glucosylate the macromolecules. These reactions, with the exception of methylation (Srinivasan & Borek, 1966; Lal & Burdon, 1967; Burdon, Martin & Lal, 1967; R. H. Burdon & R. L. P. Adams, private communication), have been described in detail only for bacterial systems.

Finally, hydrolytic enzymes must be taken into account; these include enzymes acting on the polymers (DNases and RNases) and on the monomers (phosphatases).

Similar considerations of enzyme induction apply likewise to the biosynthesis of protein; among the relevant enzymes are the aminoacyl-tRNA synthetases involved in activation of amino acids and their transfer to tRNA, and the transferases concerned in synthesis of peptide bonds at the ribosomal level.

The experimental approach then is based upon elaboration of virus-induced enzymes, and has generally proceeded in two phases: (i) demonstration of an increased level of the activity of the individual enzymes in the host cell as a consequence of viral infection; and (ii) demonstration that the activity increments are attributable to enzyme molecules different from those present in the non-infected host cell, or perhaps completely absent from the host cell. The implication is that the appearance of new enzymes is the result of direct programming from the viral genome. However, this point is difficult to prove, since the information for synthesis of new enzymes might well be coded in the genome of the host cell.

ENZYME INDUCTION IN CELLS INFECTED WITH HERPES SIMPLEX VIRUS

DNA Polymerase (DNA nucleotidyltransferase, EC 2.7.7.7)

The early experiments (Keir & Gold, 1963) were conducted using BHK 21–C 13 (baby hamster kidney) cells (Macpherson & Stoker, 1962) propagated in monolayer culture, and the HFEM strain of herpesvirus (Wildy, Russell & Horne, 1960). Comparison of soluble and partially purified fractions from control and infected cells revealed an initial decline of polymerase activity at 1–2 hr. after infection followed by an increase above the control level 2–4 hr. later. At 5 hr., the activity was increased fivefold beyond the control level. The subsequent marked fall of polymerase activity at 6–8 hr. was attributed to a concomitant increase of activity of DNase which degraded DNA template and product during polymerase assays with the infected preparations. The distribution of polymerase in the control and infected cell fractions (nuclei, small particles and soluble cell sap) was compatible with the demonstration that the progeny virus particles are assembled in the cell nucleus (Morgan, Ellison, Rose & Moore, 1954; Morgan, Rose, Holden & Jones, 1959); viral DNA is also synthesized at that site (Munk & Sauer, 1963). The site of synthesis of the induced polymerase, however, remains uncertain, but it may be significant that the small-particle fraction shows appreciable increment of the enzyme after infection (Keir & Gold, 1963). Moreover, Sydiskis & Roizman (1966) have shown that the bulk of herpes viral proteins is probably made on cytoplasmic polysomes.

Subsequent experimentation has amply confirmed these observations on intracellular location of the polymerase (Russell et al. 1964; Keir, 1965). It is not clear whether virus-induced synthesis of the enzyme is switched off at a specific time, for although these experiments suggest termination of synthesis at 6 hr. after infection, it is possible that the concomitant increase of DNase activity (Keir & Gold, 1963; Russell et al. 1964) in the infected cells interferes with the DNA-template and DNA-product during the polymerase assay. This seems quite possible because the virus-induced DNase has been shown to be a DNA-exonuclease releasing 5'-monophosphates from the 3'-end of DNA (Morrison & Keir, 1967); this would act exactly in opposition to the accepted mechanism for action of the polymerase (Mitra & Kornberg, 1966). However, conditions can be selected in which the harvested, infected cells show high activity of the polymerase and minimal DNase activity (Keir, Hay, Morrison & Subak-Sharpe, 1966); moreover, properties of the two enzymes have been determined which allow assay of one with essentially no interference from the other (Keir, Subak-Sharpe, Shedden, Watson & Wildy, 1966; Morrison & Keir, 1967).

The main question at this point is whether or not the induced polymerase activity is a new enzyme directly programmed from the viral DNA. Among the possible explanations that might account for the post-infection increase of polymerase are: (i) the existing host-cell polymerase is activated by virus-induced elaboration of an activator, or removal of an inhibitor; (ii) there is a build-up of host-cell polymerase due to increased synthesis of the enzyme or a decreased rate of its destruction; (iii) derepression of a normally inactive host gene promotes synthesis of a host-coded polymerase isoenzyme; and (iv) there is synthesized in the host-cell a new polymerase whose amino acid sequence is coded in a gene of the invading viral DNA. The following survey of the properties of the herpes-induced DNA polymerase eliminates possibilities (i) and (ii), but does not clearly distinguish between (iii) and (iv).

Carefully controlled experiments with mixtures of the control and virus-induced enzyme preparations show the calculated additive response, indicating absence of mutual inhibition or activation (Morrison, 1967).

The induced enzyme behaves differently from the control in its response to univalent cations in the reaction medium. It is optimally active at 200 mM NH_4^+, while the control enzyme shows optimal activity at 8 mM NH_4^+ and is virtually inactive at 200 mM NH_4^+ (Keir, Subak-Sharpe, Shedden, Watson & Wildy, 1966). This provides a convenient method of measuring the induced enzyme without contribution to the reaction from the presence of control enzyme in the same fraction. A further important consideration is that the virus-induced DNase and endogenous host-cell DNase are also strongly inhibited at 100–200 mM K^+, Na^+ or NH_4^+ (Morrison & Keir, 1967), and thus their actions can virtually be eliminated as complicating factors during assay of the induced polymerase.

The induced polymerase is more heat-stable than the control enzyme when subjected to heating in the presence of DNA (Keir, Hay, Morrison & Subak-Sharpe, 1966). Presumably this reflects protection of the enzyme through relatively strong binding to its template, for if heat-inactivation is conducted in the absence of DNA, the opposite effect is observed, that is, the induced polymerase is more rapidly destroyed than the host enzyme.

The host-cell polymerase is saturated with respect to template at about 40 μg. of DNA in the standard assay system; on the other hand, the induced enzyme appears not to be saturated by up to 225 μg. of template. The template concentration curves for both enzyme preparations are sigmoid in shape; however, while the sigmoid character is consistently observed with the induced enzyme, the control enzyme frequently shows this property only slightly or not at all. A possible explanation for this phenomenon together with the heat-inactivation behaviour has been suggested (Keir, 1965; Keir, Hay, Morrison & Subak-Sharpe, 1966).

The standard assay system uses thermally denatured (single-stranded) DNA as template, but double-helical DNA templates have also been used to prime the reaction. Mammalian DNA polymerases either do not accept double-helical template at all (Bollum, 1963), or, if the enzyme is not extensively purified, generally use double-helical template to a limited extent, about 10–20 % as efficiently as single-stranded DNA (Keir, 1965). The responses given by the control and herpes-induced polymerases to denatured and double-helical DNA templates are shown in Table 1 for soluble enzyme preparations obtained after subcellular fractionation in sucrose–calcium chloride medium (Morrison, 1967).

Table 1. *Priming of DNA polymerase from nuclei and cytoplasm of control and herpes-infected BHK21–C13 cells*

	Control	Infected	Infected / Control
(i) With denatured DNA-template			
Cytoplasmic	945	1166	1·23
Nuclear	297	1302	4·38
Total units	1242	2468	1·99
(ii) With double-helical DNA-template			
Cytoplasmic	326	733	2·25
Nuclear	88	310	3·52
Total units	414	1043	2·52
(iii) Ratio of denatured template: double-helical template			
Cytoplasmic	2·9	1·6	—
Nuclear	3·4	4·2	—
Total	3·0	2·4	—

The values in (i) and (ii) for the control and infected samples are the units of polymerase measured in the same amount of total soluble protein extracted. Ratios in (iii) are obtained directly from the values in (i) and (ii). Subcellular fractionation was performed in Ca^{2+}-sucrose medium. The DNA was prepared from Landschütz ascites-tumour cells.

There are clear differences between the control and infected cell samples with respect to intracellular distribution and response to condition of the template. Other experiments (Keir, Hay, Morrison & Subak-Sharpe, 1966) have indicated a preference for herpes-DNA as template in the case of the induced polymerase. A definitive interpretation of these results must await an extensive study with nuclease-free polymerase. In this context, gel filtration studies (Morrison, 1967) have shown that the DNA polymerase of both control and infected cells is only slightly retarded on Sephadex G-200 and separable from control-cell DNase which emerges later from the gel. However, a virus-induced DNase elutes with the polymerase.

The control and induced polymerases are standard in requiring all four deoxyribonucleoside 5′-triphosphates and in displaying an absolute requirement for DNA and Mg^{2+} ions. However, the virus-induced enzyme has a Mg^{2+} optimum at 20–25 mM while the host enzyme is optimally active at 8–10 mM Mg^{2+} (Keir, 1965).

The two enzyme preparations show different levels of inhibition by iodoacetamide and p-mercuribenzoate (Keir, 1965), the enzyme from infected cells being more resistant to the action of these thiol-group inhibitors.

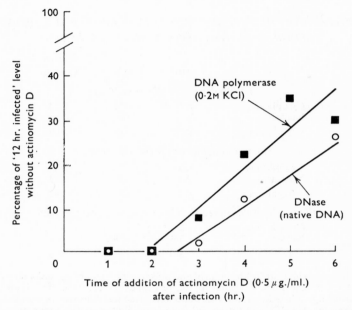

Fig. 2. Effect of time of addition of actinomycin D to infected cultures on induction of DNA polymerase and DNase after herpes-infection of BHK 21–C 13 cells. The enzymes were assayed under conditions determined to be optimal for the virus-induced activities (see text). These included assay of polymerase in the presence of 0·2 M K^+ and assay of the DNase with double-helical [^{32}P]DNA as substrate.

The induction of DNA polymerase in BHK21–C13 cells by herpes-virus requires transcription of RNA from DNA and also synthesis of protein (J. M. Morrison & H. M. Keir, unpublished observations). Addition of actinomycin D to the medium at a final concentration of 0·5 μg./ml., up to 2 hr. after infection, prevents the appearance of the enzyme, and at later times results in reduced yields (Fig. 2). The inclusion of puromycin throughout at 50 μg./ml. also eliminates the induction effect.

Immunological evidence (Keir, Subak-Sharpe, Shedden, Watson & Wildy, 1966) indicates that the induced polymerase is antigenically different from the control-cell enzyme. Antisera prepared in rabbits against antigens produced in rabbit kidney (RK) cells by herpesvirus (Watson *et al.* 1966) inactivate the polymerase induced by herpes-infection of BHK 21–C 13 cells and HEp-2 (human epithelioid carcinoma) cells, but not the polymerases of non-infected BHK 21–C 13 and HEp-2 cells (Fig. 3). Moreover, the herpes-specific antisera do not inhibit the DNA polymerase induced by infection with pseudorabies virus.

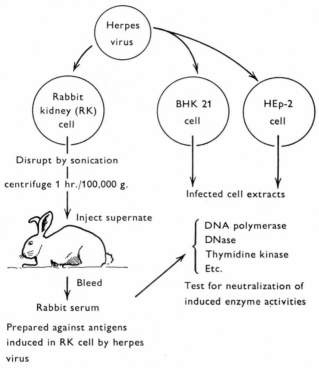

Fig. 3. The schedule for production of rabbit antisera containing herpes-specific antibodies. In the earlier experiments (Keir, Subak-Sharpe, Shedden, Watson & Wildy, 1966) whole cell sonicates of infected RK cells were injected into rabbits. In the later experiments (e.g. Fig. 4), a 100,000 *g* supernate obtained from the infected cells was injected. It contained the bulk of the virus-induced enzymes.

In addition to these immunological experiments, the other experiments described above with BHK 21–C 13 cells have been repeated with HEp-2 cells as hosts. The results were similar in all main aspects. The genetic origins of these two cell lines are quite different and therefore, since the same virus stock was used throughout, the conclusion seems inescapable that the induced polymerase is virus-coded.

However, incontrovertible proof that the information for the amino acid sequence of the induced polymerase resides in the nucleotide

sequence of the viral genome is not easy to come by. Even the immunological evidence does not eliminate the possibility that induction of the
enzyme is a consequence of derepression of a normally inactive host
gene, thus giving rise to a host-coded isoenzyme with properties somewhat different from those of the polymerase that exists in the non-
infected cell. However, it must be emphasized that the antisera used were
obtained from rabbits inoculated with herpes-specific proteins contained
in extracts of infected RK cells, and further, that the polymerase preparations that were inactivated by these antisera were obtained from infected, heterologous cells (BHK21–C13 and HEp-2). Therefore, rejection of the postulate that the induced enzyme is herpesvirus-coded
demands acceptance of the alternative, and very remote, possibility that
herpes infection induces synthesis in three different mammalian cell types
of a DNA polymerase isoenzyme with the same antigenic properties.

It seems possible that conclusive proof of the genetic origin of the
induced polymerase could be obtained by an experimental approach
based on transcription of RNA from herpes DNA, followed by demonstration that the RNA is capable of programming synthesis of the
enzyme *in vitro* in a purified, cell-free system. The technical difficulties
inherent in such a project are obvious. Even synthesis of incomplete,
virus-specific polypeptides would require prior knowledge of the amino
acid sequence of the induced enzyme for certain identification. It would
seem that such positive identification is a long way off. The most
acceptable alternative approach would consist of isolation of conditional
lethal mutants, for example, temperature-sensitive mutants, of the virus,
coupled with the demonstration that the temperature-sensitive lesion
resides in the induced polymerase molecule. This type of approach has
been successful in identifying the structural gene for DNA polymerase
in bacteriophages T4 and T5 (de Waard, Paul & Lehman, 1965; Warner
& Barnes, 1966). These studies have a most important implication: they
suggest that the DNA polymerase activity induced by infection with
bacteriophage and purified and characterized *in vitro*, is the enzyme
responsible for synthesizing the viral DNA in the infected bacteria.

At present, it is inferred that the herpes-induced DNA polymerase
functions in the synthesis of progeny viral DNA molecules. However,
direct evidence for this is lacking and must await the outcome of specific
kinetic and binding studies with purified enzyme and purified herpes
DNA.

Deoxyribonuclease

The experimental system is the same as that described in the previous
section for DNA polymerase. The early observations revealed that

DNase activity (measured at pH 7·5) increased markedly in cells infected with herpes simplex virus (Keir & Gold, 1963; Russell *et al.* 1964). The activity appeared slightly later than the polymerase (cf. Fig. 2). The intracellular distribution of the induced DNase was determined; after infection in these experiments, 17 % of the enzyme was located in the nuclear fraction, 30 % in the small particles (mitochondria + microsomes) and 52 % in the cell sap. This represented post-infection increases of 161 % in the nuclei, 105 % in the small particles and 107 % in the cell sap. Again, as with the polymerase, these are factors to be assessed in relation to the nuclear site of synthesis of progeny virions, and also, in relation to the function of the enzyme.

Whether there is precise regulation of termination of synthesis of the induced DNase activity is not certain, but it seems probable that switch-off occurred at 7–9 hr. in the earlier experiments (Keir & Gold, 1963; Russell *et al.* 1964). McAuslan, Herde, Pett & Ross (1965) and Sauer, Orth & Munk (1966) have more recently observed increases of 'alkaline' DNase following infection of monkey kidney cells and L-cells respectively.

There are several properties of the induced DNase activity which suggest that it is not present in the non-infected control cells. Chromatography of infected cell extracts on columns of DEAE-cellulose gives two peaks of DNase activity, one emerging in the non-adsorbed fractions, and a second one emerging at 0·12 M KCl during gradient elution. Control cell extracts show only the non-adsorbed peak, no DNase being detectable at the point where the virus-induced activity elutes. The chromatographic patterns on DEAE-cellulose for DNase activity from infected cell cytoplasmic and nuclear fractions are qualitatively identical (Morrison & Keir, 1967).

The control DNase is inhibited by Na^+ or K^+ above 15 mM, whereas the induced enzyme is stimulated above that level to an optimum at 50–60 mM. All control and induced DNase preparations display an absolute requirement for a bivalent cation. Mg^{2+} and Mn^{2+} fulfil this role, but while the activities of the control preparations are the same at 2 mM Mg^{2+} or 0·5 mM Mn^{2+} (the optimal concentrations), the infected-cell preparations are only half as active with Mn^{2+} as with Mg^{2+} (Morrison & Keir, 1966).

The control and induced DNases are optimally active at pH 9·1, although small differences are observed between the two enzyme activities when denatured DNA and double-helical DNA are compared as substrates (Morrison & Keir, 1966).

The heat-inactivation curves of the control and induced enzymes are quite different; the induced DNase is very sensitive to heating at 45°, while the control enzyme is virtually unaffected under the same conditions, although it begins to show a small (10–20 %) loss of activity at 50° on the same time scale (Morrison & Keir, 1966). Since the infected samples do not fall to the control levels during the heating experiments, the possibility is being considered that more than one DNase is induced by herpesvirus. The heat lability of the induced DNase is shown equally by the enzyme from nuclear and cytoplasmic fractions. A property of the induced DNase which may be related to its marked heat lability is that it must be extracted and handled in the presence of 2-mercaptoethanol at 10 mM to give maximum recoveries and activities. In contrast, the control DNase is much more resistant in the absence of 2-mercaptoethanol.

The condition of the substrate is important in comparing the control and induced enzymes. Both hydrolyse denatured DNA at a greater rate than double-helical DNA, but the induced enzyme degrades double-helical DNA to a much greater extent than does the control DNase. The DNases of control and infected cells are compared in Table 2 with respect to intracellular location and condition of the DNA substrate; the most prominent features are the large infected:control ratios given by the cytoplasmic fractions, and the marked superiority of the induced enzyme over the control enzyme in hydrolysing double-helical DNA.

Table 2. *DNase activity from nuclei and cytoplasm of control and herpes-infected BHK21–C13 cells*

	Control	Infected	Infected/Control
(i) With denatured DNA as substrate			
Cytoplasmic	65	260	4·00
Nuclear	57	81	1·42
Total units	122	341	2·80
(ii) With double-helical DNA as substrate			
Cytoplasmic	6	183	30·50
Nuclear	8	37	4·63
Total units	14	220	15·71
(iii) Ratio of double-helical DNA: denatured DNA			
Cytoplasmic	0·09	0·70	—
Nuclear	0·14	0·46	—
Total	0·11	0·64	—

The values in (i) and (ii) for the control and infected samples are the units of DNase measured in the same amount of total soluble protein extracted. Ratios in (iii) are obtained directly from the values in (i) and (ii). Subcellular fractionation was performed in Ca^{2+}-sucrose medium. The DNA was prepared from Landschütz ascites-tumour cells.

The purified, induced DNase emerges with the void volume during gel filtration through Sephadex G-200; therefore, on this basis alone a minimum molecular weight of 200,000 is indicated. The host-cell DNase is retarded on G-200 and slightly so on G-100. Neither 'acid' DNase (measured at pH 4·5 in 0·2M potassium acetate without added Mg^{2+}) nor RNase (assayed at pH 8) increases in activity after infection. The latter observation implies that the virus-induced DNase has little or no action on RNA. Increases of acid DNase have been reported for HeLa and L-cells infected with herpesvirus (Newton, 1964); the increases were not prevented by inclusion of *p*-fluorophenylalanine in the medium. The reason for this discrepancy between the two cell-virus systems is not certain but it may reflect differential fragilities of lysosomes in the different cell types (cf. Allison & Sandelin, 1963). No post-infection increase of acid DNase can be demonstrated in herpes-infected monkey kidney cells (McAuslan, 1965) or KB cells (Flanagan, 1966).

Resting cells (maintained in medium containing low concentrations of calf serum) have undetectable levels of DNase (and also of DNA polymerase). After infection of these cells the induced DNase (and the induced polymerase) appears, starting at 2 hr. and increasing to 12 hr. post-infection. These observations suggest that transcription and protein synthesis may be obligatory for elaboration of the induced DNase. Interference with transcription by addition of actinomycin D (Fig. 2), or with protein synthesis by including puromycin in the medium at 50 μg./ml., prevents the induction process (Morrison, 1967). In HeLa cells infected with herpesvirus, induction of DNase is observed starting at 8 hr. after infection and increasing by more than 600% at 18 hr.; treatment with actinomycin D 1 hr. after infection, completely and irreversibly suppresses the induction (Sauer *et al.* 1966).

Antisera, prepared against herpes-specific antigens according to the schedule illustrated in Figure 3, strongly inhibit the induced DNase but not the control enzyme (Morrison & Keir, 1967; J. M. Morrison, H. M. Keir & H. Subak-Sharpe, unpublished observations); neutral and pre-immune sera inhibit neither the induced nor the control enzyme. These effects are also observed when the enzymes separated by DEAE-cellulose chromatography are titrated against the antisera.

Control experiments using pseudorabies virus revealed that a DNase is induced after infection and has properties similar to those of the herpes-induced enzyme. However, the pseudorabies-induced DNase is not inhibited by the herpes antiserum under conditions which give virtually complete elimination of the activity of the herpes-induced DNase (Fig. 4).

The mechanism of action of the herpes-induced DNase has been determined. It is a DNA-exonuclease which catalyses release of deoxyribonucleoside 5′-monophosphates sequentially from the 3′-terminus of the DNA. The evidence for this has been derived from two main studies,

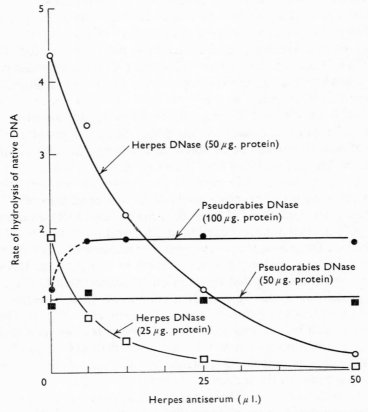

Fig. 4. Effect of herpes-specific antiserum on the DNases induced in BHK 21–C13 cells by herpes simplex virus and pseudorabies virus. DNase activity is expressed as counts per min. × 10^{-3} rendered acid-soluble from [^{32}P]DNA per mg. of protein in 30 min. under standard conditions (Morrison, 1967).

the first of which is based upon the experimental routine outlined in Figure 5. Samples of ^{32}P-labelled DNA are exposed separately to each of the enzymic treatments, chromatographed on strips of DEAE-paper using 0.75 M NH_4HCO_3 as the eluant, and the strips then scanned for ^{32}P. Non-hydrolysed DNA does not move from the origin of the strip. DNA hydrolysed by endonucleolytic attack (bovine pancreatic DNase) moves slightly according to the extent of hydrolysis (i.e. according to the size distribution of the oligodeoxyribonucleotides produced). DNA hydrolysed exonucleolytically (by snake venom 5'-phosphodiesterase) shows three radioactive bands on the chromatogram scan—macro-molecular DNA at the origin, then about half way along the strip a purine deoxyribonucleoside monophosphate band followed by a pyrimidine deoxyribonucleoside monophosphate band. Concurrent treatment of the product of action of venom diesterase with alkaline

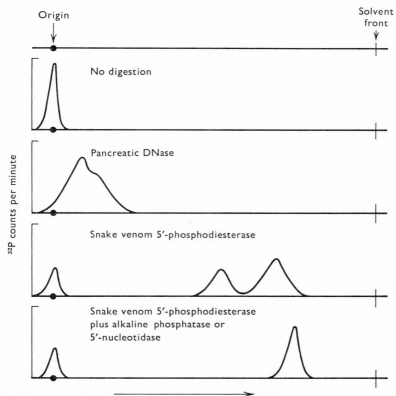

Fig. 5. Chromatography of hydrolysis products of [³²P]DNA after incubation with certain enzymes. Chromatographic system: DEAE-paper strips with 0·75 M NH₄HCO₃ as eluant, pH 8 (see text).

phosphatase or 5'-nucleotidase results in removal of ³²P-orthophosphate from the purine and pyrimidine deoxyribonucleotides, leaving one radio-active band, orthophosphate, which moves nearer the solvent front. The herpes-induced DNase behaves exactly like venom phosphodiesterase in this system, producing purine and pyrimidine deoxyribonucleoside 5'-monophosphates. Since the peak of DNA at the origin remains as sharp as unhydrolysed DNA it is inferred that either (a) the exonuclease attaches to one molecule of substrate and hydrolyses it to completion

6

before starting on a new molecule, or (b) intermediate production of enzyme-free oligodeoxyribonucleotides is not observed because they are preferred substrates and more rapidly hydrolysed than DNA. If this were not the case, a spread of oligodeoxyribonucleotides from the origin would be apparent.

The second series of experiments designed to characterize the mechanism of action of the induced DNase was conducted with a DNA substrate labelled with ^{32}P only at the 3'-end. This was achieved by herpes-induced DNA polymerase, acting under the conditions determined to be optimal for that enzyme, with 'activated' primer-DNA (pretreated briefly with pancreatic DNase and then thermally denatured) but with only one deoxyribonucleoside 5'-triphosphate present, viz. [α-^{32}P]dCTP at high specific activity. Under these circumstances a limited reaction proceeds in which the 3'-ends of the DNA become labelled with one or a few [^{32}P]dCMP residues. When this DNA is isolated and exposed to the exonucleolytic action of the herpes-induced DNase, nearly all of the ^{32}P is released as mononucleotide before substantial amounts in terms of ultraviolet-absorption are released. This indicates that the induced DNase initiates exonucleolytic attack at the 3'-end of the DNA substrate. DNA briefly treated with pancreatic DNase is more rapidly hydrolysed than non-treated DNA (Morrison, 1967); this is presumably due to an increase in the number of 3'-hydroxyterminal groups available for initiation sites for the exonuclease.

The bulk of these experiments on the induced DNase have been performed with HEp-2 cells as well as BHK 21–C 13 cells as hosts. It is particularly interesting that this type of enzyme activity is induced in both cell types by herpesvirus and by the related pseudorabies virus.

With regard to the genetic origin of the induced exonuclease, the considerations presented above for the induced DNA polymerase apply equally to the exonuclease and other induced enzymes. It seems justifiable to conclude that the enzyme is not present in the non-infected control cells. Whether its synthesis is programmed from the viral or host genome is not established, although the weight of evidence appears to favour the viral genome.

The function which the exonuclease might serve in the infective process is quite a different problem. Since the enzyme produces deoxyribonucleoside 5'-monophosphates, the obvious working hypothesis is that it degrades the host DNA to provide the monomers for kinase and polymerase action leading to the synthesis of progeny viral DNA molecules. Evidence that the host cell DNA is degraded following infection with herpesvirus has been provided by Wildy, Smith, Newton

& Dendy (1961) and Newton (1964), by labelling the host DNA with [³H]deoxythymidine for prolonged periods preceding infection; the non-infected control cells retained the incorporated radioactive label while substantial losses were observed from the infected cells. The possibility exists that some of the radio-activity lost from the host DNA might contribute to precursor pools for viral DNA synthesis; this would not be detectable in the experiments of Wildy *et al.* (1961) and Newton (1964) owing to the presence of non-radioactive deoxythymidine at 10^{-5}M in the medium during infection. However, in the absence of the high external concentration of non-radioactive deoxythymidine, it should be possible to select conditions for detection of any incorporation of radioactivity from the labelled host DNA into progeny viral DNA. Exploratory experiments (J. Hay, J. M. Morrison & H. M. Keir, unpublished observations) show that synthesis of host-cell DNA is switched off after infection and suggest that there may be some contribution from it to the precursor pool destined for incorporation into viral DNA. The experimental approach has been to label the host-cell DNA by means of exposure to [¹⁴C]deoxythymidine over an extended period, and then to infect with virus in the presence of [³H]deoxythymidine. Separation of host and viral DNA was achieved with a high degree of resolution by equilibrium centrifugation in CsCl density gradients using an angle rotor (Flamm, Bond & Burr, 1966). The preliminary observations suggest the presence of the ¹⁴C-label in the viral DNA. This result has yet to be examined rigorously both from the qualitative and quantitative points of view before conclusions can be formulated regarding the function of the DNA-exonuclease.

DNA-dependent RNA polymerase (RNA nucleotidyltransferase,
EC 2.7.7.6)

Growth of herpes simplex virus requires transcription of RNA from the viral DNA (Hay *et al.* 1966) and is prevented by actinomycin D (Sauer *et al.* 1966; Sauer & Munk, 1966). The action of DNA-dependent RNA polymerase is therefore directly implicated. In most cases, this enzyme is obtained from mammalian cells strongly bound to chromatin as an 'aggregate enzyme' readily sedimentable and not dependent on addition of exogenous DNA (Weiss, 1962; Smellie, 1963). BHK 21–C 13 cells are no exception to this; all RNA polymerase activity of both control and infected cells is located in the chromatin aggregate (H. Omura & H. M. Keir, unpublished observations), and therefore, considering that virus growth takes place in the nucleus (Morgan *et al.* 1959; Munk & Sauer, 1963) the location, at least, of the enzyme is favourable for transcription of viral DNA. But whether the host polymerase serves to perform this function is not known. Experiments conducted in this laboratory have shown that the polymerase ('aggregate enzyme') activity increased by 20 % at 1 hr. post-infection, by 30 % at 2 hr. and 90 % at 3 hr.; thereafter the activity fell to 75 % of the control level at 6 hr. post-infection. During this period, the expected increments of DNA polymerase and DNase were obtained starting at 2–3 hr. post-infection (Keir & Gold, 1963). One possible explanation for this fluctuation of RNA polymerase activity is that there is rapid stimulation of the enzyme shortly after infection; by 3 hr. the bulk of the required transcription has been completed, and so the enzyme is no longer required and its activity decays. The effect can be explained in other ways however; e.g. since the infective process produces cellular lesions which include accumulation of the chromatin at the periphery of the nucleus (Wildy *et al.* 1961), it may be that the host enzyme is rendered alternately more and less accessible to the assay system over the period 0–6 hr. after infection. Moreover, the standard assay system contains $(NH_4)_2SO_4$ which is designed to give maximal polymerase response, presumably by exposing greater stretches of template in the chromatin to the enzyme (*see* Widnell & Tata, 1966). The stimulatory effect exerted by the salt on chromatin 'aggregates' condensed to varying extents after infection is quite impossible to assess. The uncertainties inherent in the present 'aggregate' polymerase preparations are thus manifest and manifold. Nevertheless, the DNA-dependent RNA polymerase must be accorded a just measure of serious consideration. Among several hypotheses formulated to explain transcription of viral DNA, the three that follow suggest some interesting possibilities.

(i) Parental viral DNA is transcribed by the host-cell polymerase and this initiates the process leading to synthesis of the 'early proteins'. One of these is a virus-coded RNA polymerase which is concerned subsequently with transcription of progeny viral DNA for the purpose of initiating the synthesis of the 'late proteins', e.g. the viral capsid proteins. The host-cell polymerase therefore triggers off the whole process.

(ii) Parental and progeny viral DNA molecules are transcribed by the host-cell polymerase and there is no requirement for synthesis of a virus-coded RNA polymerase. This might furnish an explanation for the termination of transcription of host-cell DNA that occurs after infection (Hay *et al.* 1966) if it is supposed that the invading viral DNA has the power to pre-empt the transcribing machinery of the cell, e.g. through stripping the polymerase from the host DNA by virtue of an unusually strong capacity to bind the enzyme.

(iii) Parental viral DNA is transcribed by a virus-coded polymerase, one molecule (or a few) of which enters the host cell with each infective virus particle. This polymerase molecule, after uncoating of the viral DNA, is responsible for priming the ensuing sequence of virus-induced events, one of which is to transcribe the messenger RNA for programming synthesis of virus-coded RNA polymerase. The latter is then responsible for transcribing messenger RNA from progeny viral DNA for synthesis of the 'late proteins', and is finally itself incorporated into mature progeny virions. A most appealing feature of this hypothesis is that macromolecular specificities are preserved throughout at the enzymic level. However, highly purified preparations of herpesvirus have been checked for DNA-dependent RNA polymerase, and no activity has been detected. Of course the objections propounded earlier for chromatin and 'aggregate enzyme' apply equally to attempts to detect the enzyme in virus particles.

Definitive data are unlikely to be forthcoming unless soluble, DNA-dependent preparations of the polymerase can be obtained reproducibly and quantitatively from the 'aggregate enzyme' of control and infected cells (see Hopper, Ho & Furth, 1966). The nature of the involvement of DNA-dependent RNA polymerase in the herpesvirus system therefore remains a mystery.

dCMP deaminase (deoxy-CMP aminohydrolase, EC 3.5.4.b)

dCMP deaminase catalyses one of the reactions concerned in the de novo pathway for production of dTTP which is used almost exclusively for DNA synthesis (Fig. 1). The activity of the deaminase is generally low in non-dividing tissues but higher in dividing cells (Maley & Maley, 1960) and therefore the concept of an active role for the enzyme in DNA synthesis is supported.

Preliminary studies conducted with BHK 21–C 13 cells infected with herpes simplex virus have shown that the dCMP deaminase activity increases rapidly to a maximum in the cells at 6–8 hr. post-infection. A dramatic decline of activity ensues until the control cell level is reached at 10–12 hr. post-infection (D. J. McGeoch & H. M. Keir, unpublished observations). The effect is given by infection of either growing or resting cells, although the absolute values of deaminase activity are, of course, different for each system.

Some properties of the deaminase have been examined. The intra-cellular distribution is the same for control and infected cells. In agreement with the observations of Maley & Maley (1962) and Scarano, Geraci & Rossi (1967), the enzyme requires the presence of dCTP for full activity; all fractions tested (control and infected) gave maximal activation at 5 to 7×10^{-5} M dCTP. Heat-inactivation studies revealed differences between the deaminase fractions from control and infected cells. In one of the experiments, the enzyme from cells infected for 6 hr. lost only 20 % of its activity after heating for 90 min. at 55°, while control-cell preparations lost about 60 % of initial activity. This effect appears to be dependent on several factors including the presence of a suitable concentration of 2-mercaptoethanol in the preparative buffers and assay medium. The relationship between the heat-stability of the deaminase and the rapid decline of its activity beyond 6 hr. post-infection is not clear, but it could be explained by rapid loss of the enzyme (perhaps by leakage) from the infected cells.

The deaminase reaction is subject to complex allosteric controls (Maley & Maley, 1962; Monod, Wyman & Changeux, 1965; Scarano et al. 1967), and therefore kinetic data obtained from experiments involving a relatively crude enzyme preparation are open to question. At this stage, it is clearly impossible to express an opinion on the origin and the function of the deaminase activity induced by herpes-infection, but any view of either problem must be considered in the light of the observation that the activities of dTMP synthetase and tetrahydrofolate dehydrogenase do not rise after infection of several mammalian cell lines (Frearson, Kit & Dubbs, 1965, 1966).

Deoxythymidine kinase (EC 2.7.1.21) and dTMP kinase (EC 2.7.4.c)

The activity of deoxythymidine kinase increases in mammalian cells infected with herpes simplex virus (Kit & Dubbs, 1963a; Newton, 1964; Borman & Roizman, 1965; Prusoff, Bakhle & Sekely, 1965; Klemperer, Haynes, Shedden & Watson, 1967).

The induction effect has been extensively studied in BHK 21–C 13 cells infected with the HFEM strain of herpesvirus (Klemperer et al. 1967) and will be described in some detail here. The virus-induced increase commences at 2–3 hr. post-infection and rises to a final maximal level 20-fold higher than the non-infected controls at 8 hr. post-infection. A switch-off of enzyme synthesis appears to be quite clearly established at this time-point. The induced enzyme has a sharp pH optimum at 5·8, whereas the host-cell kinase has a more diffuse optimum in the region of pH 8·0.

The Michaelis constant (measured at pH 6 or pH 8) for the reaction catalysed by the induced enzyme is about 4×10^{-7} M with respect to deoxythymidine, while the constant for the control-cell enzyme is about $2·5 \times 10^{-6}$ M (measured at pH 8).

Heat-inactivation studies of the virus-induced kinase show that it is markedly more stable during heating at 40° than is the host-cell enzyme. Moreover, it is possible to demonstrate that this relative heat-stability is conferred upon the enzyme in soluble cell extracts largely by endogenous deoxythymidine derivatives. Removal of these small-molecular components by gel filtration results in loss of the resistance to heating, but supplementation of the gel-filtered extracts with deoxythymidine at 2 μM and dTTP at 5 μM promotes retention of the heat-stable condition. Supplementation of the host-cell preparations with these concentrations of the thymine derivatives does not significantly alter their high heat-lability, although, for another host-cell system (Kit & Dubbs, 1965) it has been observed that the host enzyme can be stabilized by higher concentrations of deoxythymidine (25 μM) and dTTP (50–100 μM). At these concentrations only a small difference in heat-stability can be demonstrated between the control and induced enzymes.

The deoxythymidine kinase of BHK 21–C 13 cells is strongly inhibited (77–86 %) by inclusion of dTTP at 50 μM in the assay medium, but the activity of the virus-induced enzyme is reduced by only 11–16 % under the same conditions.

Immunological experiments conducted essentially as illustrated in Fig. 3 demonstrate that the virus-induced enzyme is inhibited extensively by the herpes antiserum but only slightly by the pre-immune serum (and then at high serum concentrations only). The control-cell enzyme is not inhibited by either serum.

A strain of mouse fibroblast cells lacking in deoxythymidine kinase acquires the activity after infection with herpes simplex virus (Kit & Dubbs, 1963a, b; Dubbs & Kit, 1964a, 1965). The induction can be prevented by actinomycin D and by puromycin; therefore, transcription from DNA and synthesis of new protein appear to be necessary for elaboration of the enzyme. A number of mutants of herpesvirus deficient in induction of the kinase have been isolated. Although certain of them cannot induce synthesis of the kinase in the kinase-less host cells, virus multiplication is not impaired. When the normal host cells (containing the kinase) are infected with kinase-less herpes mutants, there occurs a reduction of the enzyme activity in the infected cells. When deoxythymidine kinase-less host cells are mixedly infected with the parental kinase-inducing virus and a kinase-less virus mutant, it is found that the kinase-less virus mutant has the power to inhibit the kinase induction that is normally promoted by the kinase-inducing strain (Munyon & Kit, 1965). Further the enzyme induction effect given by the kinase-inducing virus is inhibited when the mixed infection is carried out with a kinase-less mutant of vaccinia virus. Munyon & Kit (1965) have proposed several mechanisms that might account for this interesting phenomenon.

These observations, when considered in conjunction with the immunological behaviour of the virus-induced enzyme and its various properties, certainly lend very strong support to the view that the information for the amino acid sequence of a deoxythymidine kinase is coded in the nucleotide sequence of the viral DNA.

The metabolic significance of deoxythymidine kinase in general, and in the cell-virus system in particular, is quite obscure. The kinase-deficient host cells appear to function quite satisfactorily without the enzyme. Further, infection of these cells with the kinase-less herpes mutants appears to permit virus production that is essentially unimpaired. Presumably the retention by normal host and virus of the capacity to elaborate the kinase has conferred some selective advantage upon both, perhaps relating to feedback controls exercised by dTTP. Negative feedback control by dTTP is known to be exerted upon the ribonucleotide reductase reaction (Reichard, Canellakis & Canellakis, 1960; Morris & Fischer, 1960), upon dCMP deaminase (Maley & Maley, 1962; Scarano *et al.* 1967) and upon the deoxythymidine kinase reaction itself (*see* Davidson, 1965). Since the deoxythymidine kinase induced by herpes infection is resistant to feedback inhibition by dTTP (Klemperer *et al.* 1967), differential effects are implied which, at least in that cell-virus system, would seem to favour the operation of the virus-induced enzyme during the infective process. The intracellular pool of dTTP is known to be increased after infection (Newton, 1964), therefore feedback inhibition exerted on ribonucleotide reductase and dCMP deaminase presumably impedes the flow of *de novo* precursors to the DNA polymerase reaction. Also, it has been suggested earlier that host DNA may supply some precursor material for synthesis of viral DNA. Therefore, in circumstances in which, from the point of view of the virus, a high concentration of dTTP is probably desirable, the induced deoxythymidine kinase continues to supply dTMP at a high rate through this 'scavenger pathway'.

The activity of dTMP kinase increases after infection of L cells with herpesvirus (Newton, 1964). There is no comparative information on the properties of this kinase before and after infection, but the time sequence of induction is interesting. The increase commences about 2 hr. later than that of deoxythymidine kinase and somewhat earlier than the over-all DNA synthesizing reaction; the increase of the latter corresponds in time to the onset of DNA synthesis in the infected L-cells.

It has been reported that infection of monkey kidney cells with herpes-virus does not give a detectable difference in the level of dTMP kinase (Prusoff *et al.* 1965).

Other enzymes

Infection by herpes simplex virus of a strain of BHK 21 cells lacking in IMP pyrophosphorylase failed to give any indication of induction of the enzyme (Subak-Sharpe, 1965). It is therefore inferred that the herpes genome does not carry information for synthesis of this enzyme.

Flanagan (1966) was unable to detect significant alterations of the activities of β-glucuronidase, acid protease, acid RNase and acid phosphatase after infection of KB cells with herpes simplex virus.

Synthesis of DNA and RNA proceeds by polymerization of the monomer units (nucleoside monophosphate residues) derived from the proximal precursors (nucleoside triphosphates), and for each monomer incorporated into polynucleotide one molecule of inorganic pyrophosphate is eliminated. Although the rate of synthesis of RNA decreases after infection of cells with herpesvirus (Hay *et al.* 1966), large amounts of DNA are made within a relatively short time and most of it is viral DNA (Russell *et al.* 1964). It might therefore be supposed that large amounts of inorganic pyrophosphate accumulate in the infected cells, and since the DNA polymerase reaction is reversible (*see* Keir, 1965), the pyrophosphate could inhibit synthesis. However, the equilibrium could continue in favour of synthesis if a mechanism were available to eliminate the pyrophosphate. Preparations from control and herpes-infected cells were accordingly assayed for inorganic pyrophosphatase (Morrison, 1967). The level of the enzyme was very high and essentially the same in both cases. The data allow the conclusion that the pyrophosphatase activity is already sufficiently high in control cells to permit synthesis of DNA to proceed unimpeded after infection, but do not exclude the possibility that synthesis of a virus-coded pyrophosphatase concurrent with decay of the host-coded pyrophosphatase maintains a constant level of the enzyme after infection.

The genome of herpes simplex virus

It has been estimated that there are 75×10^6 molecular weight units of DNA in the herpes virion (Russell & Crawford, 1963), enough to code for about 100 proteins of molecular weight 50,000. There are indications that information for the synthesis of about five enzymes resides in the viral genome, and the virus must also code for synthesis of its own structural proteins. Since the number of different proteins in the herpes capsid is not yet known, it is impossible reliably to extend the calculation further, but if it is assumed that the number of different structural proteins is small and that only one gene codes the synthesis of each protein, then it can be concluded that the bulk of the coding potential of the DNA of herpesvirus has yet to be defined. If the number of different structural proteins is large, the bulk of the coding potential of the viral DNA may be described in terms of structural proteins plus a few key enzymes of DNA metabolism. Transcription of herpes-specific tRNA from viral DNA must, of course, be taken into account (Subak-Sharpe, Shepherd & Hay, 1966). Furthermore, it is recognized that these speculations have not taken into consideration the possibility that some of the herpes DNA, apart from that portion responsible for specifying tRNA: (i) may never be transcribed, (ii) may be transcribed but not translated, and (iii) may specify synthesis of non-enzymic, non-structural proteins, the existence of which might be difficult to demonstrate.

ENZYME INDUCTION IN CELLS INFECTED WITH PSEUDORABIES VIRUS

Rabbit kidney cells (RK) maintained at the stationary phase of growth and infected with pseudorabies virus have provided a cell-virus system accessible to analysis for virus-induced enzymes. In these resting cells, the activities of deoxythymidine kinase, dTMP kinase and DNA polymerase are very low, but they increase substantially following infection (Nohara & Kaplan, 1963; Kaplan, Ben-Porat & Kamiya, 1965; Hamada, Kamiya & Kaplan, 1966). The activities of the kinases for dAMP, dCMP and dGMP are already high in the host cells and do not undergo any quantitative alteration after infection. A DNA-exonuclease has been identified in cells infected with pseudorabies virus (J. M. Morrison, H. M. Keir & H. Subak-Sharpe, unpublished observations).

By means of an antiserum containing antibodies against 'early antigens' induced by pseudorabies infection but not against viral structural proteins (Hamada & Kaplan, 1965), it has been shown that the virus-induced deoxythymidine kinase is not immunologically related to the corresponding host-cell enzyme. On the other hand, the DNA polymerase and dAMP kinase of infected cells appear to be immunologically related to the host-cell enzymes. This observation with the polymerase merits particular attention in view of the finding (Keir, Subak-Sharpe, Shedden, Watson & Wildy, 1966) that the DNA polymerase induced in cells infected with herpes simplex virus (a virus related to pseudorabies) is apparently distinct in several properties, including inhibition by a herpes antiserum which has no inhibitory effect on the control cell polymerase.

ENZYME INDUCTION IN CELLS INFECTED WITH POXVIRUSES

The poxviruses have recently been reviewed in detail by Joklik (1966), and it would therefore be inappropriate to present here more than a brief survey of what is known.

The DNA of the poxviruses presents an even greater genetic and biochemical challenge than the herpesviruses, since the molecular weight (16×10^7) suggests a much higher coding potential (some 110 to 220 proteins). They differ from the other DNA viruses in multiplying in the cytoplasm (Cairns, 1960) where, initially, the virion is uncoated in a two-phase operation (Joklik, 1964 a, b), apparently involving the action of host-coded enzymes, but the precise nature of the enzymes has not been established.

Induction of the 'early enzymes' ensues. Three of these have been moderately well characterized, DNA polymerase (Jungwirth & Joklik, 1965), deoxythymidine kinase (Dubbs & Kit, 1964b; Kit & Dubbs, 1965; Munyon & Kit, 1965) and several DNases (Eron & McAuslan, 1966). Magee & Miller (1967) prepared from rabbits immunized with partially purified host-cell polymerase, a γ-globulin fraction that effectively inactivated the host-cell polymerase but not the virus-induced increments of the enzyme.

While the data relating to the poxvirus-induced DNases do not allow a decision to be taken on their genetic origins (viral or host), the evidence is very strong indeed that the induced polymerase and deoxythymidine kinase are virus-coded.

The position occupied by DNA polymerase in the over-all mechanism of replication of poxviruses is quite different from that for the other mammalian DNA viruses since viral replication occurs exclusively in the cytoplasm. Despite the fact that much of the DNA polymerase from non-infected mammalian cells is found in the cytoplasm or soluble fraction after cell disruption, there are indications that the enzyme may be located in the nucleus during DNA synthesis in the living cell (Keir, 1965). Therefore there would seem to be an absolute requirement for production of a cytoplasmic DNA polymerase in response to poxvirus infection; this need seems to be met in the vaccinia system (Jungwirth & Joklik, 1965; Magee & Miller, 1967).

A more intriguing problem is set by the requirement for DNA-dependent RNA polymerase which is exclusively a nuclear enzyme, with the exception of the DNA-dependent RNA synthesis that occurs in mitochondria (Luck & Reich, 1964; Wintersberger, 1964). The prior requirement for transcription from viral DNA in vaccinia-infected cells (Salzman *et al*. 1964; Becker & Joklik, 1964) and the cytoplasmic sites of vaccinia replication, suggest either that the host RNA polymerase migrates from the nucleus shortly after infection, or that another mechanism, possibly based on one of those proposed above for herpesvirus, provides RNA polymerase activity in the cytoplasm for transcription from viral DNA. Munyon & Kit (1966) have examined this problem, and have concluded that induction of synthesis of cytoplasmic RNA by vaccinia infection is not prevented in circumstances in which protein synthesis and deoxythymidine kinase induction are inhibited. It is inferred that cytoplasmic RNA synthesis is catalysed by RNA polymerase which is already present in the host, or has been brought in with the virion.

ENZYME INDUCTION IN CELLS INFECTED WITH VIRUSES OF THE POLYOMA GROUP

The DNAs of this virus group are relatively small, with molecular weights of 3×10^6, enough to code for about four to eight proteins. Thus, complete characterization of the coding potential of these small, oncogenic viruses seems to be a feasible proposition.

In terms of enzyme induction, most is known about polyoma virus and Simian virus 40 (Dulbecco, Hartwell & Vogt, 1965; Gershon, Hausen, Sachs & Winocour, 1965; Hartwell, Vogt & Dulbecco, 1965; Carp, Kit & Melnick, 1966; Frearson *et al*. 1966; Hatanaka & Dulbecco, 1966; Kit, Dubbs & Frearson, 1966; Kit, Dubbs, Frearson & Melnick, 1966; Kit, Dubbs, Piekarski, de Torres & Melnick, 1966; Sheinin, 1966; Vogt, Dulbecco & Smith, 1966; Kára & Weil, 1967; Smart, Fried & Pitts, 1967). The activities of DNA polymerase, deoxythymidine kinase, dTMP synthetase, tetrahydrofolate dehydrogenase, dCMP deaminase and dTMP kinase are clearly elevated after infection. Moreover, infection conducted under the appropriate resting condition of the host cells causes induction of synthesis of cellular DNA.

There really is insufficient evidence to assess the genetic origin of the induced enzymes, although, for a variety of reasons, the host genome seems to be favoured in most cases.

ENZYME INDUCTION IN CELLS INFECTED
WITH ADENOVIRUSES

The members of the adenovirus group show a wide range of DNA base compositions, but all appear to have molecular weights of DNA about 21 to 23×10^6 (Green, 1966). The coding potential of this group is therefore probably in the range 15 to 30 proteins, between that of the herpesviruses and poxviruses on the one hand, and polyoma virus on the other.

Green, Piña & Chagoya (1964) demonstrated that infection of KB cells with adenovirus type 2 gave no significant alteration in the activities of DNA polymerase, deoxythymidine kinase, and dAMP, dCMP, dGMP and dTMP kinases. However, this may be a function of the host cell type used to support viral replication. Using monkey kidney cells in primary culture, infected with adenovirus type 2, Ledinko (1966) detected increases in the activities of deoxythymidine kinase, dCMP deaminase, and aspartate carbamoyltransferase. No alteration was detected for DNase, RNase or DNA polymerase. The activity of DNA dependent RNA polymerase declined after infection. Kit, Dubbs, de Torres & Melnick (1965) found increased levels of deoxythymidine kinase in monkey kidney cells after infection by certain adenoviruses. The increase of aspartate carbamoyltransferase observed in HeLa cells infected with adenovirus type 5 may be attributable to activation of the host cell enzyme as a consequence of response to fluctuation of the intracellular concentrations of allosteric effector molecules (Consigli & Ginsberg, 1964a, b).

CONCLUDING REMARKS

It seems justifiable to conclude that an experimental approach to characterization of the coding potential of viral nucleic acid based upon identification of virus-induced enzymes is capable of providing relevant information. The demonstration that several new enzymes are synthesized in cells infected with members of the herpesvirus and poxvirus groups, and that they are distinct from the corresponding activities of the host cells, lends very strong support to the hypothesis that the enzymes are directly coded in the viral genome. Extensive enzyme induction also occurs in cells infected with polyoma virus and SV40, but the evidence that the information for the primary structure of the enzymes resides in the nucleotide sequence of the viral DNA is, in nearly all cases, non-existent or, at best, very tenuous. By inference, it would seem reasonable to suppose that most of the induced enzymes are host-coded; in any event, only one or two enzymes (if any) could theoretically be coded in the small amount of viral DNA available. The problem is quite the reverse with the herpesviruses and poxviruses whose DNAs accommodate some 100 to 200 genes (or even more). Less is known about the adenoviruses whose DNA is of an intermediate size.

It is clear that proteins other than enzymes must additionally be coded by viral genes. These certainly include the viral capsid proteins and possibly virus-specific antigens, e.g. the transplantation and tumour antigens elaborated by polyoma and SV40 viruses. It is conceivable that proteins with a derepressor action may also be coded in the viral DNA. Thus, one virus-specified derepressor protein could account for induction of the block of enzymes (host-coded) of DNA synthesis in cells infected with polyoma virus.

All such data contribute to defining the coding potential of viral nucleic acid, but additionally, the hypothesis should be entertained that not all of the viral DNA need specify synthesis of polypeptide. There is at least one case known in which viral genes function to produce virus-specific tRNA. Further, it is not established that all viral DNA must be transcribed.

The over-all fundamental and clinical implications are clear. Definitive information on the precise function of viral genes is a prerequisite for full understanding of the mechanism of viral replication.

ACKNOWLEDGEMENTS

I wish to thank Professor J. N. Davidson, F.R.S., Professor R. M. S. Smellie and Professor M. G. P. Stoker for their interest and support. The work with herpesvirus is part of a collaborative project and I am grateful in this respect to my colleagues for their support.

REFERENCES

ALLISON, A. C. & SANDELIN, K. (1963). Activation of lysosomal enzymes in virus-infected cells and its possible relationship to cytopathic effects. *J. exp. Med.* **117**, 879.

BECKER, Y. & JOKLIK, W. K. (1964). Messenger RNA in cells infected with vaccinia virus. *Proc. natn. Acad. Sci. U.S.A.* **51**, 577.

BENJAMIN, T. L. (1966). Virus-specific RNA in cells productively infected or transformed by polyoma virus. *J. molec. Biol.* **16**, 359.

BOLLUM, F. J. (1963). Intermediate states in enzymatic DNA synthesis. *J. cell. comp. Physiol.* **62**, Suppl. 1, p. 61.

BORMAN, G. S. & ROIZMAN, B. (1965). The inhibition of herpes simplex virus multiplication by nucleosides. *Biochim. biophys. Acta*, **103**, 50.

BURDON, R. H., MARTIN, B. T. & LAL, B. M. (1967). Molecular processes involved in the maturation of mammalian cell transfer ribonucleic acid. *Biochem. J.* **103** 69 P.

CAIRNS, J. (1960). The initiation of vaccinia infection. *Virology*, **11**, 603.

CARP, R. I., KIT, S. & MELNICK, J. L. (1966). The effect of ultraviolet light on the infectivity and enzyme-inducing capacity of papovavirus SV40. *Virology*, **29**, 503.

CONSIGLI, R. A. & GINSBERG, H. S. (1964*a*). Control of aspartate transcarbamylase activity in type 5 adenovirus-infected HeLa cells. *J. Bact.* **87**, 1027.

CONSIGLI, R. A. & GINSBERG, H. S. (1964*b*). Activity of aspartate transcarbamylase in uninfected and type 5 adenovirus-infected HeLa cells. *J. Bact.* **87**, 1034.

DAVIDSON, J. N. (1965). *The Biochemistry of the Nucleic Acids*, 5th edition, p. 191. London: Methuen and Co. Ltd.

DUBBS, D. R. & KIT, S. (1964*a*). Mutant strains of herpes simplex deficient in thymidine kinase-inducing activity. *Virology*, **22**, 493.

DUBBS, D. R. & KIT, S. (1964*b*). Isolation and properties of vaccinia mutants deficient in thymidine kinase inducing activity. *Virology*, **22**, 214.

DUBBS, D. R. & KIT, S. (1965). The effect of temperature on induction of deoxythymidine kinase activity by herpes simplex mutants. *Virology*, **25**, 256.

DULBECCO, R., HARTWELL, L. H. & VOGT, M. (1965). Induction of cellular DNA synthesis by polyoma virus. *Proc. natn. Acad. Sci. U.S.A.* **53**, 403.

ERON, L. J. & MCAUSLAN, B. R. (1966). The nature of poxvirus-induced deoxyribonucleases. *Biochem. biophys. Res. Commun.* **22**, 518.

FLAMM, W. G., BOND, H. E. & BURR, H. E. (1966). Density gradient centrifugation of DNA in a fixed-angle rotor. A higher order of resolution. *Biochim. biophys. Acta*, **129**, 310.

FLANAGAN, J. F. (1966). Hydrolytic enzymes in KB cells infected with poliovirus and herpes simplex virus. *J. Bact.* **91**, 789.

FREARSON, P. M., KIT, S. & DUBBS, D. R. (1965). Deoxythymidylate synthetase and deoxythymidine kinase activities of virus-infected animal cells. *Cancer Res.* **25**, 737.

FREARSON, P. M., KIT, S. & DUBBS, D. R. (1966). Induction of dihydrofolate reductase activity by SV40 and polyoma virus. *Cancer Res.* **26**, 1653.

GERSHON, D., HAUSEN, P., SACHS, L. & WINOCOUR, E. (1965). On the mechanism of polyoma virus-induced synthesis of cellular DNA. *Proc. natn. Acad. Sci. U.S.A.* **54**, 1584.

GREEN, M. (1966). Biosynthetic modifications induced by DNA animal viruses. *A. Rev. Microbiol.* **20**, 189.

GREEN, M., PIÑA, M. & CHAGOYA, V. (1964). Biochemical studies on adenovirus multiplication. V. Enzymes of deoxyribonucleic acid synthesis in cells infected by adenovirus and vaccinia virus. *J. biol. Chem.* **239**, 1188.

HAMADA, C., KAMIYA, T. & KAPLAN, A. S. (1966). Serological analysis of some enzymes present in pseudorabies virus-infected and noninfected cells. *Virology*, **28**, 271.

HAMADA, C. & KAPLAN, A. S. (1965). Kinetics of synthesis of various types of antigenic proteins in cells infected with pseudorabies virus. *J. Bact.* **89**, 1328.

HARTWELL, L. H., VOGT, M. & DULBECCO, R. (1965). Induction of cellular DNA synthesis by polyoma virus. II. Increase in the rate of enzyme synthesis after infection with polyoma virus in mouse kidney cells. *Virology*, **27**, 262.

HATANAKA, M. & DULBECCO, R. (1966). Induction of DNA synthesis by SV40. *Proc. natn. Acad. Sci. U.S.A.* **56**, 736.

HAY, J., KÖTELES, G. J., KEIR, H. M. & SUBAK-SHARPE, H. (1966). Herpes virus specified ribonucleic acids. *Nature, Lond.* **210**, 387.

HOPPER, D. K., HO, P. L. & FURTH, J. J. (1966). Detection of RNA and DNA polymerase in animal tissues by ammonium sulphate fractionation of cell-free extracts. *Biochim. biophys. Acta*, **118**, 648.

JOKLIK, W. K. (1964a). The intracellular uncoating of pox virus DNA. I. The fate of radioactively labelled rabbitpox virus. *J. molec. Biol.* **8**, 263.

JOKLIK, W. K. (1964b). The intracellular uncoating of poxvirus DNA. II. The molecular basis of the uncoating process. *J. molec. Biol.* **8**, 277.

JOKLIK, W. K. (1966). The poxviruses. *Bact. Rev.* **30**, 33.

JUNGWIRTH, C. & JOKLIK, W. K. (1965). Studies on early enzymes in HeLa cells infected with vaccinia virus. *Virology*, **27**, 80.

KAPLAN, A. S., BEN-PORAT, T. & KAMIYA, T. (1965). Incorporation of 5-bromo-deoxyuridine and 5-iododeoxyuridine into viral DNA and its effect on the infective process. *Ann. N.Y. Acad. Sci.* **130**, art. 1, 226.

KÁRA, J. & WEIL, R. (1967). Specific activation of the DNA-synthesizing apparatus in contact-inhibited mouse kidney cells by polyoma virus. *Proc. natn. Acad. Sci. U.S.A.* **57**, 63.

KEIR, H. M. (1965). DNA polymerases from mammalian cells. *Prog. Nucleic Acid Res. molec. Biol.* **4**, 81. Ed. J. N. Davidson and W. E. Cohn. New York: Academic Press.

KEIR, H. M. & GOLD, E. (1963). Deoxyribonucleic acid nucleotidyltransferase and deoxyribonuclease from culture and cells infected with herpes simplex virus. *Biochim. biophys. Acta*, **72**, 263.

KEIR, H. M., HAY, J., MORRISON, J. M. & SUBAK-SHARPE, H. (1966). Altered properties of deoxyribonucleic acid nucleotidyltransferase after infection of mammalian cells with herpes simplex virus. *Nature, Lond.* **210**, 369.

KEIR, H. M., SUBAK-SHARPE, H., SHEDDEN, W. I. H., WATSON, D. H. & WILDY, P. (1966). Immunological evidence for a specific DNA polymerase produced after infection by herpes simplex virus. *Virology*, **30**, 154.

KIT, S. & DUBBS, D. R. (1963a). Acquisition of thymidine kinase activity by herpes simplex infected mouse fibroblast cells. *Biochem. biophys. Res. Commun.* **11**, 55.

KIT, S. & DUBBS, D. R. (1963b). Non-functional thymidine kinase cistron in bromo-deoxyuridine resistant strains of herpes simplex virus. *Biochem. biophys. Res. Commun.* **13**, 500.

KIT, S. & DUBBS, D. R. (1965). Properties of deoxythymidine kinase partially purified from non-infected and virus-infected mouse fibroblast cells. *Virology*, **26**, 16.

KIT, S., DUBBS, D. R. & FREARSON, P. M. (1966). Enzymes of nucleic acid metabolism in cells infected with polyoma virus. *Cancer Res.* **26**, 638.

KIT, S., DUBBS, D. R., FREARSON, P. M. & MELNICK, J. L. (1966). Enzyme induction in SV40-infected green monkey kidney cultures. *Virology*, **29**, 69.

KIT, S., DUBBS, D. R., PIEKARSKI, L. J., DE TORRES, R. A. & MELNICK, J. L. (1966). Acquisition of enzyme function by mouse kidney cells abortively infected with papovavirus SV40. *Proc. natn. Acad. Sci. U.S.A.* **56**, 463.

KIT, S., DUBBS, D. R., DE TORRES, R. A. & MELNICK, J. L. (1965). Enhanced thymidine kinase activity following infection of green monkey kidney cells by Simian adenoviruses, Simian papovavirus SV40, and an adenovirus–SV40 hybrid. *Virology*, **27**, 453.

KLEMPERER, H. G., HAYNES, G. R., SHEDDEN, W. I. H. & WATSON, D. H. (1967). A virus-specific thymidine kinase in BHK21 cells infected with herpes simplex virus. *Virology*, **31**, 120.

LAL, B. M. & BURDON, R. H. (1967). Maturation of low molecular weight RNA in tumour cells. *Nature, Lond.* **213**, 1134.

LEDINKO, N. (1966). Changes in metabolic and enzymatic activities of monkey kidney cells after infection with adenovirus 2. *Virology*, **28**, 679.

LUCK, D. J. L. & REICH, E. R. (1964). DNA in mitochondria of *Neurospora crassa*. *Proc. natn. Acad. Sci. U.S.A.* **52**, 931.

MCAUSLAN, B. R. (1965). Deoxyribonuclease activity of normal and poxvirus-infected HeLa cells. *Biochem. biophys. Res. Commun.* **19**, 15.

MCAUSLAN, B. R., HERDE, P., PETT, D. & ROSS, J. (1965). Nucleases of virus-infected animal cells. *Biochem. biophys. Res. Commun.* **20**, 586.

MACPHERSON, I. A. & STOKER, M. G. P. (1962). Polyoma transformation of hamster cell clones—an investigation of genetic factors affecting cell competence. *Virology*, **16**, 147.

MAGEE, W. E. & MILLER, O. V. (1967). Immunological evidence for the appearance of a new DNA polymerase in cells infected with vaccinia virus. *Virology*, **31**, 64.

MALEY, F. & MALEY, G. F. (1960). Nucleotide interconversions. II. Elevation of deoxycytidylate deaminase and thymidylate synthetase in regenerating rat liver. *J. biol. Chem.* **235**, 2968.

MALEY, G. F. & MALEY, F. (1962). Nucleotide interconversions. IX. The regulatory influence of deoxycytidine 5'-triphosphate and deoxythymidine 5'-triphosphate on deoxycytidylate deaminase. *J. biol. Chem.* **237**, PC 3311.

MITRA, S. & KORNBERG, A. (1966). Enzymatic repair mechanisms. In *Macromolecular Metabolism*, p. 59. Proceedings of a Symposium sponsored by the New York Heart Association. London: J. and A. Churchill Ltd.

MONOD, J., WYMAN, J. & CHANGEUX, J.-P. (1965). On the nature of allosteric transitions. *J. molec. Biol.* **12**, 88.

MORGAN, C., ELLISON, S. A., ROSE, H. M. & MOORE, D. H. (1954). Structure and development of viruses in the electron microscope. I. Herpes simplex virus. *J. exp. Med.* **100**, 195.

MORGAN, C., ROSE, H. M., HOLDEN, M. & JONES, E. P. (1959). Electron microscopic observations on the development of herpes simplex virus. *J. exp. Med.* **110**, 643.

MORRIS, N. R. & FISCHER, G. A. (1960). Studies concerning inhibition of the synthesis of deoxycytidine by phosphorylated derivatives of thymidine. *Biochim. biophys. Acta*, **42**, 183.

MORRISON, J. M. (1967). Studies on mammalian viral DNA and its metabolism with special reference to herpes virus. Ph.D. Thesis, University of Glasgow.

MORRISON, J. M. & KEIR, H. M. (1966). Heat-sensitive deoxyribonuclease activity in cells infected with herpes simplex virus. *Biochem. J.* **98**, 37 C.

MORRISON, J. M. & KEIR, H. M. (1967). Characterization of the deoxyribonuclease activity induced by infection with herpes simplex virus. *Biochem. J.* **103**, 70 P.

MUNK, K. & SAUER, G. (1963). Autoradiographische untersuchungen über das verhalten der desoxyribonucleinsaure in herpes-virus-infizierten zellen. *Z. Naturf.* **18** B, 211.

MUNYON, W. & KIT, S. (1965). Inhibition of thymidine kinase formation in LM (TK⁻) cells simultaneously infected with vaccinia and a thymidine kinaseless vaccinia mutant. *Virology*, **26**, 374.

MUNYON, W. H. & KIT, S. (1966). Induction of cytoplasmic ribonucleic acid (RNA) synthesis in vaccinia-infected LM cells during inhibition of protein synthesis. *Virology*, **29**, 303.

NEWTON, A. A. (1964). Synthesis of DNA in cells infected by virulent DNA viruses. In *Acidi nucleici e Loro Funzione Biologica*, p. 109. Istituto Lombardo Accademia di Scienze e Lettere; Convegno Antonio Baselli.

NOHARA, H. & KAPLAN, A. S. (1963). Induction of a new enzyme in rabbit kidney cells by pseudorabies virus. *Biochem. biophys. Res. Commun.* **12**, 189.

PRUSOFF, W. H., BAKHLE, Y. S. & SEKELY, L. (1965). Cellular and antiviral effects of halogenated deoxyribonucleosides. *Ann. N.Y. Acad. Sci.* **130**, Art. 1, 135.

REICHARD, P., CANELLAKIS, Z. N. & CANELLAKIS, E. S. (1960). Regulatory mechanisms in the synthesis of deoxyribonucleic acid *in vitro*. *Biochim. biophys. Acta*, **41**, 558.

ROSE, J. A., REICH, P. R. & WEISSMAN, S. M. (1965). RNA production in adeno-virus-infected KB cells. *Virology*, **27**, 571.

RUSSELL, W. C. & CRAWFORD, L. V. (1963). Some characteristics of the deoxy-ribonucleic acid from herpes simplex virus. *Virology*, **21**, 353.

RUSSELL, W. C., GOLD, E., KEIR, H. M., OMURA, H., WATSON, D. H. & WILDY, P. (1964). The growth of herpes simplex virus and its nucleic acid. *Virology*, **22**, 103.

SALZMAN, N. P., SHATKIN, A. J. & SEBRING, E. D. (1964). The synthesis of a DNA-like RNA in the cytoplasm of HeLa cells infected with vaccinia virus. *J. molec. Biol.* **8**, 405.

SAUER, G. & MUNK, K. (1966). Interference of actinomycin D with the replication of the DNA of herpes virus. II. Relationship between yield of virus and time of actinomycin treatment. *Biochim. biophys. Acta*, **119**, 341.

SAUER, G., ORTH, H. D. & MUNK, K. (1966). Interference of actinomycin D with the replication of herpes virus DNA. I. Difference in behaviour of cellular and viral nucleic acid synthesis following treatment with actinomycin D. *Biochim. biophys. Acta*, **119**, 331.

SCARANO, E., GERACI, G. & ROSSI, M. (1967). Deoxycytidylate aminohydrolase. II. Kinetic properties. The activatory effect of deoxycytidine triphosphate and the inhibitory effect of deoxythymidine triphosphate. *Biochemistry*, **6**, 192.

SHEININ, R. (1966). Studies on the thymidine kinase activity of mouse embryo cells infected with polyoma virus. *Virology*, **28**, 47.

SMART, M. E., FRIED, A. M. C. & PITTS, J. D. (1967). Polyoma virus-induced DNA polymerase activity in resting mouse embryo cells. *Biochem. J.* **104**, 29 P.

SMELLIE, R. M. S. (1963). The biosynthesis of ribonucleic acid in animal systems. *Prog. Nucleic Acid Res.* **1**, 27. Ed. J. N. Davidson and W. E. Cohn. New York: Academic Press.

SRINIVASAN, P. R. & BOREK, E. (1966). Enzymic alteration of macromolecular structure. *Prog. Nucleic Acid Res. molec. Biol.* **5**, 157. Ed. J. N. Davidson and W. E. Cohn. New York: Academic Press.

SUBAK-SHARPE, H. (1965). Failure of herpes simplex virus to initiate hypoxanthine utilisation in BHK 21 cells lacking inosinic acid pyrophosphorylase activity. *Biochem. J.* **94**, 6 P.

SUBAK-SHARPE, H., SHEPHERD, W. M. & HAY, J. (1966). Studies on s-RNA coded by herpes virus. *Cold Spring Harb. Symp. quant. Biol.* **31**, 583.

SYDISKIS, R. J. & ROIZMAN, B. (1966). Polysomes and protein synthesis in cells infected with a DNA virus. *Science*, **153**, 76.

THOMAS, D. C. & GREEN, M. (1966). Biochemical studies on adenovirus multiplica-tion. XI. Evidence of a cytoplasmic site for the synthesis of viral-coded proteins. *Proc. natn. Acad. Sci. U.S.A.* **56**, 243.

VOGT, M., DULBECCO, R. & SMITH, B. (1966). Induction of cellular DNA synthesis by polyoma virus. III. Induction in productively infected cells. *Proc. natn. Acad. Sci. U.S.A.* **55**, 956.

WAARD, A. DE, PAUL, A. V. & LEHMAN, I. R. (1965). The structural gene for deoxy-ribonucleic acid polymerase in bacteriophages T4 and T5. *Proc. natn. Acad. Sci. U.S.A.* **54**, 1241.

WARNER, H. R. & BARNES, J. E. (1966). Deoxyribonucleic acid synthesis in *Esche-richia coli* infected with some deoxyribonucleic acid polymerase-less mutants of bacteriophage T4. *Virology*, **28**, 100.

WATSON, D. H., SHEDDEN, W. I. H., ELLIOT, A., TETSUKA, T., WILDY, P., BOUR-GAUX-RAMOISY, D. & GOLD, E. (1966). Virus-specific antigens in mammalian cells infected with herpes simplex virus. *Immunology*, **11**, 399.

WEISS, S. B. (1962). Biosynthesis of ribopolynucleotides. *Fedn Proc.* **21**, 120.

WIDNELL, C. C. & TATA, J. R. (1966). Studies on the stimulation by ammonium sulphate of the DNA-dependent RNA polymerase of isolated rat liver nuclei. *Biochim. biophys. Acta*, **123**, 478.

WILDY, P., RUSSELL, W. C. & HORNE, R. W. (1960). The morphology of herpes virus. *Virology*, **12**, 204.

WILDY, P., SMITH, C., NEWTON, A. A. & DENDY, P. (1961). Quantitative cytological studies on HeLa cells infected with herpes virus. *Virology*, **15**, 486.

WINTERSBERGER, E. (1964). DNA-abhangige RNA-synthese in Rattenleber-mitochondrien. *Z. physiol. Chem.* **336**, 285.

THE REPLICATION OF VIRAL DNA

R. L. SINSHEIMER

*Division of Biology, California Institute of Technology,
Pasadena, California*

The central act in the complex, multifaceted drama of viral infection is the replication, many-fold, of the viral genetic material—for the viruses to be considered here, DNA. Each replicated DNA, appropriately packaged within the progeny virus particle, contains within itself both the specific information and the operational potential, such that once introduced within a susceptible host cell it can subvert the synthetic machinery of that cell and redirect it toward the specific production of more virus particles.

It has become clear that the replication of DNA is a considerably more complex process than was perhaps naïvely envisioned a few years ago and requires considerably more than nucleoside triphosphate precursors and a polymerase. The manner of (apparent) simultaneous synthesis of two DNA strands of opposite polarity (Cairns, 1963) remains unsolved. The initiation of DNA synthesis is evidently under some form of control (Lark, 1966). The unwinding of the DNA helix, always a thorny problem, has been further complicated by the recognition that at least some replicating DNAs are topologically circular. It seems likely that a rather complex machinery, of which the known polymerases are only components, is required to perform this intricate task. If this is so, it is then plausible to propose that the replication of viral DNAs—at least those of the smaller viruses—are less autonomous processes than have formerly been considered and instead make considerable use of pre-existing cellular machinery for DNA replication—with, of course, appropriate modifications. With the discovery of cytoplasmic (mitochondrial) DNA and DNA replication (Reich & Luck, 1966; Van Bruggen *et al.* 1966), even the replication of the DNA of those viruses known to be localized to the cytoplasm may make use of a cellular apparatus. The study of viral DNA replication may then provide a valuable avenue of approach to the analysis of the cellular DNA replication machinery. The advantages of this approach—the synthesis of DNAs of defined size and function which can be initiated at desired and defined times—may be of crucial importance.

It has also become clear that viral DNAs illustrate in a variety of instances potentials of DNA structure such as unusual bases, single-strandedness, and oddities of tertiary structure that have not, at least as yet, been observed in cellular DNA. Presumably these vagaries of structure provide each viral DNA with an 'edge' within its ecological niche. These variations provide us with a better framework within which to view the inherent potential of the basic genetic substance.

SPECIAL STRUCTURAL FEATURES OF VIRAL DNAs

While some viral DNAs (T1, T7) contain no known structural features by which they could be distinguished from the cellular DNAs of their hosts, many other viral DNAs do possess unusual structural features. These features must originate in the structure of the parental DNA of each virus and must be conserved by means of specialized processes in the course of the replication of that DNA. Though the existence of the former group would seem to imply that these structural variations are not a viral *sine qua non*, it can be inferred that they do have an adaptive value, each for its particular virus in its particular ecology. It is a measure of our ignorance that we are in almost all instances unable to discern this adaptive value.

Unusual bases

It is well known that in the T-even bacterial viruses (T2, T4, T6) cytosine is entirely replaced by 5-hydroxymethyl cytosine (Hershey, Dixon & Chase, 1953). The hydroxymethyl group in turn is variously substituted with α-glucose, β-glucose or the disaccharide, gentiobiose (Lehman & Pratt, 1960; Kuno & Lehman, 1962). If by growth of the virus in a UDP glucose-deficient host the viral DNA is encapsulated without the glucose substitution, the resultant virus particles are defective in the normal host (subsequent to infection the viral DNA will be degraded), although they can still successfully infect certain atypical related bacterial strains (Hattman & Fukasawa, 1963). Evidently the glucosylation is essential—for unknown reasons—in the normal host. The enzymes necessary for the synthesis of 5-hydroxymethyldeoxy-cytidylic acid (Flaks & Cohen, 1959) and for its glucosylation after DNA synthesis are encoded in the viral DNA—as are also enzymes which by degrading deoxycytidine triphosphate to deoxycytidine monophosphate (Warner & Barnes, 1966) and further degrading this to deoxy-uridine monophosphate (Fleming & Bessman, 1965) insure that no cytosine is incorporated into the viral DNA.

In cells infected with T-even phage, the host DNA is degraded *pari passu* with viral DNA synthesis (Hershey *et al.* 1953). Wiberg has suggested that this is accomplished by a viral-encoded enzyme which specifically recognizes DNA containing cytosine residues (Wiberg, 1967). The use of such a mode of DNA recognition might at least in part account for the use of 5-hydroxylmethylcytosine in the viral DNA.

In certain of the *Bacillus subtilis* phages the distinctive base of DNA, thymine, has been replaced by uracil (PBS-1) (Marmur & Cordes, 1963) or by hydroxymethyl uracil (SPO-1, SP-8, φe, SP5C) (Roscoe & Tucker, 1966). Again special viral-encoded enzymes are synthesized to produce specifically the appropriate, unusual DNA precursors (Roscoe & Tucker, 1966).

The *Serratia marcescens* phage η has been reported to contain an unusual and unknown base replacing a portion of its guanine (Pons, 1966).

Methylated bases

The presence of small and varied amounts of methylated bases (5-methyl-cytosine and 6-methylaminopurine) is a distinctive feature of cellular DNAs. The function of such methylation is unclear, but it is known that DNA methylation is produced by enzymes of high specificity (Gold, Hausmann, Maitra & Hurwitz, 1964; Hausmann & Gold, 1966). Variations in specificity of DNA methylation have been postulated to underlie enzymatic restriction of cellular or viral DNAs introduced into incompatible hosts (Arber, 1965; Klein & Sauerbier, 1965).

Some viral DNAs are completely unmethylated, as is that of phage T3 (Gefter, Hausmann, Gold & Hurwitz, 1966; Hausmann, 1967). This is a consequence of the synthesis during infection with this phage of an enzyme to destroy S-adenosylmethionine, the methyl precursor. Polyoma DNA is similarly unmethylated, although methyl groups are incorporated into host DNA during polyoma infection (Kaye & Winocour, 1967).

In phage T2 infection a new methylase for 6-methylaminopurine formation appears in considerable activity. This enzyme appears to have a specificity distinct from the analogous enzyme normally present in the host cell (Hausmann & Gold, 1966; Sellin, Srinivasan & Borek, 1966).

These instances suggest that differences in methylation may provide a subtle but important distinction between viral and cellular DNA.

Single-stranded DNA

Single-stranded DNA has been demonstrated to be the genetic substance of two classes of bacterial viruses (Fiers & Sinsheimer, 1962; Marvin & Schaller, 1966; Takeya & Amako, 1966) and at least one species of animal virus (Crawford, 1966). In each instance the DNA appears to be in a ring form.

Single-stranded DNA is not normally observed in cells and in the instances studied the single-stranded viral DNA has been rapidly converted to a double-stranded form (the replicative form) within the infected cell and shown to replicate as such (Sinsheimer, Starman, Nagler & Guthrie, 1962; Ray, Preuss & Hofschneider, 1966). At a later stage single-stranded DNA for progeny particles is produced. While (see below, p. 113) the replication of these viral DNAs can be brought within the general framework of DNA replication, the adaptive basis for a single-stranded DNA in the life cycle of these viruses is unclear.

Circular DNA

In addition to the nucleic acid of the single-stranded DNA viruses the DNAs of many double-stranded DNA viruses (polyoma, SV 40, papilloma) have been shown to be rings (Vinograd et al. 1965; Crawford & Black, 1964; Crawford, 1965). In addition, the DNA of some viruses (λ), although known to be in a linear form while in the virus particle, has been shown to enter into a circular form during some stages of viral replication (Young & Sinsheimer, 1964; Bode & Kaiser, 1965; Ogawa & Tomizawa, 1967; Lipton & Weissbach, 1966; Weissbach, Lipton & Lisio, 1966). Indeed it is possible to make an argument that all DNA replication involves the participation of a circular form.

When the double-stranded circular DNA form (both strands closed) of these viral DNAs is isolated from the particle or host cell, it is invariably found in a 'super-coiled' form (Vinograd et al. 1965). This twisted form appears to indicate that the winding number of the DNA at the time of closure of the second strand is inadequate to provide the number of turns required for the Watson–Crick double-helical form assumed in free solution. This deficiency amounts to some 12–20 turns (Crawford & Waring, 1967; J. Vinograd, personal communication) in the case of polyoma DNA.

Single-strand nicks

In many viral DNAs, either linear or circular, the single DNA strands have been shown to be continuous covalently linked polynucleotide chains (Studier, 1965; Tomizawa & Anraku, 1965; Young & Sinsheimer, 1965; Davison & Freifelder, 1966; Abelson & Thomas, 1966). Of course single-strand breaks—'nicks'—even if present would not in general be inactivating (phage α appears to be an exception (Freifelder, 1966)) as they could be repaired by the action of host enzymes (Olivera & Lehman, 1967; Becker, Gefter & Hurwitz, 1967) such as polynucleotide ligase or, in some instances (T4), by the analogous phage-induced enzyme (Weiss & Richardson, 1967).

Abelson & Thomas (1966) have shown that the DNA of bacteriophage T5 is nicked specifically at four sites, three in one chain and one in the other. It is plausible to suppose that these nicks are functional. One may, for instance, define the extent of T5 injected into the cell during the early phase of T5 infection (McCorquodale & Lanni, 1964; Lanni, McCorquodale & Wilson, 1964).

Single-stranded ends

The DNAs of certain viruses (λ, ϕ80, P2) have short (approximately 16–18 nucleotides) single-stranded segments at each 5′ end which are complementary in composition. By annealing these ends the DNA can be transformed into a hydrogen-bonded ring (Hershey, Burgi & Ingraham, 1963; Baldwin et al. 1966; Mandell, 1967).

Terminal redundancy

The DNAs of a number of viral species (T2, T3, T7, P22) appear to contain terminal redundancies—i.e. the DNA sequence at one end of the molecule is repeated at the other end (Ritchie, Thomas, MacHattie & Wensink, 1967). The length of the redundant segment may range from a few hundred to several thousands of nucleotides. As a consequence of this terminal redundancy, if the 3′ ends of the two polynucleotide strands are partially digested away, the exposed 5′ ends—now single-stranded—will be at least partially complementary. Hence if such partially digested viral DNA molecules are annealed, at high dilution, hydrogen-bonded viral DNA rings will be formed.

Circular permutation

The DNA molecules of some viral species (T2, P22) appear to be both terminally redundant and circularly permuted (MacHattie, Ritchie, Thomas & Richardson, 1967). By circular permutation is meant that while each DNA molecule contains a full genome (plus a terminal repetition) the DNA molecule in one virus particle does not in general begin with the same gene as does that of another particle (as abcd...xyzab and defg...zabcde). If a collection of such permuted molecules is denatured and then annealed, the random pairing of permuted polynucleotide chains will result in formation of a large number of hydrogen-bonded DNA rings.

INTERMEDIATES IN VIRAL DNA REPLICATION

Analyses of the DNA extracted from infected cells during viral replication have revealed a variety of structures, many as yet only incompletely characterized. The association of these DNA structures with viral DNA has been demonstrated in several ways, as by incorporation of parental virus label, by a precursor relationship to progeny viral DNA, by incorporation of the label under circumstances wherein host DNA synthesis is known to be blocked, and by infectivity *per se* in an appropriate assay system.

The role of DNA circularity

Almost *a priori* it would appear to be advantageous that a DNA which is to be rapidly replicated many-fold will have a circular topology, so the second cycle of replication could commence immediately upon completion of the first. Such a circular topology has indeed been found in at least some stages of the DNA replication of all viruses in which the intermediates have been well characterized, as well as in the cellular DNA of *Escherichia coli* (Cairns, 1963) and PPLO (Riggs, 1966).

It should be pointed out that for this purpose a hydrogen-bonded ring, which might be much more difficult to detect, could in principle serve as well as a covalently linked ring.

The polynucleotide chains of the single-stranded DNA viruses are rings. In both ϕX174 and M13 infections it has been shown that this DNA can be found within the cell in a double-stranded ring, present in two forms: (I) with both strands covalently closed, or (II) with one strand open and one strand covalently closed (Kleinschmidt, Burton & Sinsheimer, 1963; Burton & Sinsheimer, 1965; Ray, Preuss & Hofschneider, 1966; Jaenisch, Hofschneider & Preuss, 1966; Pouwels, Jansz, van Rotterdam & Cohen, 1966). It has been shown in ϕX infection that while the latter form (II) plays a special role in DNA replication, the two forms are in fact interconvertible. Both forms are infective to bacterial spheroplasts; the analogous forms of the double-stranded DNA of polyoma virus (Vinograd *et al.* 1965; Hirt, 1967) are similarly infective to cultured mammalian cells. While the DNA of bacteriophage λ is known to be linear when in the virus particle, it has been shown to be present in forms I and II (Young & Sinsheimer, 1964; Bode & Kaiser, 1965; Ogawa & Tomizawa, 1967) within the bacterial cell. (This is also true of bacteriophage P22.) Again both forms are infective to bacterial spheroplasts (Young, 1967).

In ϕX (Denhardt & Sinsheimer, 1965), polyoma (Hirt, 1966) and λ (Young, 1967) infection it has been demonstrated that the replication of the double-stranded DNA rings is semiconservative. Recalling that ring form I is infective, the topology of the ring then requires that at least one of the polynucleotide strands must be opened to permit strand separation. It seems necessary to postulate an unknown specific enzyme for this purpose, as a part of the replicative machinery.

While a ring form has been proposed as an intermediate in the replication of the large DNA-containing virus T4 (Kozinski & Lin, 1965) the evidence is not as yet convincing. The frequent genetic recombination occurring in this system (Stahl, Edgar & Steinberg, 1964) with the consequent likely existence at any time of single-stranded DNA regions or hydrogen-bonded branched forms may make clear recognition of such a form much more difficult. (The possibility of multiple replication points on one DNA molecule must also be considered.)

The circular permutation of the DNAs of the T-even phages (Mac-Hattie *et al.* 1967)—which it should be emphasized can arise within a single burst—would appear to be most easily explained by a mechanism which—perhaps starting from a circular form—generates long chains of DNA in which the full genome is continually repeated. If then, DNA segments of fixed length somewhat greater than one full genome are successively cut off and packaged into virus particles, the terminal redundancy in circular permutation would be expected. The observation that deletion of a segment of the T4 genome results in a greater region of terminal redundance is in full accord with this hypothesis (Streisinger, Emrich & Stahl, 1967).

For those viruses in which the particle DNA is in a ring form, the intracellular form can be simply packaged as such. In the case of bacteriophage λ (the only instance so far studied) in which the viral DNA form is linear, it appears that the viral form is not produced by a simple scission of the intracellular rings but rather that it is generated *de novo* in an unknown manner (with its unusual ends) by a process in which the ring form serves primarily as a template (Young, 1967).

Dependent upon the mechanism of recombination and the uniformity of permutation of the genome, both circular and linear DNAs can in principle give rise to circular or linear maps (Stahl & Steinberg, 1964). It is reported that the genetic map of phage S13 (Tessman, 1965; Tessman & Tessman, 1966; Tessman, Ishiwa, Kumar & Baker, 1967) (a single-stranded DNA ring) is circular (I. Tessman, personal communication). The genetic map of the circularly permuted T4 phage is circular (Epstein *et al.* 1963; Edgar & Wood, 1966). The genetic map of vegetative λ phage however is linear (Campbell, 1961; Campbell, 1962; Protass & Korn, 1966).

Genetic recombination by breakage and fusion between DNA rings would result in double-size rings unless it is always accompanied by a second recombination event. If this second event were always in a specific site in the rings, the resultant map would be linear, despite the topology.

That recombination can occur between DNA rings is demonstrated by the recovery of recombinant forms of the ϕX RF from mixedly infected cells, but the details of this recombination are unknown.

'Concatenates'

A second class of intermediate termed 'concatenate' has been observed in infections with T-even phages (Frankel, 1966a, b, c) and with λ (Smith & Skalka, 1966). This DNA form is relatively shear-sensitive and has a sedimentation rate up to two or more times that of the viral DNA. If this form were homologous with the (linear) viral DNA, such a sedimentation rate would imply a molecular weight up to eight times that of the viral DNA. However, the possible presence of single-strand nicks or short single-stranded regions which would permit greater molecular flexibility and the possibility of branching at replication points makes such numerical arguments quite questionable. F. R. Frankel (personal communication) has recently demonstrated that pieces of single-stranded DNA of sedimentation coefficient greater than that of T-even viral DNA single strands can be obtained from these forms by alkaline denaturation.

A heterogeneous, pulse-labelled DNA component has also been observed in T5 infection (Smith & Burton, 1966) with sedimentation rate up to twice that of the viral DNA.

It should be noted that not all groups have observed this form in λ infection. Young and Sinsheimer instead observed, during the later phase of λ infection, at the time of progeny viral DNA synthesis, a form 'X', heterogeneous in sedimentation rate and extending in S from that of RF form II to somewhat greater than that of RF form I (Young, 1967). Studies of the alkaline denaturation of 'X' revealed that it contains, as one component, a ring form of λ DNA.

ROLE OF THE HOST IN THE REPLICATION OF VIRAL DNA

The autonomy of viruses as regards the replication of their DNA is quite varied. It might be expected that the larger viruses, with more information content, will be the more autonomous (T4 DNA contains

200,000 nucleotide pairs (Rubenstein, Thomas & Hershey, 1961); λDNA, 50,000 (Burgi & Hershey, 1963; MacHattie & Thomas, 1964); φX DNA, 5500 nucleotides (Sinsheimer, 1959)), requiring from the host only a supply of deoxynucleoside triphosphate precursors and perhaps even augmenting that supply. Clearly, in those viral infections in which the viral DNA contains an unusual nucleotide, the virus must encode specifications for enzymes to produce the appropriate triphosphate precursor. It has been shown that the T-even viruses can multiply in the absence of added thymine in cells normally requiring thymine (Barner & Cohen, 1954; also, T5, Crawford, 1958), and that in such infections the rate of DNA synthesis can become ten times that found in normal cell growth (Hershey & Melechen, 1957). T4 mutants which lack the ability to induce the formation of thymidylate synthetase (Simon & Tessman, 1963) are at a selective disadvantage, at least in dilute culture (Mathews, 1965). Viral coded enzymes catalyse the necessary methylation and kinase activities. Animal viruses such as herpes simplex (Kit & Dubbs, 1963), vaccinia (Kit & Dubbs, 1964), and possibly polyoma (Sheinin, 1966; but see Kit, Dubbs & Frearson, 1966) appear to specify a new thymidine kinase activity.

Simpler viruses may rely not only on precursors provided by the host, but also on host polymerases and even host sites for DNA replication.

Several approaches can be taken to attempt to ascertain the particular contributions of host and virus toward viral DNA replication:

(1) Bacterial mutants are known in which the capacity to synthesize DNA is temperature-sensitive (Bonhoeffer & Schaller, 1965). While the specific labile synthetic step in such strains is, in general, not known, it is established that it is not at the precursor triphosphate level. The ability or inability of viruses to grow in such strains at the restrictive (high) temperature is then an indication of a role of the host DNA synthetic machinery in the viral DNA system. Thus, in a mutant of this type isolated by Bonhoeffer (1966), host DNA synthesis is blocked at 42°, but DNA synthesis of such phages as T4, T1, T7 or P1 is un-impeded (F. Bonhoeffer, personal communication). However, the replication of less autonomous viruses such as λ or φX174 and fd (both of which contain single-stranded DNA) is blocked in such mutants at 42°. In the case of φX174 it has been shown that not only is the first step of DNA replication (conversion from single-stranded to double-stranded DNA) blocked at the high temperature, but if the infection is allowed to proceed at the low temperature until progeny virus are being produced, then a rise in temperature at that time will block further synthesis of progeny (D. T. Denhardt, personal communication). Thus

in such an infection there is a continuing need for participation of a host function for DNA synthesis.

(2) Certain bacterial mutants carry 'mutator genes' which by influencing, in presently unknown ways, DNA replication, cause these strains to have abnormally high rates of mutation (Yanofsky, Cox & Horn, 1966; Zamenhof, 1966). Demonstration of a similar effect on the mutation frequency of a virus growing in such a strain would implicate participation of this particular step of host DNA synthesis in the viral DNA replication. As might perhaps be expected, studies of the mutation rate of T4 growth in mutator strains revealed no effect of the mutant locus, confirming the relative autonomy of T4 replication. It is more surprising, however, that only small effects have been found of these loci on the mutation rate of ϕX174 or S13. The Treffers locus (Yanofsky *et al.* 1966) has been shown to have a marked effect on the mutation rates of λ and of T7, but negligible effects in infections with T1, T3, and T5 (E. C. Cox, personal communication).

(3) The synthesis of new DNA polymerases has been demonstrated during infection with several viruses (T2 (Aposhian & Kornberg, 1962); T4 (de Waard, Paul & Lehman, 1965); T5 (de Waard *et al.* 1965; Orr, Herriott & Bessman, 1965); Poxvirus (Jangwirth & Joklik, 1965); Herpes simplex (Keir *et al.* 1966)); no new polymerase activity could be detected during lambda phage infection (Pricer & Weissbach, 1964). It is significant that most of the viral-coded polymerases characterized to date will use only denatured or single-stranded DNA as a template. Thus they may well not be the true viral DNA replication device, but may instead be repair enzymes which fill in polynucleotide gaps left by recombination or other events. That the polymerases of T4 infection do play a catalytic role *in vivo* has been convincingly proven by the demonstration that viral infections with mutant strains, known to carry mutations which give rise to temperature-sensitive polymerases, will, when performed at an intermediate temperature, result in an abnormally high rate of production of viral particles, mutant in various loci (Speyer, 1965). This result is also an important indication that the polymerase itself plays some role in the precision of DNA replication.

It is also relevant that although the host DNA polymerase activity persists throughout T4 infection (de Waard *et al.* 1965), cells infected with T4 *ts* mutants of gene 43, which are known to produce a temperature-sensitive polymerase, are unable to synthesize DNA at 37°.

(4) Another approach to the question of the autonomy of viral DNA replication is to inquire into the number of potential sites of such replication within a cell. If the virus is fully autonomous and able to

generate new sites as it reproduces, no sharp restriction will be expected. Thus, in T4 infection, Werner (1967) has shown that the number of sites of DNA replication increases during the infection and at a late stage can be as high as 20, or one per viral DNA equivalent in the multiplying DNA pool. On the contrary, in ϕX infection it has been shown by a variety of approaches that the number of sites at which ϕX DNA replication can occur is dependent upon the physiology of the cell and in previously starved cells is most often only one (Yarus & Sinsheimer, 1967). A simple, but indirect, experimental approach to this number is to infect cells with a multiplicity of three or four each of four distinguishable mutant virus strains and then to inquire how many different strains appear in the progeny of single cell bursts. With previously starved cells the result is that most often only one strain appears, occasionally two, never more. With log phase cells as many as four strains are observed in single bursts, although not as frequently as would be expected, assuming random statistics. More direct physical analysis of intracellular viral DNA has confirmed the result that only one DNA is actively replicating within the specially grown cells (Stone, 1967).

It has been postulated that in the case of ϕX infection the number of available sites for viral DNA replication is the number of active sites of DNA synthesis in the cell prior to infection.

(5) It might be expected that if the replication of a viral DNA makes use of host enzymes or machinery, the synthesis of host DNA would thereby be affected. However, the converse is not necessarily true and the relationship of viral DNA synthesis to host DNA synthesis appears to be complex.

In T-even phage infection, which we have seen appears to be quite autonomous as regards DNA replication, host DNA synthesis is halted abruptly upon infection (Tomizawa & Sunakawa, 1956) and the host DNA is at least partially degraded (Hershey et al. 1953). These effects require the synthesis of viral proteins (Nomura, Witten, Mantei & Echols, 1966). Similarly in T5 infection, the host DNA synthesis ceases and the host DNA is extensively degraded prior to viral DNA synthesis (Murray & Whitfield, 1953; Pfefferkorn & Amos, 1958). In T7 infection the host DNA is extensively degraded and the degradation products re-used in the synthesis of the new viral DNA (Putnam, Miller, Palm & Evans, 1952).

In both λ lytic and lysogenic infections, host DNA synthesis undergoes a complex pattern of suppression; in a lytic infection, it is not completely halted until late in infection, while in a lysogenic infection it is depressed and then resumes at a normal rate at a later time (Smith & Levine, 1964; Young, 1967). After λ induction, host DNA synthesis (and mRNA synthesis) continue at a normal rate until close to lysis (Joyner, Isaacs, Echols & Sly, 1966; Dove, 1966).

In ϕX infection there are two distinct phases with respect to viral and host DNA synthesis. In the early phase, accumulation of double-stranded viral DNA proceeds at a low rate, while host DNA synthesis continues at nearly its normal rate (Lindqvist & Sinsheimer, 1967). Then, abruptly, host DNA synthesis is completely blocked and the synthesis of progeny viral single-stranded DNA begins at a rate five to ten times that of early viral DNA synthesis, and nearly equal to the previous rate of cellular DNA synthesis. The host DNA is not degraded, and a simple explanation of these observations is to suggest that at the time of transition, the host DNA is displaced from the replicating site by the viral DNA. The 'site' for the early stage of low level viral DNA synthesis, however, is unclear.

In fd or M13 viral infection the progeny virus particles are extruded from the cell without lysis. The cells remain viable and divide and continue DNA synthesis indefinitely, although at a somewhat lesser rate than is normal for the particular growth medium (Ray, Bscheider & Hofschneider, 1966; Hoffmann-Berling & Mazé, 1964). However, these cells, which are initially F$^+$, lose their male character (C. Dowell, personal communication). It is possible that the viral DNA replication takes place at the normal episomal DNA replication site.

(6) Interference between unrelated viruses infecting the same cell is a common phenomenon. Many of the instances are readily explained as a consequence of nuclease activity (Lesley, French, Graham & van Rooyen, 1951), selective inhibition of messenger RNA synthesis (Hayward & Green, 1965), change in absorption properties of the cell surface, lysis which is premature for one viral type, etc. However, in some instances, such as in cells mixedly infected with ϕX and λ viruses, the interference, which is marked (W. Salivar, personal communication), appears to be more subtle and can be interpreted as a competition within the cell for the limited number of potential sites of DNA synthesis (Yarus & Sinsheimer, 1967).

THE ROLE OF VIRAL CODED PROTEINS

The role of viral coded polymerases has already been discussed. However, there is convincing evidence for roles of uncertain nature to be played by other viral coded proteins.

Lark has presented evidence for the participation of two proteins in the initiation of each new round of DNA synthesis in *E. coli* (Lark,

1966). The synthesis of one of these proteins is surprisingly resistant to inhibition by chloramphenicol. A similar chloramphenicol-resistant protein has been demonstrated to be necessary for initiation of viral DNA synthesis in both ϕX and S13 infection (Tessman, E. S., 1966; Sinsheimer, Hutchison & Lindqvist, 1967) and λ infection (Young, 1967). This protein is made in 30 μg./ml. chloramphenicol but not appreciably in 100 μg./ml. By virtue of this property these initiator proteins have been identified in both ϕX and λ infections (A. Levine, personal communication); the two proteins appear to have very similar properties. Viral mutants which do not make these proteins under restrictive conditions and thus cannot initiate DNA synthesis are known.

New deoxyribonucleases have been demonstrated in infections with T-even viruses (Hershey *et al.* 1953; Stone & Burton, 1962; Short & Koerner, 1965), λ (Radding, 1964), T5 (Paul & Lehman, 1966) and poxvirus (McAuslan & Kates, 1966). The function of these enzymes in the infective process is unknown.

In poxvirus infection (Kates & McAuslan, 1967) (and possibly in phage T5 infection, Pfefferkorn & Amos, 1958; Crawford, 1959) it has been shown that a block to protein synthesis at almost any time during infection will quickly result in a cessation of viral DNA synthesis. The nature of the protein involved is not known, although in the case of poxvirus it has been shown that this essential protein can accumulate if viral DNA synthesis is blocked. The necessary protein is not the poxvirus-induced polymerase.

Gene 30 of phage T4 codes for an enzyme, polynucleotide-ligase, which repairs single-strand breaks in DNA (Weiss & Richardson, 1967). It is of interest that T4 mutants, temperature-sensitive in gene 30, synthesize almost no viral DNA at the restrictive temperature (R. S. Edgar, personal communication), indicating an essential if unexplained role for this enzyme.

THE ROLE OF THE HOST IN VIRAL GENETIC RECOMBINATION

Genetic recombination between viral genomes—or as in lysogenization between viral and host genomes—appears to be a consequence of a DNA breakage and fusion process (Meselson, 1964). This process presumably requires the participation of several enzymes (Howard-Flanders & Boyce, 1966). Bacterial mutants lacking in one or another of these recombination enzymes are known (Clark & Margulies, 1965). The effects of such mutations on the recombination of viral DNAs during mixed infection again depend on the degree of autonomy of the

8

particular virus. Thus recombination among T4 (or T7) strains is unaffected by such host mutations (Brooks & Clark, 1967) (viral genes affecting T4 recombination have been demonstrated (Tomizawa, Anraku & Iwama, 1966)). However, recombination of S13 strains is reduced by a significant factor—although not to zero—by host cell mutations affecting recombination (Tessman, I., 1966).

The effect of such host mutations upon λ viral infections is instructive. Recombination deficient hosts can be lysogenized by λ, but these lysogens are then resistant to induction by ultraviolet radiation (Fuerst & Siminovitch, 1965; Brooks & Clark, 1967). In non-lysogenic infections, recombination between λ genomes is not affected by the host mutation (Brooks & Clark, 1967). An analysis of the situation suggests that the λ virus itself encodes specifications for two recombination enzymes, a specific recombinase which promotes incorporation of the λ genome into the host DNA at a specific site and a second, nonspecific recombinase. Recombination between λ genomes can be prevented only if both the host and the viral enzymes are defective (E. Signer, personal communication).

REPLICATION OF A SINGLE-STRANDED DNA VIRUS

Several of the theses presented in this paper may be illustrated by the summary of the current status of our knowledge of the replication of the DNA of the ϕX174 virus, which has been extensively studied in our laboratory (Fig. 1).

The DNA of this virus is a single-stranded ring containing 5500 nucleotides. Immediately after its introduction into the bacterial cell, it is converted to a double-stranded DNA ring (Sinsheimer et al. 1962), the replicative form (RF). As this conversion can take place in the absence of protein synthesis (in high concentrations of chloramphenicol, or in auxotrophs in the absence of a required amino acid) it is presumably accomplished by pre-existing host enzymes. Just which enzyme is not known, although it is known that the DNA polymerase isolated by Kornberg can produce a second strand, given the ϕX DNA ring a sa template, and that this second strand can be closed into a ring by the action of the previously mentioned polynucleotide ligase (M. Goulian, personal communication).

The RF is found within the cell in two forms: I, in which both strands are covalently closed and II, in which one strand is open and one covalently closed. These forms appear to be interconvertible within the cell although the equilibrium is normally well toward form I.

After formation, the RF becomes associated with a pre-existent site within the bacterial cell (Yarus & Sinsheimer, 1967). This site can be localized with the 'membrane fraction' of the cell. The number of such sites depends upon the physiology of the cell and in previously starved cells is most often only one. Only an RF associated with such a site can be replicated. Replication of the RF also requires the participation of a

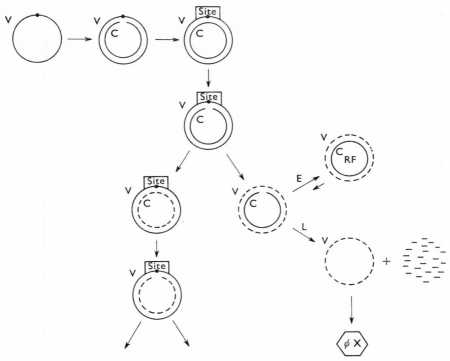

Fig. 1. A model for the replication of bacteriophage ϕX174 DNA.
V = viral strand; C = complementary strand; E = early; L = late.

viral coded protein, the synthesis of which is chloramphenicol-resistant Tessman, E. S., 1966; Sinsheimer *et al.* 1967). RF replication is a semi-conservative process in which the input parental strand remains associated with the site, exchanging partners at each replication. To reproduce semi-conservatively one of the two strands of the ring must be opened to allow them to separate. The RF on the site is therefore most often found in form II. The nascent RF which comes free from the site is also found as form II, but is quickly converted to form I.

If the cell has more than one site, a daughter RF may secure a second site and commence, in turn, to replicate.

The rate of replication is about two RF molecules per site per minute. This process continues from about 3 to about 12 min. after infection during which time some 15–20 RF molecules are produced. During this time host DNA synthesis continues at nearly its normal rate.

At about 12 min. after infection host DNA synthesis is terminated (Lindqvist & Sinsheimer, 1967). Net RF synthesis also ceases. The rate of viral DNA synthesis increases some five- to tenfold at this time. The only DNA made in net amount thereafter is single-stranded viral DNA which is immediately packaged into progeny virus particles.

However, during this period of single-stranded progeny synthesis, the immediate product of DNA replication is not single-stranded DNA, but RF form II DNA. The semi-conservative replication of the RF on the site appears to continue at an enhanced rate; the distinction from the early phase of replication is that the daughter RF II molecules are not converted to RF form I, but instead serve as precursors for the single strands of the viral DNA. The fate of the complementary strands of these RF form II molecules is not as yet certain but it appears possible that they are simply degraded.

The evidence given in various parts of this paper suggests that the replication of the small single-stranded DNAs is largely accomplished by means of the host DNA synthetic machinery—first, very likely, the host DNA repair system to form the RF and then the host apparatus to replicate double-stranded DNA. Viral proteins are necessary to accomplish this molecular subversion and in the later stages to achieve the conversion of the double-stranded DNA product to the single-stranded DNA of the progeny particles.

CONCLUSION

In no instance is the replication of a viral DNA fully understood. The nucleotide (and thus the information) content of viral DNAs spans a range of two orders of magnitude. Gross and subtle distinctions of composition and tertiary structure are observed among various species. The extent of autonomy of the viral DNA replication is varied and new approaches are becoming available for the analysis of the steps of reproduction and the interactions with the host. Two or three temporally distinct stages can be defined in the replication of several viral DNAs; these involve different structures and functions. Genetic, physiological, biochemical and biophysical approaches are available to permit further exploration of these problems.

REFERENCES

ABELSON, J. N. & THOMAS, C. A., JUN. (1966). The anatomy of the T5 bacteriophage DNA molecule. *J. molec. Biol.* **18**, 262.

APOSHIAN, H. V. & KORNBERG, A. (1962). Enzymatic synthesis of deoxyribonucleic acid. IX. The polymerase formed after T2 bacteriophage infection of *Escherichia coli*: A new enzyme. *J. biol. Chem.* **237**, 519.

ARBER, W. (1965). Host specificity of DNA produced by *Escherichia coli*. V. The role of methionine in the production of host specificity. *J. molec. Biol.* **11**, 247.

BALDWIN, R. L., BARRAND, P., FRITSCH, A., GOLDTHWAIT, D. A. & JACOB, F. (1966). Cohesive sites on the deoxyribonucleic acids from several temperate coliphages. *J. molec. Biol.* **17**, 343.

BARNER, H. D. & COHEN, S. S. (1954). The induction of thymine synthesis by T2 infection of a thymine requiring mutant of *Escherichia coli*. *J. Bact.* **68**, 80.

BECKER, A., GEFTER, M. & HURWITZ, J. (1967). Reactions at termini of DNA. *Fedn Proc. Fedn Am. Socs exp. Biol.* **26**, 395.

BODE, V. C. & KAISER, A. D. (1965). Changes in the structure and activity of λ DNA in a superinfected immune bacterium. *J. molec. Biol.* **14**, 399.

BONHOEFFER, F. (1966). DNA transfer and DNA synthesis during bacterial conjugation. *Z. VererbLehre*, **98**, 141.

BONHOEFFER, F. & SCHALLER, H. (1965). A method for selective enrichment of mutants based on the high UV sensitivity of DNA containing 5-bromouracil. *Biochem. biophys. Res. Commun.* **20**, 93.

BROOKS, K. & CLARK, A. J. (1967). Behaviour of λ bacteriophage in a recombination deficient strain of *Escherichia coli*. *J. Virol.* **1**, 283.

BURGI, E. & HERSHEY, A. D. (1963). Sedimentation rate as a measure of molecular weight of DNA. *Biophys. J.* **3**, 309.

BURTON, A. & SINSHEIMER, R. L. (1965). The process of infection with bacteriophage φX174. VII. Ultracentrifugal analysis of the replicative form. *J. molec. Biol.* **14**, 327.

CAIRNS, J. (1963). The bacterial chromosome and its manner of replication as seen by autoradiography. *J. molec. Biol.* **6**, 208.

CAMPBELL, A. (1961). Sensitive mutants of bacteriophage lambda. *Virology*, **14**, 22.

CAMPBELL, A. M. (1962). Episomes. *Adv. Genet.* **11**, 101.

CLARK, A. J. & MARGULIES, A. D. (1965). Isolation and characterization of recombination deficient mutants of *Escherichia coli* K12. *Proc. natn. Acad. Sci. U.S.A.* **53**, 451.

CRAWFORD, L. V. (1958). Thymine metabolism in strains of *Escherichia coli*. *Biochim. biophys. Acta*, **30**, 428.

CRAWFORD, L. V. (1959). Nucleic acid metabolism in *Escherichia coli* infected with phage T5. *Virology*, **7**, 359.

CRAWFORD, L. V. (1965). A study of human papilloma virus DNA. *J. molec. Biol.* **13**, 362.

CRAWFORD, L. V. (1966). A minute virus of mice. *Virology*, **29**, 605.

CRAWFORD, L. V. & BLACK, P. H. (1964). The nucleic acid of simian virus 40. *Virology*, **24**, 388.

CRAWFORD, L. V. & WARING, M. J. (1967). Supercoiling of polyoma virus DNA measured by its interaction with ethidium bromide. *J. molec. Biol.* **25**, 23.

DAVISON, P. F. & FREIFELDER, D. (1966). Lability of single-stranded deoxyribonucleic acid to hydrodynamic shear. *J. molec. Biol.* **16**, 490.

DENHARDT, D. T. & SINSHEIMER, R. L. (1965). The process of infection with bacteriophage φX174. IV. Replication of the viral DNA. *J. molec. Biol.* **12**, 647.

DE WAARD, A., PAUL, A. V. & LEHMAN, I. R. (1965). The structural gene for deoxyribonucleic acid polymerase in bacteriophages T4 and T5. *Proc. natn. Acad. Sci. U.S.A.* **54**, 1241.

DOVE, W. F. (1966). Action of the lambda chromosome. I. Control of functions late in bacteriophage development. *J. molec. Biol.* **19**, 187.

EDGAR, R. S. & WOOD, W. B. (1966). Morphogenesis of bacteriophage T4 in extracts of mutant-infected cells. *Proc. natn. Acad. Sci. U.S.A.* **55**, 498.

EPSTEIN, R. H., BOLLE, A., STEINBERG, C. M., KELLENBERGER, E., BOY DE LA TOUR, E., CHEVALLEY, R., EDGAR, R. S., SUSMAN, M., DENHARDT, G. H. & LIELAUSIS, A. (1963). Physiological studies of conditional lethal mutants of bacteriophage T4D. *Cold Spring Harb. Symp. quant. Biol.* **28**, 375.

FIERS, W. & SINSHEIMER, R. L. (1962). The structure of the DNA of bacteriophage ϕX174. III. Ultracentrifugal evidence for a ring structure. *J. molec. Biol.* **5**, 424.

FLAKS, J. G. & COHEN, S. S. (1959). Virus-induced acquisition of metabolic function. II. Studies on the origin of the deoxycytidylate hydroxymethylase of bacteriophage-infected *E. coli. J. biol. Chem.* **234**, 1507.

FLEMING, W. H. & BESSMAN, M. J. (1965). The enzymology of virus-infected bacteria. VIII. The deoxycytidylate deaminase of T6-infected *Escherichia coli. J. biol. Chem.* **240**, PC 4108.

FRANKEL, F. R. (1966a). The absence of mature phage DNA molecules from the replicating pool of T-even-infected *Escherichia coli. J. molec. Biol.* **18**, 109.

FRANKEL, F. R. (1966b). Studies on the nature of replicating DNA in T4-infected *Escherichia coli. J. molec. Biol.* **18**, 127.

FRANKEL, F. R. (1966c). Studies on the nature of replicating DNA in *Escherichia coli* infected with certain amber mutants of T4. *J. molec. Biol.* **18**, 144.

FREIFELDER, D. (1966). Inactivation of phage α by single-strand breakage. *Virology*, **30**, 328.

FUERST, C. R. & SIMINOVITCH, L. (1965). Characterization of an unusual defective lysogenic strain of *Escherichia coli* K12 (λ). *Virology*, **27**, 449.

GEFTER, M., HAUSMANN, R., GOLD, M. & HURWITZ, J. (1966). The enzymatic methylation of ribonucleic acid and deoxyribonucleic acid. X. Bacteriophage T3-induced S-adenosylmethionine cleavage. *J. biol. Chem.* **241**, 1995.

GOLD, M., HAUSMANN, R., MAITRA, U. & HURWITZ, J. (1964). The enzymatic methylation of RNA and DNA. VIII. Effects of bacteriophage infection on the activity of the methylating enzymes. *Proc. natn. Acad. Sci. U.S.A.* **52**, 292.

HATTMAN, S. & FUKASAWA, T. (1963). Host-induced modification of T-even phages due to defective glucosylation of their DNA. *Proc. natn. Acad. Sci. U.S.A.* **50**, 297.

HAUSMANN, R. (1967). Synthesis of an S-adenosylmethionine-cleaving enzyme in T3-infected *Escherichia coli* and its disturbance by co-infection with enzymatically incompetent bacteriophage. *J. Virol.* **1**, 57.

HAUSMANN, R. & GOLD, M. (1966). The enzymatic methylation of ribonucleic and deoxyribonucleic acid. IX. Deoxyribonucleic acid methylase in bacteriophage-infected *Escherichia coli. J. biol. Chem.* **241**, 1985.

HAYWARD, W. S. & GREEN, M. H. (1965). Inhibition of *Escherichia coli* and bacteriophage lambda messenger RNA synthesis by T4. *Proc. natn. Acad. Sci. U.S.A.* **54**, 1675.

HERSHEY, A. D., BURGI, E. & INGRAHAM, L. (1963). Cohesion of DNA molecules isolated from phage lambda. *Proc. natn. Acad. Sci. U.S.A.* **49**, 748.

HERSHEY, A. D., DIXON, J. & CHASE, M. (1953). Nucleic acid economy in bacteria infected with bacteriophage T2. I. Purine and pyrimidine composition. *J. gen. Physiol.* **36**, 777.

HERSHEY, A. D. & MELECHEN, N. E. (1957). Studies of phage-precursor nucleic acid in the presence of chloramphenicol. *Virology*, **3**, 207.

HIRT, B. (1966). Evidence for semiconservative replication of circular polyoma DNA. *Proc. natn. Acad. Sci. U.S.A.* **55**, 997.

HIRT, B. (1967). Selective extraction of polyoma DNA from infected mouse cell cultures. *J. molec. Biol.* **26**, 365.

HOFFMANN-BERLING, H. & MAZÉ, R. (1964). Release of male-specific bacteriophages from surviving host bacteria. *Virology*, **22**, 305.

HOWARD-FLANDERS, P. & BOYCE, R. P. (1966). DNA repair and genetic recombination: Studies on mutants of *Escherichia coli* defective in these processes. *Radiat. Res.*, Suppl. 6, p. 156.

JAENISCH, R., HOFSCHNEIDER, P. H. & PREUSS, A. (1966). Über infektiose substrukturen aus *Escherichia coli* bakteriophagen. VIII. On the tertiary structure and biological properties of ϕX174 replicative form. *J. molec. Biol.* **21**, 501.

JANGWIRTH, C. & JOKLIK, W. K. (1965). Studies on 'early' enzymes in HeLa cells infected with vaccinia virus. *Virology*, **27**, 80.

JOYNER, A., ISAACS, L. N., ECHOLS, H. & SLY, W. S. (1966). DNA replication and messenger RNA production after induction of wild-type λ bacteriophage and λ mutants. *J. molec. Biol.* **19**, 174.

KATES, J. R. & MCAUSLAN, B. R. (1967). Relationship between protein synthesis and viral deoxyribonucleic acid synthesis. *J. Virol.* **1**, 110.

KAYE, A. M. & WINOCOUR, E. (1967). On the 5-methylcytosine found in the DNA extracted from polyoma virus. *J. molec. Biol.* **24**, 475.

KEIR, H. M., SUBAK-SHARPE, H., SHEDDEN, W. I. H., WATSON, D. H. & WILDY, P. (1966). Immunological evidence for a specific DNA polymerase produced after infection by herpes simplex virus. *Virology*, **30**, 154.

KIT, S. & DUBBS, D. R. (1963). Acquisition of thymidine kinase activity by herpes simplex mouse fibroblast cells. *Biochem. biophys. Res. Commun.* **11**, 55.

KIT, S. & DUBBS, D. R. (1964). Acquisition of thermostable thymidine-deoxyuridine kinase in vaccinia-infected cells. *Fedn Proc. Fedn Am. Socs exp. Biol.* **23**, 382.

KIT, S., DUBBS, D. R. & FREARSON, P. M. (1966). Enzymes of nucleic acid metabolism in cells infected with polyoma virus. *Cancer Res.* **26**, 638.

KLEIN, A. & SAUERBIER, W. (1965). Role of methylation in host-controlled modification of phage T1. *Biochem. biophys. Res. Commun.* **18**, 440.

KLEINSCHMIDT, A., BURTON, A. & SINSHEIMER, R. L. (1963). Electron microscopy of the replicative form of the DNA of bacteriophage ϕX174. *Science*, **142**, 961.

KOZINSKI, A. W. & LIN, T. H. (1965). Early intracellular events in the replication of T4 phage DNA. I. Complex formation of replicative DNA. *Proc. natn. Acad. Aci. U.S.A.* **54**, 273.

KUNO, S. & LEHMAN, I. R. (1962). Gentiobiose, a constituent of deoxyribonucleic acid from coliphage T6. *J. biol. Chem.* **237**, 1266.

LANNI, Y. T., MCCORQUODALE, D. J. & WILSON, C. M. (1964). Molecular aspects of DNA transfer from phage T5 to host cells. II. Origin of first-step-transfer DNA fragments. *J. molec. Biol.* **10**, 19.

LARK, K. G. (1966). Regulation of chromosome replication and segregation in bacteria. *Bact. Rev.* **30**, 3.

LEHMAN, I. R. & PRATT, E. A. (1960). On the structure of the glucosylated hydroxymethyl cytosine nucleotides of coliphages T2, T4 and T6. *J. biol. Chem.* **235**, 3254.

LESLEY, S. M., FRENCH, R. C., GRAHAM, A. F. & VAN ROOYEN, C. E. (1951). Studies on the relationship between virus and host cell. II. The breakdown of T2r+ bacteriophage upon infection of its host, *Escherichia coli*. *Can. J. med. Sci.* **29**, 128.

LINDQVIST, B. H. & SINSHEIMER, R. L. (1967). The process of infection with bacteriophage φX174. XIV. Studies on macromolecular synthesis during infection with a lysis-defective mutant. *J. molec. Biol.* **28**, 87.

LIPTON, A. & WEISSBACH, A. (1966). The appearance of circular DNA after lysogenic induction in *Escherichia coli* CR34 (λ). *J. molec. Biol.* **21**, 517.

MACHATTIE, L. A. & THOMAS, C. A., JUN. (1964). DNA bacteriophage lambda: molecular length and conformation. *Science*, **144**, 1142.

MACHATTIE, L. A., RITCHIE, D. A., THOMAS, C. A., JUN. & RICHARDSON, C. C. (1967). Terminal repetition in permuted T2 bacteriophage DNA molecules. *J. molec. Biol.* **23**, 355.

MANDELL, M. cited in WU, R. & KAISER, A. D. (1967). Mapping the 5′ terminal nucleotides of the DNA of bacteriophage λ and related phages. *Proc. natn. Acad. Sci. U.S.A.* **57**, 170.

MARMUR, J. & CORDES, S. (1963). Studies on the complementary strands of bacteriophage DNA. In *Symposium on Informational Macromolecules*, pp. 79–87. Ed. H. J. Vogel, V. Bryson and J. O. Lampen. New York: Academic Press.

MARVIN, D. A. & SCHALLER, H. (1966). The topology of DNA from the small filamentous bacteriophage fd. *J. molec. Biol.* **15**, 1.

MATHEWS, C. K. (1965). Phage growth and deoxyribonucleic acid synthesis in *Escherichia coli* infected by a thymine-requiring bacteriophage. *J. Bact.* **90**, 648.

MCAUSLAN, B. R. & KATES, J. R. (1966). Regulation of virus-induced deoxyribonucleases. *Proc. natn. Acad. Sci. U.S.A.* **55**, 1581.

MCCORQUODALE, D. J. & LANNI, Y. T. (1964). Molecular aspects of DNA transfer from phage T5 to host cells. I. Characterization of first-step-transfer material. *J. molec. Biol.* **10**, 10.

MESELSON, M. (1964). On the mechanism of genetic recombination between DNA molecules. *J. molec. Biol.* **9**, 734.

MURRAY, R. G. E. & WHITFIELD, J. F. (1953). Cytological effects of infection with T5 and some related phages. *J. Bact.* **65**, 715.

NOMURA, M., WITTEN, C., MANTEI, N. & ECHOLS, H. (1966). Inhibition of host nucleic acid synthesis by bacteriophage T4: effect of chloramphenicol at various multiplicities of infection. *J. molec. Biol.* **17**, 273.

OGAWA, H. & TOMIZAWA, J. (1967). Bacteriophage lambda DNA with different structures found in infected cells. *J. molec. Biol.* **23**, 265.

OLIVERA, B. A. & LEHMAN, I. R. (1967). Linkage of polynucleotides through phosphodiester bonds by an enzyme from *Escherichia coli*. *Proc. natn. Acad. Sci. U.S.A.* **57**, 1426.

ORR, C. W. M., HERRIOTT, S. T. & BESSMAN, M. J. (1965). The enzymology of virus-infected bacteria. VII. A new deoxyribonucleic acid polymerase induced by bacteriophage T5. *J. biol. Chem.* **240**, 4652.

PAUL, A. V. & LEHMAN, I. R. (1966). The deoxyribonucleases of *Escherichia coli*. VII. A deoxyribonuclease induced by infection with phage T5. *J. biol. Chem.* **241**, 3441.

PFEFFERKORN, E. & AMOS, H. (1958). Deoxyribonucleic acid breakdown and resynthesis in T5 bacteriophage infection. *Virology*, **6**, 299.

PONS, F. W. (1966). Untersuchung der DNS einiger serratiastämme und derer phage. *Biochem. Z.* **346**, 26.

POUWELS, P. H., JANSZ, H. S., VAN ROTTERDAM, J. & COHEN, J. A. (1966). Structure of the replicative form of bacteriophage φX174. Physicochemical studies. *Biochim. biophys. Acta*, **119**, 289.

PRICER, W. E., JUN. & WEISSBACH, A. (1964). The effect of lysogenic induction with mitomycin C on the deoxyribonucleic acid polymerase of *Escherichia coli* K12 λ. *J. biol. Chem.* **239**, 2607.

PROTASS, J. J. & KORN, D. (1966). Function of the N cistron of bacteriophage lambda. *Proc. natn. Acad. Sci. U.S.A.* **55**, 1089.

PUTNAM, F. W., MILLER, D., PALM, L. & EVANS, E. A., JUN. (1952). Biochemical studies of virus reproduction. X. Precursors of bacteriophage T7. *J. biol. Chem.* **199**, 177.

RADDING, C. M. (1964). Nuclease activity in defective lysogens of phage λ. II. A hyperactive mutant. *Proc. natn. Acad. Sci. U.S.A.* **52**, 965.

RAY, D. S., BSCHEIDER, H. & HOFSCHNEIDER, P. H. (1966). Replication of the single-stranded DNA of the male-specific bacteriophage M13. Isolation of intracellular forms of phage-specific DNA. *J. molec. Biol.* **21**, 473.

RAY, D. S., PREUSS, A. & HOFSCHNEIDER, P. H. (1966). Replication of the single-stranded DNA of the male-specific bacteriophage M13. Circular forms of the replicative DNA. *J. molec. Biol.* **21**, 485.

REICH, E. & LUCK, D. L. (1966). Replication and inheritance of mitochondrial DNA. *Proc. natn. Acad. Sci. U.S.A.* **55**, 1600.

RIGGS, A. D. (1966). Studies on Mycoplasma gallisepticum. Ph.D. Thesis, California Institute of Technology.

RITCHIE, D. A., THOMAS, C. A., JUN., MacHATTIE, L. A. & WENSINK, P. C. (1967). Terminal repetition in non-permuted T3 and T7 bacteriophage DNA molecules. *J. molec. Biol.* **23**, 365.

ROSCOE, D. H. & TUCKER, R. G. (1966). The biosynthesis of 5-hydroxymethyl-deoxyuridylic acid in bacteriophage-infected *Bacillus subtilis*. *Virology*, **29**, 157.

RUBENSTEIN, I., THOMAS, C. A., JUN. & HERSHEY, A. D. (1961). The molecular weights of T2 bacteriophage DNA and its first and second breakage products. *Proc. natn. Acad. Sci. U.S.A.* **47**, 1113.

SELLIN, H., SRINIVASAN, P. R. & BOREK, E. (1966). Studies of a phage-induced DNA methylase. *J. molec. Biol.* **19**, 219.

SHEININ, R. (1966). Studies on the thymidine kinase activity of mouse embryo cells infected with polyoma virus. *Virology*, **28**, 47.

SHORT, E. C., JUN. & KOERNER, J. F. (1965). Separation of deoxyribo-oligonucleotidases induced by infection with bacteriophage T2. *Proc. natn. Acad. Sci. U.S.A.* **54**, 595.

SIMON, E. H. & TESSMAN, I. (1963). Thymidine-requiring mutants of phage T4. *Proc. natn. Acad. Sci. U.S.A.* **50**, 526.

SINSHEIMER, R. L. (1959). A single-stranded deoxyribonucleic acid from bacteriophage φX174. *J. molec. Biol.* **1**, 43.

SINSHEIMER, R. L., HUTCHISON, C. A., III. & LINDQVIST, B. (1967). Bacteriophage φX174: viral functions. In *Molecular Biology of Viruses* (Symposium). Ed. J. S. Colter. New York: Academic Press.

SINSHEIMER, R. L., STARMAN, B., NAGLER, C. & GUTHRIE, S. (1962). The process of infection with bacteriophage φX174. I. Evidence for a 'replicative form'. *J. molec. Biol.* **4**, 142.

SMITH, H. O. & LEVINE, M. (1964). Two sequential repressions of DNA synthesis in the establishment of lysogeny by phage P22 and its mutants. *Proc. natn. Acad. Sci. U.S.A.* **52**, 356.

SMITH, M. G. & BURTON, K. (1966). Fractionation of deoxyribonucleic acid from phage-infected bacteria. *Biochem. J.* **98**, 229.

SMITH, M. & SKALKA, A. (1966). Some properties of DNA from phage-infected bacteria. *J. gen. Physiol.* (Suppl.), **49**, 127.

SPEYER, J. F. (1965). Mutagenic DNA polymerase. *Biochem. biophys. Res. Commun.* **21**, 6.

STAHL, F. W., EDGAR, R. S. & STEINBERG, J. (1964). The linkage map of bacteriophage T4. *Genetics*, **50**, 539.

STAHL, F. W. & STEINBERG, C. M. (1964). The theory of formal phage genetics for circular maps. *Genetics*, **50**, 531.

STONE, A. B. (1967). Some factors which influence the replication of the replicative form of bacteriophage φX174. *Biochem. biophys. Res. Commun.* **26**, 247.

STONE, A. B. & BURTON, K. (1962). Studies on the deoxyribonucleases of bacteriophage-infected *Escherichia coli. Biochem. J.* **85**, 600.

STREISINGER, G., EMRICH, J. & STAHL, M. M. (1967). Chromosome structure in phage T4. III. Terminal redundancy and length determination. *Proc. natn. Acad. Sci. U.S.A.* **57**, 292.

STUDIER, F. W. (1965). Sedimentation studies of the size and shape of DNA. *J. molec. Biol.* **11**, 373.

TAKEYA, K. & AMAKO, K. (1966). A rod-shaped Pseudomonas phage. *Virology*, **28**, 163.

TESSMAN, E. S. (1965). Complementation groups in phage S13. *Virology*, **25**, 303.

TESSMAN, E. S. (1966). Mutants of bacteriophage S13 blocked in infectious DNA synthesis. *J. molec. Biol.* **17**, 218.

TESSMAN, I. (1966). Genetic recombination of phage S13 in a recombination-deficient mutant of *Escherichia coli* K12. *Biochem. biophys. Res. Commun.* **22**, 169.

TESSMAN, I., ISHIWA, H., KUMAR, S. & BAKER, R. (1967). Bacteriophage S13: A seventh gene. *Science*, **156**, 824.

TESSMAN, I. & TESSMAN, E. S. (1966). Functional units of phage S13: Identification of two genes that determine the structure of the phage coat. *Proc. natn. Acad. Sci. U.S.A.* **55**, 1459.

TOMIZAWA, J. & ANRAKU, N. (1965). Molecular mechanisms of genetic recombination in bacteriophage. IV. Absence of polynucleotide interruption in DNA of T4 and λ phage particles with special reference to heterozygosis. *J. molec. Biol.* **11**, 509.

TOMIZAWA, J., ANRAKU, N. & IWAMA, Y. (1966). Molecular mechanisms of genetic recombination in bacteriophage. VI. A mutant defective in the joining of DNA molecules. *J. molec. Biol.* **21**, 247.

TOMIZAWA, J. & SUNAKAWA, S. (1956). The effect of chloramphenicol on deoxyribonucleic acid synthesis and the development of resistance to ultraviolet irradiation in *E. coli* infected with bacteriophage T2. *J. gen. Physiol.* **39**, 553.

VAN BRUGGEN, E. F. J., BORST, P., RUTTENBERG, G. J. C. M., GRUBER, M. & KROON, A. M. (1966). Circular mitochondrial DNA. *Biochim. biophys. Acta*, **119**, 437.

VINOGRAD, J., LEBOWITZ, J., RADLOFF, R., WATSON, R. & LAIPIS, P. (1965). The twisted circular form of polyoma DNA. *Proc. natn. Acad. Sci. U.S.A.* **53**, 1104.

WARNER, H. R. & BARNES, J. E. (1966). Evidence for a dual role for the bacteriophage T4-induced deoxycytidine triphosphate nucleotidohydrolase. *Proc. natn. Acad. Sci. U.S.A.* **56**, 1233.

WEISS, B. & RICHARDSON, C. C. (1967). Enzymatic breakage and joining of deoxyribonucleic acid. I. Repair of single-strand breaks in DNA by an enzyme system from *Escherichia coli* infected with T4 bacteriophage. *Proc. natn. Acad. Sci. U.S.A.* **57**, 1021.

WEISSBACH, A., LIPTON, A. & LISIO, A. (1966). Intracellular forms of λ deoxyribonucleic acid in *Escherichia coli* infected with clear or virulent mutants of bacteriophage λ. *J. Bact.* **91**, 1489.

WERNER, R. (1967). Replicating points in T4 RNA. *Carnegie Institution Year Book*, **65**, 567.

WIBERG, J. S. (1967). Mutants of bacteriophage T4 unable to cause breakdown of host DNA. *Proc. natn. Acad. Sci. U.S.A.* **55**, 614.

YANOFSKY, C., COX, E. C. & HORN, V. (1966). The unusual mutagenic specificity of an *E. coli* mutator gene. *Proc. natn. Acad. Sci. U.S.A.* **55**, 274.

YARUS, M. & SINSHEIMER, R. L. (1967). The process of infection with bacteriophage ϕX174. XIII. Evidence for an essential bacterial 'site'. *J. Virol.* **1**, 135.

YOUNG, E. T., II. (1967). Structure and synthesis of bacteriophage lambda DNA. Ph.D. Thesis, California Institute of Technology.

YOUNG, E. T., II. & SINSHEIMER, R. L. (1964). Novel intra-cellular forms of lambda DNA. *J. molec. Biol.* **10**, 562.

YOUNG, E. T., II. & SINSHEIMER, R. L. (1965). A comparison of the initial actions of spleen deoxyribonuclease and pancreatic deoxyribonuclease. *J. biol. Chem.* **240**, 1274.

ZAMENHOF, P. J. (1966). A genetic locus responsible for generalized high mutability in *Escherichia coli*. *Proc. natn. Acad. Sci. U.S.A.* **56**, 845.

THE REPLICATION OF VIRAL RNA

L. MONTAGNIER

Institut du Radium-Biologie, Faculté des Sciences,
91, Orsay, France

According to current concepts of molecular biology, functions of the two nucleic acids in a cell are clearly separated: DNA is the depository of genetic information, RNA is involved in the expression of this information, either at the transcription level of DNA or in the translation process of messages into proteins. DNA is a self-replicating molecule, which exists normally as a double helix, RNA is synthesized along a DNA strand and is usually single-stranded.

Viral RNA is an exception to this scheme, since it can perform both functions in the host-cell: it possesses the abilities to direct protein synthesis and to replicate. This view has progressively arisen from numerous studies made during the past 12 years over a wide range of virus–host systems, and particularly from the following.

First, the RNA component of several groups of viruses, including plant, animal and bacterial viruses, was shown to be infective, hence to carry information for its own synthesis and the synthesis of capsidal proteins (Gierer & Schramm, 1956; Colter, Bird & Brown, 1957; Davis, Strauss & Sinsheimer, 1961). Confirmation that TMV–RNA contained the information for its coat protein was provided by the induction of mutants with modified protein following chemical alteration of the bases of infecting RNA (Wittmann, 1960). More direct evidence came from the synthesis in a cell-free system of polypeptides of the coat protein of phage f2 directed by its own RNA (Nathans, Notani, Schwartz & Zinder, 1962).

Then, with the use of specific inhibitors in cultured cells, it was demonstrated that RNA replication was not dependent on the integrity of cell DNA (Reich & Franklin, 1961), and did not require its transcription (Reich, Franklin, Shatkin & Tatum, 1962), or the synthesis of a new DNA (Simon, 1961). In contrast, continuous protein synthesis was needed for viral RNA replication (Levintow, Thoren, Darnell & Hooper, 1962).

These results suggested that viral RNA might induce a new kind of replicating enzyme which would use it as template. Indeed, complexes of RNA-polymerase bound to an RNA template were found in cells infected with animal viruses (Baltimore & Franklin, 1962) and bacteriophages

(Weissmann, Simon & Ochoa, 1963). In two phage systems (MS2 and Qβ), the polymerases were isolated free from their endogenous template (Haruna, Nozu, Ohtaka & Spiegelman, 1963; Haruna & Spiegelman, 1965): they showed in the test-tube a preferential requirement for homologous viral RNA as template, and, in the case of Qβ phage, infectivity of the product confirmed its identity to the RNA model (Spiegelman et al. 1965).

Viral RNA could thus be defined as a self-replicating polycistronic messenger, so that viral multiplication would depend on the host-cell only for the translation machinery and energy sources. This picture is probably oversimplified: multiplication of large RNA viruses is in some way dependent on cell information, and it will be seen that even in the case of small RNA viruses casual participation of cellular enzymes cannot be excluded.

THE ROLE OF BASE PAIRING IN RNA REPLICATION

In the approach to the replication mechanism, the first question to be answered was whether specific base pairing by hydrogen bonds was involved or not.

Unlike DNA, the double-stranded structure of which immediately suggests a model for replication (Watson & Crick, 1953), the answer to this question was not obvious for single-stranded viral RNA.

However, the finding of a double-stranded 'replicative form' in bacteria infected with the single-stranded DNA phage φX174 indicated that a single-stranded nucleic acid might pass through a double-stranded state while replicating (Sinsheimer, Starman, Nagler & Guthrie, 1962). Furthermore, there was no physical impediment to the existence of double helices of RNA held by pairing between complementary bases (adenine–uracil and guanine–cytosine). Double-stranded RNA could be produced artificially (Geiduschek, Moohr & Weiss, 1962) and was found to occur naturally in some viruses (Gomatos & Tamm, 1963).

Search for a double-stranded 'replicative form' (RF) of single-stranded viral RNA was successful in cells infected with picornaviruses (Montagnier & Sanders, 1963b; Baltimore, Becker & Darnell, 1964), phages (Kelly & Sinsheimer, 1964; Weissmann et al. 1964) and plant viruses (Shipp & Haselkorn, 1964; Mandel, Matthews, Matus & Ralph, 1964). Virus-specific double-stranded RNA has been found in practically all the systems where it has been looked for, with the possible exception of oncogenic RNA viruses.

Complete demonstration that base pairing is involved in RNA replication requires that two conditions should be satisfied:

(1) The structure of the 'replicative form' should be that of a double helix made of two complementary chains, one of which should be identical to viral RNA.

(2) It should be an obligatory intermediate of replication, and not a pathological product of cell metabolism.

Conversely, failure to find a virus-specific double-stranded RNA in infected cells does not necessarily mean that base pairing is not involved, but simply that it is too transient to be detectable.

As regards the first condition, the double-stranded structure of the 'RF' has been proved in most cases indirectly, owing to the small quantities available, by comparison with naturally occurring double-stranded RNAs. Helical structures of reovirus RNA (Langridge & Gomatos, 1963) and rice dwarf virus RNA (Sato *et al.* 1966) studied by X-ray diffraction do not differ fundamentally from DNA double helices; some minor differences, arising from the possibility of extra hydrogen bonds with the OH of the ribose, may explain the greater thermostability of double-stranded RNA (Arnott *et al.* 1966).

The X-ray diffraction pattern of the MS2 phage 'RF' is identical to that of reovirus RNA (Langridge *et al.* 1964), thus providing direct evidence of its double-stranded structure.

Other common characteristics of 'RF' and other double-stranded RNAs are the following: the molar ratio of adenine to uracil and that of guanine to cytosine are equal to one; the helical structure melts sharply when temperature increases and the separated strands can anneal to give rise again to a double-stranded molecule (Weissmann *et al.* 1964); double-stranded RNA is resistant to ribonuclease in moderate and high ionic strength. This last property has been particularly useful in allowing separation of the replicative form from the bulk of cellular and viral single-stranded RNAs (Montagnier & Sanders, 1963*b*).

Double-stranded RNA molecules behave in sedimentation like flexible rods, and can be visualized in electron microscopy with techniques used for DNA (Kleinschmidt *et al.* 1964) (see Plates 1 and 2).

Several other properties (solubility in concentrated salt solutions, density in caesium salts, lack of reactivity with formaldehyde, chromatographic behaviour) make this kind of RNA clearly distinct from single-stranded RNA.

Evidence for the viral origin of 'RF' includes the following:

(1) That it contains the viral ('plus') strand is shown by its infectivity (animal viruses), or the appearance of infectivity, after melting and

release of the viral strand (bacteriophages) (Ammann, Delius & Hof-schneider, 1964).

(2) The molecular weight as deduced from its sedimentation rate or its molecular length is twice that of viral RNA (Plates 1 and 2).

(3) Melting and reannealing experiments indicate that nearly half of the strands separated from the duplexes (the complementary or 'minus' strands) can re-form a double-stranded structure with an excess of RNA extracted from purified virus (Weissmann *et al.* 1964).

The possibility that this RNA would be a product of cellular enzymes irrelevant to viral RNA replication can be considered now as very un-likely, in the light of recent *in vitro* studies: viral double-stranded RNA is an early product of the purified virus-induced polymerase (Mills, Pace & Spiegelman, 1966).

Thus, base pairing involvement in viral RNA replication is now widely accepted. However, the exact role of double-stranded RNA in the replication mechanism is still a matter of discussion. It is not clear yet if the extracted double-stranded RNA exists in the same state in in-fected cells and is used as such for replication, or if it is a part of a more complex macromolecular structure, or if it is an artifact resulting from the pairing after extraction of pre-existing free 'plus' and 'minus' strands.

In this connection, it must be kept in mind that the ionic and macro-molecular environment, the hydration conditions in cells and especially in the RNA replication sites, may differ fundamentally from the media used for studying RNA molecules after their extraction, so that the isolated structures might be poor reflections of the actual ones.

Besides, RNA replication is a dynamic, very rapid process, and the methods currently used tend to select the most stable stages of replica-tion, which are not necessarily the most active.

These remarks apply not only to double-stranded RNA but also to the other structures found in RNA replication.

MECHANISM OF SINGLE-STRANDED VIRAL
RNA REPLICATION

Most of our knowledge of viral RNA replication has come from studies on two groups of viruses, the RNA bacteriophages (MS 2, R 17, f 2, M 12, Qβ) and the small animal RNA viruses (picorna-) (polio, EMC, Mengo, ME, FMDV).* The mechanism to be presented concerns mainly these groups, although available data suggest it may also apply to the arboviruses (lipid-containing) group and to some plant viruses (TMV,

* An explanation of the abbreviations used is given on p. 143.

TYMV). The RNA component of these viruses behaves in solution as a single chain, of a molecular weight ranging from 1×10^6 daltons to 3×10^6 daltons (picorna- and arboviruses).

Where studies of single infection cycles have been possible (phages and animal viruses), these systems have shown some common features:

(1) The infection cycle is short (less than 60 min. in bacteria; a few hours in animal cells) and ends in the lysis of the infected cells and the release of virions in the incubation medium.

(2) The site of synthesis of the virus-specific components is cytoplasmic, and is usually insensitive to actinomycin D, at doses which markedly inhibit cellular RNA synthesis.

(3) Viral RNA synthesis begins shortly after virus adsorption and uncoating of the infecting RNA. It usually reaches its maximal rate before the appearance of mature virions in the infected cells.

To study how this synthesis occurs, two main types of experimentation have been used, either *in vivo* or *in vitro*: in the first, the fate of parental RNA is followed by its previous labelling with radioactive precursors or by its infectivity. In the second, the progeny RNA is studied, by labelling the infected cells with radioactive precursors in presence of actinomycin D or by its infectivity.

Results of these investigations have led to distinguishing two steps in the replication mechanism. The first step corresponds to the initiation of the replication: after having directed the synthesis of one or several polymerase molecules, parental RNA is used by them as template for the synthesis of complementary 'minus' strands, and consequently appears upon extraction in a double-stranded structure. In the second step, the 'minus' strands of the duplexes serve as template for the synthesis of viral 'plus' strands.

First step

If initiation of the replication depends on the synthesis of an enzyme coded by viral RNA, one should find the infecting molecules, immediately after being uncoated, to be bound to the cell machinery of protein synthesis. This has proved to be so in several systems, and the case of phage MS2, object of detailed studies, will be taken as example (Godson & Sinsheimer, 1967).

Immediately after infection of *E. coli* cells (between 1 and 4 min.) MS2 parental RNA, as traced by radioisotope labelling, is found to be associated with 30S ribosomal subunits, then with polysomes. Between 4 and 10 min. (infection of bacterial cells is not synchronized), a fraction (up to 14%) of parental RNA is found in a double-stranded

9

(ribonuclease-resistant after phenol extraction) structure, which sediments before deproteinization, associated with polysomes or in a complex possibly formed with polymerase molecules. If inhibitors of protein synthesis are added from the beginning of infection, appearance of polysomes and double-stranded RNA is prevented. Parental RNA remains in the complex as an intact molecule, as shown by experiments using labelling with heavy isotopes (Kelly & Sinsheimer, 1967).

Although *in vitro* conditions differ from the *in vivo* situation, studies with purified phage-induced polymerases (replicases) have also confirmed enzymatic synthesis of double-stranded RNA on single-stranded RNA templates (Shapiro & August, 1965; Mills *et al.* 1966). In the latter work, conversion of infective Qβ phage RNA into a complex having a double-stranded structure was followed by isotope labelling and by changes in biological activity of the template: the double-stranded complex was not infective, but became infective after melting of its secondary structure, which released the viral 'plus' strand. The work of Feix, Slor & Weissmann (1967) on the same system indicates that the double-stranded character of the complex results to a large extent from its deproteinization: before phenol extraction, only 10 % of the complex is ribonuclease-resistant, while this figure reaches 73 % after phenol treatment. The authors have concluded that the largest part of complementary RNA being synthesized is not paired to the template strand. The double-stranded structure would be thus an artifact of phenol extraction. Alternatively, this structure may actually exist but would be disrupted in many points, each corresponding to a replication point. The latter hypothesis will be developed in the model proposed below.

In the case of picornaviruses, *in vivo* studies of the fate of parental RNA have been obscured owing to the difficulty of tracing the molecules actually involved in replication: the ratio of infective virus particles to physical particles is usually low in these systems (less than one to ten), and the non-infective particles remain attached to cell membranes, so that their RNA can be found in cell extracts as well as the RNA participating in replication.

Mengovirus parental RNA has been found associated with polysomal structures (Tobey, 1964b), but no conversion into a ribonuclease-resistant structure could be detected (Tobey, 1964a; Homma & Graham, 1965). Several explanations of this failure can be advanced: the double-stranded stage may be very transient and coupled with the second step of replication, or its occurrence in membranous structures could make its extraction difficult. Our own work on the EMC virus system suggests that formation of minor amounts of double-stranded RNA takes place from parental labelled RNA, but it is not clear whether this material arises from intact parental molecules or from nucleotides resulting from the breakdown of the latter.

Second step

All data show that the majority of RNA molecules synthesized during infection are identical to the parental 'plus' strands, and that only small quantities of complementary 'minus' strands are formed. For example, the base composition of the RNA synthesized after poliovirus infection in presence of actinomycin D resembles that of RNA extracted from purified virions (Darnell, 1962). Likewise, most, if not all, of the RNA molecules synthesized after MS2 phage infection, are of the 'plus' type, when analysed by a specific annealing test (Weissmann *et al.* 1964).

The *size* of the synthesized RNA is also usually that of the parental RNA, although some single-stranded RNAs with higher and smaller sedimentation rates have been repeatedly found in some cases, especially in picorna- and arbovirus systems (Montagnier & Sanders, 1963*a*; Brown & Cartwright, 1964; Friedman, Levy & Carter, 1966; Sonnabend, Martin & Mécs, 1967). These RNAs may represent varying configurations of molecules of the same length, or definite fragments and aggregates of viral RNA. In this connection, it must be pointed out that the behaviour of viral RNA as single chain may only be apparent: the possibility has to be considered that some regions of the RNA chain are held together by non-covalent bonds. This situation seems to occur in some large RNA viruses. In the case of small RNA viruses, its relevance to the replication mechanism would be only a matter of speculation at the present time and will not be taken into consideration.

A second feature of this step is that the number of templates for the synthesis of 'plus' strands (namely the 'minus' strands in double-stranded structures), is comparatively small: for example, the fraction of double-stranded RNA synthesized 4 hr. after infection by EMC virus, as estimated by its ribonuclease resistance after extraction, is only 0·1 % of the total viral RNA being made (Montagnier & Sanders, 1963*b*). Moreover, the growth of viral RNA is exponential, at least in the first half of a cycle, and the amount of double-stranded RNA increases in parallel with viral RNA (Baltimore & Girard, 1966). At the late part of the infection cycle, double-stranded RNA accumulates in the cell; it cannot be considered then as a precursor, but rather as a by-product, since most of the viral RNA has been incorporated into virions by this time.

These facts are best interpreted by assuming that 'minus' strands of the double-stranded RNA serve many times as template for the synthesis of 'plus' strands. Two mechanisms are possible. One is a semi-conservative mechanism: the parental 'plus' strand of the initial double-stranded template is displaced by the new 'plus' strand and becomes again available for a new cycle of replication (Fig. 1*a*), and so on. The other is conservative, somewhat analogous to the *in vivo* transcription of DNA: the duplex remains conserved when the polymerase is working, and its secondary structure is disrupted at the replication point to allow the copying to proceed along the 'minus' strand (Fig. 1*b*). The new 'plus' strand may again enter the two-step cycle.

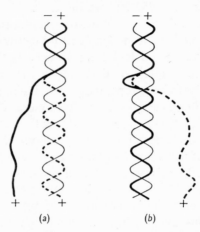

Fig. 1. Two possible structures of the 'replicative intermediate': (*a*) in a semi-conservative mechanism; (*b*) in a conservative mechanism. The dotted lines represent the 'plus' strand being synthesized.

Kinetic studies of incorporation of pulses of radioactive precursors during virus infection as well as studies on the fate of parental RNA have shown that the double-stranded structures isolated during the active phase of replication are associated with fragments of single-stranded RNA molecules, provided the ribonuclease treatment is omitted during the isolation procedure. These complexes were first described in R17 phage-infected cells (Fenwick, Erikson & Franklin, 1964), and have been found also in poliovirus-infected cells (Baltimore & Girard, 1966). As the radioactive precursors are first incorporated into them (Darnell *et al.* 1967), these structures are good candidates to represent the actual intermediates of replication and have been called 'replicative intermediates' (RI) to differentiate them from the double-stranded 'replicative form' (Erikson, Fenwick & Franklin, 1964, 1965).

The existence of the 'RI' could fit in with both of the possible mechanisms, but its molecular structure should differ in the two cases:

in a semi-conservative mechanism, the double-stranded 'core' of 'RI' is made from one 'minus' strand and two or more fragments of 'plus' strands: the old one, and, where replication has taken place, the newly synthesized ones. The single-stranded 'tails' are the parts of old 'plus' strands which have been displaced (Fig. 1 a).

In a conservative mechanism, the double-stranded molecule is made from the 'plus' and 'minus' old strands, the single-stranded tails are the growing chains of 'plus' strands (Fig. 1 b). At the present time, none of the studies made on the structure of the 'RI' allows us to exclude one of these two models. Results of labelling kinetics are best explained by a semi-conservative mechanism: they show that the 'plus' strand of the double-stranded component turns over and becomes single-stranded again (Hausen, 1965; Billeter, Libonati, Vinuela & Weissmann, 1966). However, studies on the duplex containing the parental 'plus' strand have given contradictory results: in experiments using a temperature-sensitive mutant of phage f2, a chase of the parental strand from the duplex was observed (Lodish & Zinder, 1966c); some other work on phage MS2 suggests that the parental strand remains in the duplex while the complementary 'minus' strand is displaced (Kelly & Sinsheimer, 1967). In any case, not all the movable strands are displaced at each replication. Hence the possibility remains in these experiments that part of the duplexes are involved in a conservative replication.

Two other findings are more easily explained in a conservative mechanism. Firstly, parental RNA has never been found incorporated in the progeny virions, as it should be if it was chased from the double-stranded structures (Davis & Sinsheimer, 1963). Secondly, upon denaturation by heat, phage double-stranded RNA directly purified from infected cells (Francke & Hofschneider, 1966), or obtained by ribonuclease treatment of the 'RI' (Erikson, Erikson & Gordon, 1966) yields infective molecules of the size of intact viral RNA. Unless one admits that 'minus' strands are also infective, this result cannot be understood if the 'plus' strand of the duplex is in several pieces. Likewise, solvent denaturation of poliovirus 'RF' releases single-stranded molecules of the same size as viral RNA (Katz & Penman, 1966) and no smaller pieces. The only way to explain these results, in terms of the semi-conservative model, is to assume that the strands of the 'RI' rearrange themselves after phenol extraction, so that the displaced intact 'plus' strand will reform a duplex with the 'minus' strand. But this rearrangement can also be invoked in the conservative model to explain the apparent displacement of strands from the duplexes.

One or two enzymes?

From their studies of RNA replication in temperature-sensitive and 'amber' mutants of phage f2, Lodish & Zinder (1966*a*, *b*) have suggested that each step of replication, respectively defined as the synthesis of double-stranded RNA from the parental 'plus' strand and the synthesis of 'plus' strands on double-stranded RNA, was performed by a different enzyme. *In vitro* studies seemed at first to correlate well with this hypothesis, since two types of enzymes could be isolated, some bound to an endogenous template and synthesizing mostly 'plus' strands (Weissmann, 1965), some using added single-stranded RNA to make a double-stranded product (Shapiro & August, 1965). However, the replicase isolated from Qβ phage-infected cells is endowed with both abilities, since it converts first the parental RNA into a double-stranded complex, then reads off the 'minus' strand of the complex to synthesize infective 'plus' strands (Mills *et al.* 1966). Therefore, at least in this system, the hypothesis of two virus-induced enzymes is no longer necessary.

Binding sites

An important characteristic of replicases is their high specificity towards the RNA template, since they work correctly only with homologous viral RNA. This suggests the existence on viral RNA of a specific site to which the enzyme binds. Indeed, Banerjee, Eoyang, Hori & August (1967) have shown that the first nucleotide of a 'plus' strand to be synthesized by the Qβ replicase is guanylic acid, and mentioned that the RNA template must contain cytosine in order to be copied. Thus the 'minus' strand is likely to have at its 3' OH end a binding site beginning with cytosine. As the enzyme is also capable of using 'plus' strands as template (first step), 'plus' strands should also have a specific binding site at their 3' OH ends. If the two binding sites were identical, both 'plus' and 'minus' strands should be terminated at their 5' end by a nucleotide or a nucleotide sequence complementary to the binding site (Fig. 2). Assuming that the two chains of a duplex are antiparallel, double-stranded RNA should also have self-complementary sequences at both ends, and form circles in certain ionic conditions.

So far, no evidence of circular molecules has been produced in phage RNA replication: examination of phage 'RF' has revealed only linear configurations (Granboulan & Franklin, 1966; see also Pl. 1). This would suggest that binding sites of 'plus' and 'minus' strands are different or too short to form stable bonds with their complementary counterpart. In the case of EMC-virus RNA 'RF', while most molecules appear in convoluted configurations with free ends (Pl. 2b), a minor, more rapidly sedimenting, fraction of molecules is in ring-like structures (Pl. 2c). Ribonuclease treatment does not affect these configurations (Granboulan & Montagnier, in preparation). As linear forms are infectious as well as ring-like structures, this possible circularity does not seem to be necessary for the initiation of replication.

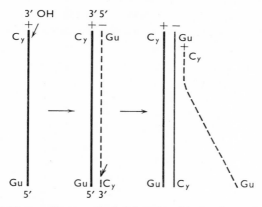

Fig. 2. How one initiation site on each viral 'plus' and 'minus' strand can order a single enzyme to carry out both steps of replication. Cy (cytidylic) is the initiation site (or its beginning), Gu (guanylic) is its complementary counterpart. The site used for each step is indicated by the short arrows. Newly made strands are drawn in dotted lines. (Other explanations in the text.)

A tentative model for RNA replication

This model is intended to give a synthetic picture accounting for the main facts previously reported, particularly from the *in vitro* results in the Qβ system. It is based on the existence of an initiation site on each of the 'plus' and 'minus' strands, and on the assumption that there should always be several polymerase molecules working on a conserved double-stranded template. This situation is likely to occur in *in vitro* experiments, where a large excess of enzyme has been used.

Let us consider first what could be the sequence of events in a Qβ-like system, in the *in vitro* conditions. A and B are respectively the binding sites of 'plus' and 'minus' strands.

The first polymerase molecule attaches at the site A of a 'plus' strand and starts synthesizing a 'minus' strand by moving along the template (Fig. 3a). Then a second enzyme molecule binds to the site A, being now in a double-stranded structure, and makes a second 'minus' strand, which does not remain attached to the double-stranded template (Fig. 3b).

When the second enzyme has moved away sufficiently, a third enzyme binds to A and so on. The whole complex becomes what is pictured in Figure 3c: a duplex made of the 'plus' strand and of the first 'minus' strand, with several enzymes working on it and growing 'minus' strands at the replication points. It is clear that this structure would be largely sensitive to ribonuclease, sensitive regions including the free 'minus' tails and the area of the duplex engaged in replication. Upon deproteinization, the whole length of the duplex becomes double-stranded, some of the single 'minus' strands are released in the medium, others remain attached to the duplex to form a 'replicative intermediate'.

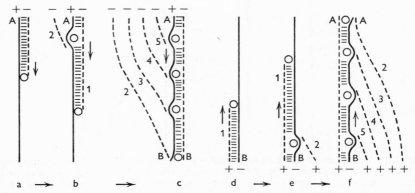

Fig. 3. A possible conservative model for RNA replication, in a, b, c: the first step; in d, e, f: the second step from one of the 'minus' strands made in the first step. Template strands are drawn in heavy lines, newly synthesized strands (numbered in chronological order) in dotted lines. Circles represent polymerase molecules. (Other explanations in the text.)

Second step: as soon as the 'minus' strands are completed, their ends will constitute new binding sites (B) for the enzyme molecules, which can thus leave the duplex. If there is still an excess of enzyme present in the medium, the process of replication on each 'minus' strand will be exactly symmetrical to that described in the first step and lead to large amounts of 'plus' strands (Fig. 3d, e, f). The structure pictured in Figure 3f would give rise, upon deproteinization, to a 'replicative intermediate' similar to that of the first step, but where single-stranded tails are 'plus' instead of being 'minus'.

As regards the duplexes made in the first step, it is possible that their 'minus' component would serve as template, while the opposite 'plus' strand is still used as template in the first step process. This situation may give rise to complete disruption of the double-stranded structure, which would not re-form after deproteinization.

In this model, polymerase molecules are acting on double-stranded templates, providing that the secondary structure of the latter has been partially disrupted by a previous enzyme molecule. When the number of available enzyme molecules decreases, the reaction will stop and give rise to inert double-stranded RNA. This is in agreement with the observations that added double-stranded RNA cannot be used as template by the Qβ replicase (Mills *et al.* 1966; Feix *et al.* 1967).

In vivo situation: the first step and the beginning of the second step will rely on the messenger activity of parental RNA. However, the environmental conditions may allow a single enzyme molecule to use double-stranded RNA as template, so that the requirements for an excess of enzyme molecules would not be as strict as *in vitro*.

The regulation of the second step will depend on the fate of the newly synthesized 'plus' strands: a large fraction will be involved in messenger functions (synthesis of all virus-specific proteins) or incorporated into virions, and consequently will not re-enter the replication cycle. In contrast, 'minus' strands can only serve for replication. This situation may explain by itself the large excess of synthesis of 'plus' strands: the number of duplexes where a 'minus' strand is used as template will be greater than that derived from 'plus' strands.

Participation of cellular enzymes?

Even though a single virus-induced enzyme can achieve the whole cycle of replication, eventual participation of cellular enzymes existing prior to infection cannot be excluded: as previously mentioned, double-stranded RNAs found after picornavirus infection are infective (Montagnier & Sanders, 1963b; Pons, 1964; Brown & Cartwright, 1964) and their biological activity is not due to single-stranded molecules attached to them. On the other hand, it is known from *in vitro* studies (Miura & Muto, 1965) that viral double-stranded RNA has no messenger activity in a cell-free system of protein synthesis. Therefore, it is likely that, *in vivo*, a cellular enzyme is capable of splitting the double-stranded RNA molecules or eventually using them as template to make single strands. Of special interest, in this connection, is the recent observation that infectivity of poliovirus double-stranded RNA is decreased by actinomycin D treatment of the recipient cells (Koch, Quintrell & Bishop, 1967).

Another example of possible participation of cellular polymerases is that given by some temperature-sensitive mutants of Qβ phage (Tsuchida, Nonoyama & Ikeda, 1966): after incubation at high temperature of cells infected with these mutants, no viral single-stranded RNA is made and no phages are produced, but the phage genome is perpetuated even after many divisions of the infected bacteria: the growth of phage genome parallels the bacterial growth, suggesting that replication of the former (presumably in a double-stranded state) is coupled with cell division.

REPLICATION OF VIRAL DOUBLE-STRANDED RNA

The RNA component of three viruses (reovirus, wound tumour virus, rice dwarf virus) is double-stranded (Gomatos & Tamm, 1963; Miura, Kimura & Suzuki, 1966). The molecular weight of whole molecules of reovirus RNA is around 12×10^6 daltons, as deduced from their length in electron microscopy (Granboulan & Niveleau, 1967). Phenol extraction yields smaller pieces in a non-random manner, and recent work suggests that there are three pieces inside the viral particle, held together by non-covalent bonds (Watanabe & Graham, 1967).

Investigation of reovirus RNA replication has been made possible by its relative insensitivity to doses of actinomycin D which reduce the level of cellular RNA synthesis (Kudo & Graham, 1965). Results can be summarized as follows:

(1) Replication of reovirus RNA does not require DNA synthesis (Loh & Soergel, 1967).

(2) 25–40 % of the newly made RNA is double-stranded and has the same sedimentation properties as virus RNA (Kudo & Graham, 1965).

(3) The remainder is single-stranded, and found associated with poly-ribosomes (Prevec & Graham, 1966); it can be separated into three size classes, each specified by one of the three pieces of double-stranded viral RNA (Watanabe & Graham, 1967).

(4) Synthesis of both RNAs depends on protein synthesis, but, whereas the requirement is continuous for double-stranded RNA production, it is only transient for single-stranded RNA. At later times, if protein synthesis is blocked, the single-stranded RNA accumulates in the cells.

(5) Single-stranded RNA does not seem to be a precursor of double-stranded RNA (Watanabe, Kudo & Graham, 1967).

It seems, therefore, that double-stranded RNA synthesis is mediated by a short-lived enzyme, synthesized in the presence of actinomycin D, hence presumably virus-coded. This replication may be semi-conservative, as for double-stranded DNA. On the other hand, a fraction of double-stranded RNA serves as template for the synthesis of single-stranded messenger RNAs. The latter function is probably mediated by a second, more stable, enzyme. This enzyme has to be of cellular origin at least for the initiation of replication, which requires transcription of the parental double-stranded RNA cistron coding for the viral polymerase.

REPLICATION OF THE RNAs OF MYXO- AND RELATED VIRUSES

In these viruses, the RNA, in a single-stranded form, is included in a filamentous nucleocapsid, enclosed in a lipid-containing envelope. Until very recently, almost nothing was known of the size, structure and replication of myxovirus RNAs. By the use of specific inhibitors, it could be shown that viral RNA replication displayed various degrees of dependence on cell information. From this point of view, three classes have to be considered: (1) The para-influenza group (Newcastle disease virus, Sendaï virus); (2) the influenza group (influenza, fowl plague virus); (3) the group of oncogenic viruses (Rous sarcoma virus, avian leukosis viruses, murine leukemia viruses, mouse mammary cancer virus).

Replication of Newcastle disease virus RNA

The RNA from purified preparations of NDV has recently been extracted in a single piece of a molecular weight of approximately 7×10^6 daltons (Duesberg & Robinson, 1965).

NDV RNA synthesis takes place in the cytoplasm and is relatively insensitive to actinomycin D, so that it could be studied in infected cells. Results of these studies have revealed some peculiar features: several RNA components with differing sedimentation rates appear during infection, all of smaller size than the RNA extracted from virions. They are mostly single-stranded, and their base composition is complementary to that of viral RNA; annealing experiments have confirmed this complementarity. Infected cells contain only small amounts of viral 'plus' RNA and double-stranded RNA (Kingsbury, 1966; Bratt & Robinson, 1967). Yet no RNA-polymerase activity has been found in the infected cells.

The finding of double-stranded and complementary RNAs suggests that the replication mechanism does not differ fundamentally from that previously mentioned. The peculiarity of this system might lie in the extensive reading of 'plus' strands presumably in double-stranded structures, which would give rise to the observed excess of complementary 'minus' strands. 'Plus' strands must also be made in order to be incorporated in virions, but only small amounts can be detected in this state, since mature virions do not accumulate in cytoplasm and are immediately expelled by membrane budding.

As suggested by Kingsbury (1966), the 'minus' strand may have a messenger function for the synthesis of viral proteins. It may be assumed, for instance, that the 'plus' strand primes an early replicase, whereas the 'minus' strand directs synthesis of the coat proteins and of some other enzymes needed for virus maturation.

Influenza group

The RNA extracted from influenza virus in the presence of divalent ions has a molecular weight of a little more than 3×10^6 daltons (Pons, 1967; Duesberg & Robinson, 1967). If divalent ions are removed, the molecule dissociates into several fragments (perhaps 4, as deduced from their size) (Duesberg & Robinson, 1967). It is therefore likely that such fragments pre-exist in the virions, held together by non-covalent bonds.

Actinomycin D, mitomycin C and u.v. irradiation inhibit multiplication of fowl plague and influenza viruses at doses which inhibit cellular RNA synthesis. These effects are maximal at the beginning of infection or just before infection, and disappear when RNA synthesis begins (Barry, 1964; Rott, Saber & Scholtissek, 1965; White & Cheyne, 1966). The resistance to u.v. light does not develop if protein synthesis is inhibited. These facts suggest that an early transcription of cell DNA is required for the initiation of viral RNA replication.

Furthermore, a nucleotidyl-transferase activity with properties differing from the cellular transcriptase has been found in infected chorio-allantoic membranes by Ho & Walters (1966). This activity is not inhibited by actinomycin D *in vitro*, but its appearance *in vivo* is prevented by treatment of the infected cells with actinomycin D.

Finally, virus-specific double-stranded RNA has been isolated from influenza virus infected chick embryo cells. Its size is that expected of a template for the viral RNA subunits. No excess of complementary single-stranded RNA has been found (Pons, 1967; Duesberg & Robinson, 1967).

Apart from the early cell-dependent event, the replication of this viral RNA does not seem to differ from the general picture. It seems likely that the early event concerns the synthesis of a viral RNA replicase, which thus would depend on cellular DNA transcription. In order to explain this dependence, it is not necessary to assume that cell DNA instead of viral RNA contains information for the enzyme; viral RNA messenger activity might require nuclear synthesis of ribosomal RNA precursors, which are known to be markedly affected by actinomycin D and u.v. light. It should be noted that, in contrast to viruses of the NDV group, viral nucleocapsids are produced in the nucleus (Schäfer, 1963).

Replication of oncogenic virus RNAs

In this group, viral RNA is larger (10^7 daltons) and has been recently extracted as a single piece (Robinson, Pitkanen & Rubin, 1965; Harel, Huppert, Lacour & Harel, 1965; Galibert, Bernard, Chenaille & Boiron, 1965). Structure of the virions is complex (de Thé & O'Connor, 1966) and the outer envelope contains cell components. Little information is available regarding viral RNA replication, most of it derived from the Rous sarcoma virus system.

RSV–RNA replication has two main aspects, corresponding to two types of cell-virus interaction:

In avian cells, infection leads to an active and continuous virus production with concomitant cell transformation. Some virus strains are defective for virus production, but defectiveness seems to affect the synthesis of viral envelope components rather than RNA replication (Hanafusa, Hanafusa & Rubin, 1964).

In mammalian cells, virus production is very low or non-existent, whichever viral strain is used. However, viral RNA is perpetuated indefinitely in the transformed cells, since contact of these cells with avian cells leads to virus production by the latter (Šimkovič, 1964; Svoboda, 1964).

Investigations of the effect of specific inhibitors on virus production in chick embryo cells have brought the following results:

(1) As in the influenza group, viral multiplication requires an early cell DNA transcription: this is shown by the irreversible inhibition of virus growth with actinomycin D early after infection or following pre-treatment of the cells with mitomycin C or small doses of u.v. light (Rubin & Temin, 1959; Temin, 1963; Vigier & Goldé, 1964; Bader, 1966).

(2) In RSV, in contrast to viruses of the influenza group, actinomycin D and u.v. light inhibit virus production at later times after infection, when RNA replication is likely to have already started. This effect is fully reversible (Temin, 1963; Vigier & Goldé, 1964; Bader, 1966). It suggests that another cell-dependent event is constantly required for the production of virions.

(3) DNA synthesis is required at a very early stage of infection, and not later: virus production is prevented by treatment of the cells at the time of infection with specific inhibitors of DNA synthesis (IUdR, BUdR, Ara-C) (Bader, 1965).

This requirement has led Temin (1964) to postulate that a DNA intermediate was involved in RSV–RNA replication and was transmitted as a 'provirus' in multiplying transformed cells.

At the present time, there is no convincing evidence that the newly made DNA is virus specific and not cell DNA. Temin's report (1964) that a fraction of RSV–RNA could be hybridized with DNA extracted from RSV-transformed cells cannot be taken as evidence for a proviral DNA, since some homology exists also between RSV–RNA and uninfected cell DNA (Harel, Harel, Goldé & Vigier, 1966).

A possible explanation for the latter homology would be the incorporation of cellular RNA in virus particles. The requirement for cell DNA synthesis may simply mean that initiation of viral RNA replication depends upon cellular enzymes or structural components present in cells only when DNA is replicating.

On the other hand, finding of virus-specific double-stranded RNA would be good evidence that RSV–RNA replication is based on the patterns established for other RNA-viruses.

We have found some small amounts of double-stranded RNA in RSV-transformed mammalian and avian cells, but a similar material exists also in *uninfected* cells, the significance of which is unknown (Montagnier, Vigier & Goldé, in preparation). It is still possible that a fraction of the double-stranded RNA found in transformed cells is virus-specific.

Attempts to detect an RNA-replicase activity in RSV-infected cells have so far failed (Wilson & Bader, 1965). However, in AMV-infected cells, a replicase using preferentially AMV–RNA as template has been recently isolated (Watson, Haruna & Beaudreau, 1967).

To conclude this short review, it can be said that we are far from knowing exactly how viral RNA is replicated. In the most studied systems, phages and picornaviruses, we know the general pattern of the replication mechanism, but the details are still a matter of discussion. In the case of myxoviruses and oncogenic viruses, we can only guess the general pattern. It seems that the more complex the virus, the greater the dependence of its replication on cell information. This dependence is also correlated with the appearance of more sophisticated forms of virus-host interaction: whereas phage and picornavirus infection is lytic for the host-cell, myxoviruses are produced in a way which does not always lead to cell death. Finally, oncogenic RNA viruses do not kill the host cell, the viral genome multiplies indefinitely, and its partial expression induces specific alteration of some cellular components. Hence, one is tempted to extrapolate and suggest that some self-replicating RNAs may have become part of cell genetic information.

List of abbreviations used in the text

TMV	tobacco mosaic virus	NDV	Newcastle disease virus
EMC	encephalomyocarditis virus	RSV	Rous sarcoma virus
		AMV	avian myeloblastosis virus
ME	Maus–Elberfeld virus	IUdR	iododeoxyuridine
FMDV	foot-and-mouth disease virus	BUdR	bromodeoxyuridine
		Ara-C	cytosine arabinoside

REFERENCES

AMMANN, J., DELIUS, H. & HOFSCHNEIDER, P. H. (1964). Isolation and properties of an intact phage-specific replicative form of RNA phage M 12. *J. molec. Biol.* **10**, 557.

ARNOTT, S., HUTCHINSON, F., SPENCER, M., WILKINS, M. H. F., FULLER, W. & LANGRIDGE, R. (1966). X-ray diffraction studies of double helical RNA. *Nature, Lond.* **211**, 227.

BADER, J. P. (1965). The requirement for DNA synthesis in the growth of Rous sarcoma and Rous-associated virus. *Virology*, **26**, 253.

BADER, J. P. (1966). Metabolic requirement for infection by Rous sarcoma virus. II. The participation of cellular DNA. *Virology*, **29**, 452.

BALTIMORE, D., BECKER, Y. & DARNELL, J. E. (1964). Virus-specific double-stranded RNA in poliovirus-infected cells. *Science*, **143**, 1034.

BALTIMORE, D. & FRANKLIN, R. M. (1962). Preliminary data on a virus-specific enzyme system responsible for the synthesis of viral RNA. *Biochem. biophys. Res. Commun.* **9**, 388.

BALTIMORE, D. & GIRARD, M. (1966). An intermediate in the synthesis of poliovirus RNA. *Proc. natn. Acad. Sci. U.S.A.* **56**, 741.

BANERJEE, A. K., EOYANG, L., HORI, K. & AUGUST, J. T. (1967). Replication of RNA viruses. IV. Initiation of RNA synthesis by the Qβ RNA polymerase. *Proc. natn. Acad. Sci. U.S.A.* **57**, 986.

BARRY, R. D. (1964). The effects of actinomycin D and ultraviolet irradiation on the production of fowl plague virus. *Virology*, **24**, 563.

BILLETER, M. A., LIBONATI, M., VINUELA, E. & WEISSMANN, C. (1966). Replication of viral ribonucleic acid. X. Turnover of virus-specific double-stranded ribonucleic acid during replication of phage MS 2 in *E. coli. J. biol. Chem.* **241**, 4750.

BRATT, M. A. & ROBINSON, W. S. (1967). Ribonucleic acid synthesis in cells infected with Newcastle disease virus. *J. molec. Biol.* **23**, 1.

BROWN, F. & CARTWRIGHT, B. (1964). Virus specific ribonucleic acids in baby hamster kidney cells infected with foot-and-mouth disease virus. *Nature, Lond.* **204**, 855.

COLTER, J. S., BIRD, H. H. & BROWN, R. A. (1957). Infectivity of ribonucleic acid from Ehrlich ascites tumor cells infected by Mengo encephalitis virus. *Nature, Lond.* **179**, 859.

DARNELL, J. E. (1962). Early events in poliovirus infection. *Cold Spring Harb. Symp. quant. Biol.* **27**, 149.

DARNELL, J. E., GIRARD, M., BALTIMORE, D., SUMMERS, D. F. & MAIZEL, J. V. (1967). The synthesis and translation of poliovirus RNA. In *Proc. Symp. Molec. Biol. Viruses, Edmonton, Canada*, 1966. New York: Academic Press (in the Press).

DAVIS, J. E. & SINSHEIMER, R. L. (1963). The replication of bacteriophage MS2. I. Transfer of parental nucleic acid to progeny phage. *J. molec. Biol.* **6**, 203.

DAVIS, J. E., STRAUSS, J. H. & SINSHEIMER, R. L. (1961). Bacteriophage MS2: another RNA phage. *Science*, **134**, 1427.

DE THÉ, G. & O'CONNOR, T. E. (1966). Structure of a murine leukemia virus after disruption with tween-ether and comparison with two myxoviruses. *Virology*, **28**, 713.

DUESBERG, P. H. & ROBINSON, W. S. (1965). Isolation of the nucleic acid of Newcastle disease virus (NDV). *Proc. natn. Acad. Sci. U.S.A.* **54**, 794.

DUESBERG, P. H. & ROBINSON, W. S. (1967). On the structure and replication of influenza virus. *J. molec. Biol.* **25**, 383.

ERIKSON, R. L., ERIKSON, E. & GORDON, J. A. (1966). Structure and function of bacteriophage R17 replicative intermediate RNA. I. Studies on sedimentation and infectivity. *J. molec. Biol.* **22**, 257.

ERIKSON, R. L., FENWICK, M. L. & FRANKLIN, R. M. (1964). Replication of bacteriophage RNA: studies on the fate of parental RNA. *J. molec. Biol.* **10**, 519.

ERIKSON, R. L., FENWICK, M. L. & FRANKLIN, R. M. (1965). Replication of bacteriophage RNA: some properties of the parental-labelled replicative intermediate. *J. molec. Biol.* **13**, 399.

FEIX, G., SLOR, H., & WEISSMANN, C. (1967). Replication of viral RNA. XIII. The early product of phage RNA synthesis *in vitro*. *Proc. natn. Acad. Sci. U.S.A.* **57**, 1401.

FENWICK, M. L., ERIKSON, R. L. & FRANKLIN, R. M. (1964). Replication of the RNA of phage R17. *Science*, **146**, 527.

FRANCKE, B. & HOFSCHNEIDER, P. H. (1966). Über infektiöse Substrukturen auf *E. coli* Bakteriophagen. VII. Formation of a biologically intact replicative form in ribonucleic acid bacteriophage (M12)-infected cells. *J. molec. Biol.* **16**, 544.

FRIEDMAN, R. M., LEVY, H. B. & CARTER, W. B. (1966). Replication of Semliki forest virus: three forms of viral RNA produced during infection. *Proc. natn. Acad. Sci. U.S.A.* **56**, 440.

GALIBERT, F., BERNARD, P., CHENAILLE, P. & BOIRON, M. (1965). Acide ribonucléique de haut poids moléculaire isolé du virus leucémogène de Rauscher. *C. r. hebd. Séanc. Acad. Sci., Paris*, **261**, 1771.

GEIDUSCHEK, E. P., MOOHR, J. W. & WEISS, S. B. (1962). The secondary structure of complementary RNA. *Proc. natn. Acad. Sci. U.S.A.* **48**, 1078.

GIERER, A. & SCHRAMM, G. (1956). Infectivity of ribonucleic acid from Tobacco Mosaic Virus. *Nature, Lond.* **177**, 702.

GODSON, G. N. & SINSHEIMER, R. L. (1967). The replication of bacteriophage MS2. VI. Interaction between bacteriophage RNA and cellular components in MS2-infected *Escherichia coli*. *J. molec. Biol.* **23**, 495.

GOMATOS, P. J. & TAMM, I. (1963). Animal and plant viruses with double-helical RNA. *Proc. natn. Acad. Sci. U.S.A.* **50**, 878.

GRANBOULAN, N. & FRANKLIN, R. M. (1966). Electron microscopy of viral RNA, replicative form and replicative intermediate of the bacteriophage R17. *J. molec. Biol.* **22**, 173.

GRANBOULAN, N. & NIVELEAU, A. (1967). Étude au microscope électronique du RNA du reovirus. *J. Microscopie*, **6**, 23.

HANAFUSA, H., HANAFUSA, T. & RUBIN, H. (1964). Analysis of the defectiveness of Rous sarcoma virus. II. Specification of RSV antigenicity by helper virus. *Proc. natn. Acad. Sci. U.S.A.* **51**, 41.

HAREL, J., HAREL, L., GOLDÉ, A. & VIGIER, P. (1966). Homologie entre génome du virus du sarcome de Rous et génome cellulaire. *C. r. hebd. Séanc. Acad. Sci., Paris*, **263**, 745.

HAREL, J., HUPPERT, J., LACOUR, F. & HAREL, L. (1965). Mise en évidence d'un acide ribonucléique de haut poids moléculaire dans le virus de la myelo-blastose aviaire. *C. r. hebd. Séanc. Acad. Sci., Paris*, **261**, 2266.

HARUNA, I., NOZU, K., OHTAKA, Y. & SPIEGELMAN, S. (1963). An RNA 'replicase' induced by and selective for a viral RNA: isolation and properties. *Proc. natn. Acad. Sci. U.S.A.* **50**, 905.

HARUNA, I. & SPIEGELMAN, S. (1965). Specific template requirements of RNA replicases. *Proc. natn. Acad. Sci. U.S.A.* **54**, 579.

HAUSEN, P. (1965). Studies on the occurrence and function of virus-induced double-stranded RNA in the ME-virus cell system. *Virology*, **25**, 523.

HO, P. K. & WALTERS, P. (1966). Influenza virus-induced ribonucleic acid nucleo-tidyltransferase and the effect of actinomycin D on its formation. *Biochemistry*, **5**, 231.

HOMMA, M. & GRAHAM, A. F. (1965). Intracellular fate of Mengo virus ribonucleic acid. *J. bact.* **89**, 64.

KATZ, L. & PENMAN, S. (1966). The solvent denaturation of double-stranded RNA from poliovirus infected HeLa cells. *Biochem. biophys. Res. Commun.* **23**, 557.

KELLY, R. B. & SINSHEIMER, R. L. (1964). A new RNA component in MS2-infected cells. *J. molec. Biol.* **8**, 602.

KELLY, R. B. & SINSHEIMER, R. L. (1967). The replication of phage MS2. VII. Non-conservative replication of double-stranded RNA. *J. molec. Biol.* **26**, 169.

KINGSBURY, D. W. (1966). Newcastle disease virus RNA. II. Preferential synthesis of RNA complementary to parental viral RNA in chick embryo cells. *J. molec. Biol.* **18**, 204.

KLEINSCHMIDT, A. K., DUNNEBACKE, T. H., SPENDLOVE, R. S., SCHAFFER, F. L. & WHITCOMB, R. F. (1964). Electron microscopy of RNA from reovirus and wound tumor virus. *J. molec. Biol.* **10**, 282.

KOCH, G., QUINTRELL, N. & BISHOP, J. M. (1967). Differential effect of Actino-mycin D on the infectivity of single- and double-stranded poliovirus RNA. *Virology*, **31**, 388.

KUDO, G. & GRAHAM, A. F. (1965). Synthesis of reovirus ribonucleic acid in L-cells. *J. Bact.* **90**, 936.

LANGRIDGE, R., BILLETER, M. A., BORST, P., BURDON, R. H. & WEISSMANN, C. (1964). The replicative form of MS2 RNA: An X-ray diffraction study. *Proc. natn. Acad. Sci. U.S.A.* **52**, 114.

LANGRIDGE, R. & GOMATOS, P. (1963). The structure of RNA. *Science*, **141**, 694.

LEVINTOW, L., THOREN, M. M., DARNELL, J. E. & HOOPER, J. L. (1962). Effect of *p*-fluorophenylalanine and puromycin on the replication of poliovirus. *Virology*, **16**, 220.

LODISH, H. F. & ZINDER, N. D. (1966*a*). Replication of the RNA of bacteriophage f2. *Science, N.Y.* **152**, 372.

LODISH, H. F. & ZINDER, N. D. (1966*b*). Mutants of the bacteriophage f2. VIII. Control mechanisms for phage-specific syntheses. *J. molec. Biol.* **19**, 333.

LODISH, H. F. & ZINDER, N. D. (1966*c*). Semiconservative replication of bacterio-phage f2 RNA. *J. molec. Biol.* **21**, 207.

LOH, P. C. & SOERGEL, M. (1967). Macromolecular synthesis in cells infected with reovirus type 2 and the effect of Ara-C. *Nature, Lond.* **214**, 622.

MANDEL, H. G., MATTHEWS, R. E. F., MATUS, A. & RALPH, R. K. (1964). Replica-tive form of plant viral RNA. *Biochem. biophys. Res. Commun.* **16**, 604.

MILLS, D. R., PACE, N. R. & SPIEGELMAN, S. (1966). The *in vitro* synthesis of a non-infectious complex containing biologically active viral RNA. *Proc. natn. Acad. Sci. U.S.A.* **56**, 1778.

MIURA, K. I., KIMURA, I. & SUZUKI, N. (1966). Double-stranded ribonucleic acid from rice dwarf virus. *Virology*, **28**, 571.

MIURA, K. I. & MUTO, A. (1965). Lack of messenger RNA activity of a double-stranded RNA. *Biochim. biophys. Acta*, **108**, 707.

MONTAGNIER, L. & SANDERS, F. K. (1963a). Sedimentation properties of infective ribonucleic acid extracted from encephalomyocarditis virus. *Nature, Lond.* **197**, 1178.

MONTAGNIER, L. & SANDERS, F. K. (1963b). Replicative form of encephalomyocarditis virus ribonucleic acid. *Nature, Lond.* **199**, 664.

NATHANS, N., NOTANI, G., SCHWARTZ, J. H. & ZINDER, N. D. (1962). Biosynthesis of the coat protein of coliphage f2 by *E. coli* extracts. *Proc. natn. Acad. Sci. U.S.A.* **48**, 1424.

PONS, M. (1964). Infectious double-stranded poliovirus RNA. *Virology*, **24**, 467.

PONS, M. W. (1967). Studies on influenza virus ribonucleic acid. *Virology*, **31**, 523.

PREVEC, L. & GRAHAM, A. F. (1966). Reovirus-specific polyribosomes in infected L-cells. *Science*, **154**, 522.

REICH, E. & FRANKLIN, R. M. (1961). Effect of mitomycin C on the growth of some animal viruses. *Proc. natn. Acad. Sci. U.S.A.* **47**, 1212.

REICH, E., FRANKLIN, R. M., SHATKIN, A. J. & TATUM, B. L. (1962). Action of actinomycin D on animal cells and viruses. *Proc. natn. Acad. Sci. U.S.A.* **48**, 1238.

ROBINSON, W. S., PITKANEN, A. & RUBIN, H. (1965). The nucleic acid of the Bryan strain of Rous sarcoma virus: purification of the virus and isolation of the nucleic acid. *Proc. natn. Acad. Sci. U.S.A.* **54**, 137.

ROTT, R., SABER, S. & SCHOLTISSEK, C. (1965). Effect on myxovirus of mitomycin C, actinomycin D and pretreatment of the host cell with ultra-violet light. *Nature, Lond.* **205**, 1187.

RUBIN, H. & TEMIN, H. M. (1959). A radiological study of cell-virus interaction in the Rous sarcoma. *Virology*, **7**, 75.

SATO, T., KYOGOKU, Y., HIGUCHI, S., MITSUI, Y., IITAKA, Y., TSUBOI, M. & MIURA, K. (1966). A preliminary investigation on the molecular structure of rice dwarf virus ribonucleic acid. *J. molec. Biol.* **16**, 180.

SCHÄFER, W. (1963). Structure of some animal viruses and significance of their components. *Bact. Rev.* **27**, 1.

SHAPIRO, L. & AUGUST, J. T. (1965). Replication of RNA viruses. II. The RNA product of a reaction catalysed by a viral RNA-dependent RNA polymerase. *J. molec. Biol.* **11**, 272.

SHIPP, W. & HASELKORN, R. (1964). Double-stranded RNA from tobacco leaves infected with TMV. *Proc. natn. Acad. Sci. U.S.A.* **52**, 401.

ŠIMKOVIČ, D. (1964). Interaction between mammalian tumor cells induced by Rous virus and chicken cells. *17th Natn. Cancer Inst. Monogr. Avian tumor viruses*, p. 351.

SIMON, E. H. (1961). Evidence for the non-participation of DNA in viral RNA synthesis. *Virology*, **13**, 105.

SINSHEIMER, R. L., STARMAN, B., NAGLER, C. & GUTHRIE, S. (1962). The process of infection with bacteriophage ϕX174. I. Evidence for a 'replicative form'. *J. molec. Biol.* **4**, 142.

SONNABEND, J. A., MARTIN, E. M. & MÉCS, E. (1967). Viral specific RNAs in infected cells. *Nature, Lond.* **213**, 365.

SPIEGELMAN, S., HARUNA, I., HOLLAND, I. B., BEAUDREAU, G. & MILLS, D. (1965). The synthesis of self-propagating and infectious nucleic acid with a purified enzyme. *Proc. natn. Acad. Sci. U.S.A.* **54**, 919.

SVOBODA, E. H. (1964). Malignant interaction of Rous virus with mammalian cells *in vivo* and *in vitro*. *17th natn. Cancer Inst. Monogr. Avian tumor viruses*, p. 277.

TEMIN, H. M. (1963). The effects of actinomycin D on growth of Rous sarcoma virus *in vitro*. *Virology*, **20**, 577.

TEMIN, H. M. (1964). Homology between RNA from Rous sarcoma virus and DNA from Rous sarcoma virus-infected cells. *Proc. natn. Acad. Sci. U.S.A.* **52**, 323.

TOBEY, R. A. (1964*a*). Mengo virus replication. I. Conservation of virus RNA. *Virology*, **23**, 10.

TOBEY, R. A. (1964*b*). Mengo virus replication. II. Isolation of polyribosomes containing the infecting viral genome. *Virology*, **23**, 23.

TSUCHIDA, N., NONOYAMA, M. & IKEDA, Y. (1966). Perpetuation of genome of a temperature-sensitive mutant of RNA bacteriophage in growing cells of *E. coli* at high temperature. *J. molec. Biol.* **20**, 575.

VIGIER, P. & GOLDÉ, A. (1964). Effects of actinomycin D and of mitomycin C on the development of Rous sarcoma virus. *Virology*, **23**, 511.

WATANABE, Y. & GRAHAM, A. F. (1967). Structural units of reovirus RNA and their possible functional significance. *J. Virol.* **1**, 665.

WATANABE, Y., KUDO, H. & GRAHAM, A. F. (1967). Selective inhibition of reovirus ribonucleic acid synthesis by cycloheximide. *J. Virol.* **1**, 36.

WATSON, J. D. & CRICK, F. H. C. (1953). The structure of DNA. *Cold Spring Harb. Symp. quant. Biol.* **18**, 123.

WATSON, K. R., HARUNA, I. & BEAUDREAU, G. S. (1967). Purification of an RNA-dependent RNA polymerase from leukemic cells. 58th Annual Meeting. *Proc. Am. Ass. Cancer Res.* **8**, 71.

WEISSMANN, C. (1965). Replication of viral RNA. VII. Further studies on the enzymatic replication of MS2 RNA. *Proc. natn. Acad. Sci. U.S.A.* **54**, 202.

WEISSMANN, C., BORST, P., BURDON, R. H., BILLETER, M. A. & OCHOA, S. (1964). Replication of viral RNA. III. Double-stranded replicative form of MS2 phage RNA. *Proc. natn. Acad. Sci. U.S.A.* **51**, 682.

WEISSMANN, C., SIMON, L. & OCHOA, S. (1963). Induction by an RNA phage of an enzyme catalyzing incorporation of ribonucleotides into ribonucleic acid. *Proc. natn. Acad. Sci. U.S.A.* **49**, 407.

WHITE, D. O. & CHEYNE, I. M. (1966). Early events in the eclipse phase of influenza and parainfluenza virus infection. *Virology*, **29**, 49.

WILSON, R. G. & BADER, J. P. (1965). Viral ribonucleic acid polymerase: chick embryo cells infected with vesicular stomatitis virus or Rous-associated virus. *Biochim. biophys. Acta*, **103**, 549.

WITTMANN, H. G. (1960). Comparison of the tryptic peptides of chemically induced and spontaneous mutants of tobacco mosaic virus. *Virology*, **12**, 609.

EXPLANATION OF PLATES

PLATE 1

Single-stranded RNA, replicative form and replicative intermediate of RNA bacteriophage R 17. (By courtesy of R. M. Franklin and N. Granboulan.) (*a*) Single-stranded RNA from bacteriophage R 17, spread in presence of urea. (Length: 1·06 μ.) Magnification: × 60,000. (This picture is taken from Granboulan & Franklin (1966) by permission of the authors and Academic Press.) (*b*) Double-stranded replicative form. (Length: 1·05 μ.) Magnification: × 60,000. (*c*) Replicative intermediate. Arrow indicates a single-stranded branch attached to the double-stranded 'backbone' molecule. Magnification: × 60,000.

PLATE 2

Single-stranded RNA and replicative form of EMC virus. (*a*) Single-stranded RNA from EMC virus, spread in presence of urea. (Length: 2·60 μ.) Magnification: × 46,800. (*b*) Double-stranded replicative form: linear molecules. (Length: 2·57 μ.) Magnification: × 46,800. (*c*) and (*d*) Double-stranded replicative form: two examples of molecules with 'ring-like' aspect. Magnification: × 45,000.

PLATE I

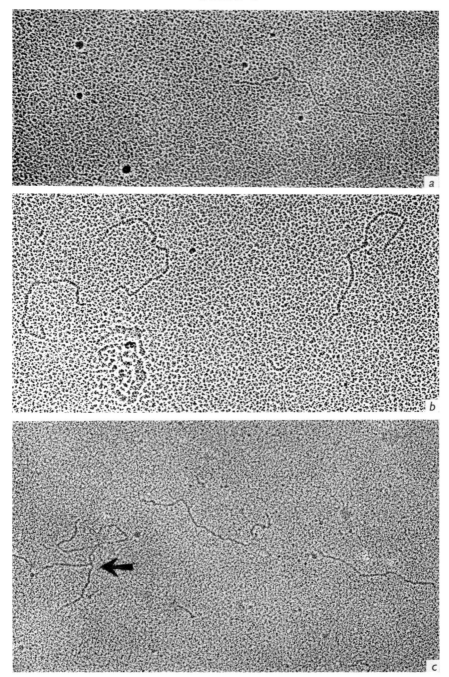

PLATE 2

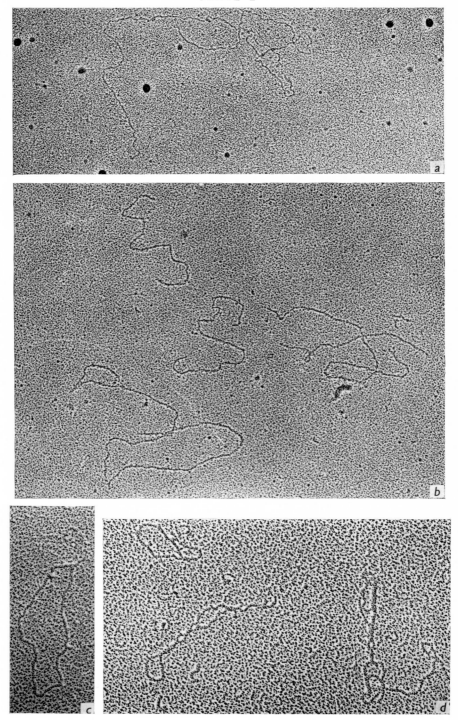

CONSIDERATIONS ON VIRUS-CONTROLLED FUNCTIONS*

JAMES E. DARNELL, JUN.†

Department of Biochemistry and Cell Biology, Albert Einstein College of Medicine, Bronx, New York, U.S.A.

Over the past decade research on the events which occur in virus infected cells has been channelled into two rather separate areas: (1) synthetic processes leading to progeny virus production, and (2) description of changes in host cell macromolecule synthesis or host cell behaviour incident to virus infection. Although there are interconnections between these two streams of research, a case can be made that by and large virus manufacture is not dependent on prior cell damage, but that the two sets of events are separate. Thus, on the one hand, viruses which replicate without killing cells are recognized, e.g. male-specific bacteriophages, SV5 myxoviruses, and Rous associated viruses (Hoffman-Berling & Mazé, 1964; Choppin, 1964; Rubin, 1962) and cell killing without virus multiplication has been frequently described, e.g. phage ghosts and u.v. inactivated vaccinia particles kill their host cells (Herriott, 1951; Joklik, 1966).

The purpose of this article will be to survey a variety of types of virus-induced events, dealing mainly with those topics which our laboratory has investigated.

NATURE OF VIRAL FUNCTIONS

Synthesis of virus materials

During the multiplication of viruses the investigator has offered to him a great variety of processes to study. Most effort in recent years has been concentrated on viral nucleic acid and protein studies and virology has in fact played a significant role in developing and advancing our concepts of gene expression and regulation.

Both the synthesis of progeny RNA and DNA, as well as the generation and translation of virus mRNA, have been widely studied. Two particularly good examples of the demonstration in virus-infected cells of generally important aspects of macromolecular synthesis can be

* This work was supported by the grants from the National Institutes of Health [CA 07861–03] and the National Science Foundation (GB 4565).
† Career Scientist of Health Research Council of City of New York.

mentioned. First, the initial demonstration of mRNA as a new class of short-lived RNA, which is derived from the DNA by base-pair copying, and which directs new protein synthesis by host ribosomes, was made in bacterial cells infected with T-even phages (Brenner, Jacob & Meselson, 1961). Subsequently convincing evidence was obtained for the existence of a class of cellular RNA which shared characteristics with phage mRNA.

Secondly, in animal cells, it was suspected on the basis of the longevity of many cell types that the mRNA might be substantially more stable than in bacterial cells. Evidence that this was the case was obtained when prolonged protein synthesis was observed after cells were treated with actinomycin D (Penman, Scherrer, Becker & Darnell, 1963), a drug which stops further RNA synthesis (Reich, Franklin, Shatkin & Tatum, 1961). An important type of experiment confirming the longevity of mRNA for particular enzyme proteins has been performed with vaccinia-infected cells. Here it was shown that enzymes induced during normal vaccinia infection (thymidine kinase and DNA polymerase) would continue to be synthesized for many hours after actinomycin D addition, although subsequent events of infection were blocked (McAuslan, 1963; Jungwirth & Joklik, 1965). Even more significant was the observation that while the accumulation of these enzymes was normally halted within a few hours after induction, the actinomycin-treated cells continued to accumulate enzyme past the normal 'shut-off' point. These and further experiments demonstrated that protein synthesis in animal cells can and in many instances must be regulated by proteins which interrupt the ongoing translation of mRNA.

In addition to following the separate synthesis and control of viral proteins and nucleic acids, studies have been performed both in infected cells and more recently in extracts or lysates of infected cells, the aim of which is to understand in molecular terms the underlying events in the morphogenesis of virus particles (Edgar & Wood, 1966; Maizel, Phillips & Summers, 1967). The success of studies of this type can be logically expected to lead to a better definition of how nucleoprotein structures are assembled in uninfected cells. Animal viruses have a particularly important role to play in studies relating to the assembly of complex structures. Many animal viruses are enveloped, that is, the nucleic acid and protein core is covered with a membrane composed of viral lipoprotein. The study of the synthesis and assembly of this type of structure should be especially rewarding since our knowledge of how lipoprotein structures are assembled is so scanty.

Viral functions which damage or change cells

Viruses are generally recognized by their ability to destroy cells, yet only meagre progress has been made in understanding, on the molecular level, the basis of virus-induced cell death. The chief cause for this is that the understanding of normal cell function and the ability to perform sensitive and sophisticated experiments on integrated cell functions was lacking, and necessarily, therefore, the points in cell metabolism where viruses act could not be well defined.

Again because of their central role in cell growth and division, cellular nucleic acid and protein synthesis in virus-infected cells have been popular areas for study.

In accord with the observations that viruses destroy cells, it has been repeatedly documented that various viruses, in bacteria and animal cells, stop host cell RNA, DNA and protein synthesis. The details of how this is accomplished are also beginning to be discovered. For example, many of the damaging effects of viruses are only expressed if the viral genome is allowed to produce products within an infected cell (Franklin & Baltimore, 1962; Penman & Summers, 1965). Thus in many cases it is not the simple entry of the virus particle which is responsible for the virus killing. On the other hand, some instances of cell killing by virus particles which themselves are inactivated are well documented. The further analysis of both kinds of virus-induced cell death is certain to be rewarding.

FUNCTIONS OF SMALL RNA VIRUSES

One of the most popular animal viruses for studies on virus macro-molecule synthesis is poliovirus. Among the reasons for this choice are: (1) the intrinsic thermal and chemical stability of the virion and consequently the ease with which it can be purified; and (2) the fact that the virus has a small genome which might ultimately allow definition of most or all of its gene products.

In addition, two fortunate experimental situations exist which allow the study of virus macromolecule synthesis without interference by continued host-cell synthesis. We will first be concerned with viral RNA synthesis and then discuss viral protein synthesis.

Viral RNA synthesis

Indispensable in studies of poliovirus synthesis is the antibiotic actino-mycin D, which completely blocks host-cell RNA synthesis while still allowing viral RNA replication (Reich *et al.* 1961). Thus incorporation of RNA precursors by poliovirus-infected actinomycin-treated cells is exclusively into viral RNA molecules (Darnell, 1962). This has provided a very sensitive means of studying viral RNA replication (Baltimore, Girard & Darnell, 1966).

New progeny RNA appears within the first hour of infection and accumulates exponentially for about 2·5–3 hr. RNA synthesis and accumulation after this time is approximately linear for an additional hour at the rate of about 2,000–3,000 viral RNA molecules per infected cell per minute. Viral RNA formed early in infection either never enters whole virus or does so very slowly while late in the developmental cycle (e.g. after 50 % of the viral RNA to be formed has been made) the newly formed molecules enter virions within a few minutes.

The enzymic process by which viral RNA is formed probably involves a newly induced viral RNA polymerase (Baltimore & Franklin, 1962) which operates by copying a strand of RNA complementary (according to Watson–Crick base-pairing characteristics) to that found in mature virions. In poliovirus infected cells it has been estimated that each template is simultaneously used by approximately four enzyme molecules for the generation of new viral RNA. The complex of RNA molecules which includes nascent viral RNA chains as well as the complementary template chain has been termed replicative intermediate (RI). (For review see Luria & Darnell, 1967, and Montagnier, this Symposium.)

Another form which is probably not active in synthesis (at least at the time it was extracted from its host cell) is a complete double-stranded molecule termed the replicative form (RF). RI predominates early in infection, whereas RF is found to a greater extent after viral RNA synthesis begins to slow down.

Interest in our laboratory has recently been focused on studies aimed at localizing by cell fractionation techniques the complexes which are active in RNA synthesis (Girard, Baltimore & Darnell, 1967). This has been accomplished by exposing cells to a radioactive RNA precursor for a period of time sufficiently short to label predominantly molecules identifiable as RI. Examination of extracts of cells labelled in this way has revealed that all of the RI is contained in a structure that sediments very rapidly (Fig. 1). The integrity of the structure, termed the *replication complex*, is destroyed by the protease, pronase, without affecting the RNA molecules in the complex (Fig. 2). Treatments designed to remove any ribosomes from the complex (e.g. EDTA treatment, or exposure of cells to puromycin before isolation of the complex) fail to change its sedimentation properties (Figs. 1 and 3). The replication complex contains all of the RNA synthesizing activity present in cell-free

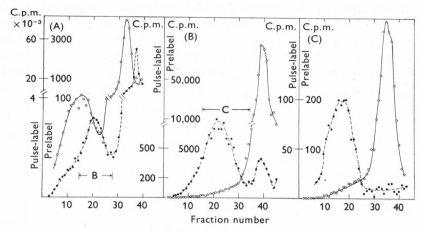

Fig. 1. Replication complex: Structural site of poliovirus RNA synthesis. Cells were labelled with ³H uridine (–O–) for a generation so that ribosomes would bear the majority of RNA label and then were infected with poliovirus. Other cells were very briefly labelled with ³H uridine (–●–, 2 min. exposure) during infection with poliovirus so that the structures bearing new virus RNA would be labelled. Extracts of both batches of cells were examined by sucrose gradient zonal sedimentation analysis (direction of sedimentation right to left in this and subsequent figures). A, Control extracts, DOC treated; B, EDTA-treated extracts, DOC treated; C, section from Part B re-examined again in EDTA buffer (Girard, Baltimore & Darnell, 1967).

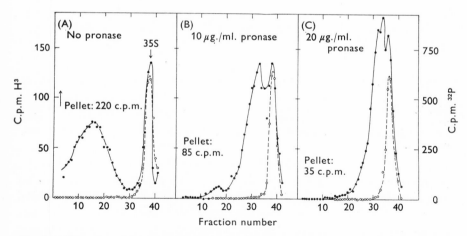

Fig. 2. Effect of pronase on replication complex. ³H-labelled (–●–) replication complex demonstrated as in Fig. 1 A. Pronase treatment destroys the complex without affecting the sedimentation behaviour of an added ³²P viral RNA (–O–).

extracts. It is therefore concluded that this complex represents in at least partially purified form, the functional site of RNA synthesis which existed inside the cell. One further important point about the replication complex is that it is observed in cell extracts only after treatment with the detergent, sodium desoxycholate. Thus, it appears that the complex may be part of some more highly organized lipid containing structure within infected cells (Penman, Becker & Darnell, 1964).

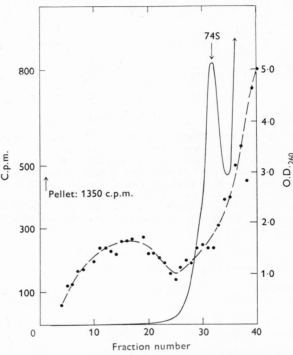

Fig. 3. Effect of puromycin on replication complex. ³H-labelled replication complex, demonstrated as in Figure 1 A, is not affected by treatment of cells with puromycin before cell fractionation (-●-). Polyribosomes, as judged by O.D.₂₆₀ (——), are entirely removed.

Viral protein synthesis

Another aspect of viral macromolecule synthesis in poliovirus-infected cells which has received wide attention has been the manufacture of viral proteins. The structures involved in viral protein synthesis are large units composed of 20–40 host-cell ribosomes simultaneously attached to and translating a complete strand of viral RNA (Summers, Maizel & Darnell, 1967). These viral polyribosomes, like the replicative complex, are not observed in cell extracts without sodium deoxycholate addition, again supporting the proposal that all virus synthetic events are localized in a lipid-associated structure (Penman *et al.* 1964). Electron microscopic examination of infected cells has not provided clear-cut evidence of such structures although the large virus polyribosomes do appear to be in somewhat closer proximity than normal cell polysomes to the irregularly sized pieces of endoplasmic reticulum seen in cultured cells (Dales, Eggers, Tamm & Palade, 1965).

A great step forward in the study of virus protein synthesis has been possible because of the improved technology with which polypeptides can be separated by acrylamide gel electrophoresis (Maizel, 1963, 1966). Also, in the study of poliovirus protein synthesis advantage can be taken of the fact that host protein synthesis is arrested in infected cells, even though viral RNA replication is completely prevented by treating infected cells with guanidine (Penman & Summers, 1965). Thus it becomes possible to suspend all host protein synthesis in the infected cell, and then remove the cells from the presence of guanidine, thus allowing RNA replication and the synthesis of large amounts of viral protein in the absence of interference by host-cell protein synthesis (Fig. 4) (Summers, Maizel & Darnell, 1965). In this situation, when radioactive amino acids are introduced into the culture medium, only virus-directed proteins become labelled. Analysis of extracts of cells labelled in this way has revealed some important characteristics of virus protein synthesis (Fig. 5).

First, when cell extracts are treated with detergent plus urea in the presence of mercaptoethanol all protein molecules are reduced to their constituent polypeptide chains, the ultimate product of a cistron. Electrophoretic analysis of such treated extracts from labelled infected cells reveals that at least ten to twelve separate polypeptides are formed in infected cells, four of which are capsid polypeptides and six to eight of which are non-capsid virus polypeptides. Poliovirus RNA is about 6,000 nucleotides long, and therefore has the capacity to code for 2,000 amino acids of ten peptide chains of average molecular weight 20,000 daltons. The pattern seen in an electrophoretic analysis therefore indicates that probably all the gene products of polio are expressed simultaneously. They are not expressed equally however, for there are clearly varying total amounts of radioactivity in different polypeptide chains which cannot be accounted for on the basis of variation in chain length. Also it can be seen that this variation in amounts is not confined to either capsid or non-capsid proteins. The capsid proteins are composed of four chains, two of which are synthesized much more extensively than the other two, and considerable variation in total amount of the non-capsid chains is also observed.

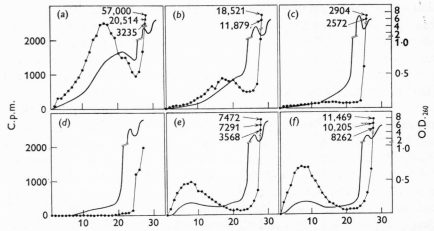

Fig. 4. Effect of poliovirus infection on polyribosome distribution in guanidine-treated HeLa cells. Cells infected with poliovirus in the presence of 3 mm guanidine were labelled with ^{14}C amino acids 60 and 120 min. after virus infection (a and b). Part of the infected culture was then centrifuged and the cells resuspended in guanidine-free medium. Samples were labelled with ^{14}C amino acids 60 (c), 120 (d), 135 (e) and 150 (f) min. later. Cytoplasmic extracts of all samples were then examined for polyribosomes by sucrose zonal sedimentation analysis (Penman et al. 1963; Summers et al. 1965). Solid line, O.D.$_{260}$; connected dots, radioactivity. (a) 60 min. post-infection, plus guanidine. (b) 120 min. post-infection, plus guanidine. (c) 180 min. post-infection, guanidine removed for 60 min. (d) 240 min. post-infection, guanidine removed for 120 min. (e) 255 min. post-infection, guanidine removed for 135 min. (f) 300 min. post-infection, guanidine removed for 180 min. (This figure is taken from Summers et al. (1965), with the permission of the National Academy of Sciences.)

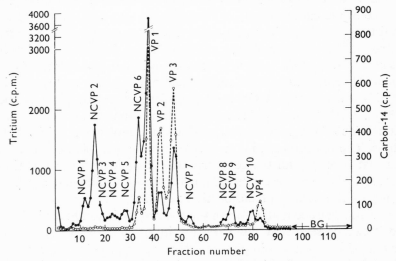

Fig. 5. Acrylamide gel electrophoresis of poliovirus proteins. Purified poliovirus labelled with ^{14}C amino acids ($-\bigcirc-$) was added to the cytoplasmic extract of cells treated as in Fig. 4 and labelled with ^{3}H amino acids ($-\bullet-$) through virus growth after removal of guanidine. Acrylamide gel analysis of the mixture was carried out according to Summers et al. (1965).

Another important aspect of the results on poliovirus protein synthesis is that the pattern of peptides formed is similar whether the examination is carried out 'early or late' in infection, that is, between the times when viral RNA replication is from 5–95% complete. The times referred to as 'early and late' are, of course, functionally different, e.g. little or no virus is formed within the time required to reach a level of viral RNA replication equivalent to 5% of the maximum yield, while virions are formed later. It should be recognized, however, that all experiments thus far have examined the proteins directed by progeny virus molecules, and *not* the original infecting strands. The interesting question of what proteins the translation of these original strands produces remains unanswered.

One final point in connection with the translation of poliovirus RNA late in infection should be pointed out. The average number of ribosomes per strand of viral mRNA declines late in infection by a factor of about two, yet as previously pointed out, the pattern of proteins synthesized is unchanged (Summers *et al.* 1967).

It seems reasonable to conclude therefore that there is no changing state of regulation of translation of this polycistronic mRNA but there is, rather, a built-in programme which automatically generates graded numbers of copies of various genes (Darnell *et al.* 1967).

Functions of RNA viruses which damage or change cells

Poliovirus infection appears to interrupt both host protein and RNA synthesis as primary events in infection (Zimmerman, Heeter & Darnell, 1963). Both effects come early in infection, with protein synthesis declining before RNA synthesis. Available evidence indicates that a virus-induced product(s) is responsible for both events: (1) u.v.-killed virus is completely inactive in harming cell function (Penman & Summers, 1965); (2) virus penetration and viral RNA release occur under circumstances where viral inhibition is not expressed (Mandel, 1967); and (3) most important, if virus expression is prevented by inhibitors of protein synthesis then no suppressive effect on host functions is observed until after protein synthesis has been restored for 15–30 min. (Penman & Summers, 1965; Franklin & Baltimore, 1962).

The mechanism of action of the proposed inhibitory viral product(s) remains unknown. For derepression of protein synthesis evidence has been described which is consistent with an attack of some subtle type on the initiating end of mRNA. Thus, as host-cell polyribosomes disappear, the average polysome size and the rate of function of individual ribosomes in polysomes remains the same (Penman *et al.* 1963; Willems &

Penman, 1966; Summers & Maizel, 1967). If the suppressing event caused a 'jamming' of the ribosomes in polysomes, the polysomes should increase in size; if the defect were a limitation of chain initiation then the polysomes should get smaller. The picture observed is most compatible with an all or none event at the beginning of the mRNA chain. Thus, according to this model, as soon as translation by the ribosomes already past the stage of initiation was completed, an event which normally requires only a minute, the affected mRNA molecule would be unable to support protein synthesis and a polysome would be destroyed.

The molecular basis for the block in RNA synthesis is equally obscure. It is apparent, however, that the major effects of virus-induced derangement in RNA synthesis are along the pathway of ribosome maturation (Darnell et al. 1967). Ribosomal precursor RNA synthesis is depressed within an hour of infection and, in addition, ribosomal maturation becomes faulty soon after infection.

These two events, depression of host protein synthesis and interruption of ribosome maturation, might be expected to be catastrophic events for the infected cell, and cell death would be certain to ensue in such a case.

Virus changes in cells at the cell surface

One of the more interesting properties of animal viruses is their ability to transform their host cells. Important changes are known to occur at the surface of cells which have been transformed by viruses (Rubin, 1962; Stoker, 1964). It seemed therefore that if the technology developed in the study of poliovirus, a cytolytic agent, could be used in a study of the phenomenon of cell surface changes a great deal of useful information might be forthcoming. As a model system for use in studying virus-induced changes of cellular surfaces we have chosen to study the arbovirus, Sindbis virus. This virus, like polio, is an RNA virus with a small genome. Sindbis virus, as shown by Pfefferkorn's very important work, possesses protein in its envelope, which is *newly made after infection* (Pfefferkorn & Hunter, 1963). This suggests that a viral product becomes the protein portion of the lipoprotein of the virus envelope. Examination of purified Sindbis virus reveals that it contains only two protein chains (Fig. 6; Strauss, Burge, Pfefferkorn & Darnell, in preparation). Treatment of the virus with low concentrations of deoxycholate, a lipid solubilizing agent, disrupts the whole infectious particle, releasing one of the two proteins present (Fig. 7). The other protein remains with the virus RNA in the form of a core particle (Fig. 8). Since the lipid (non-RNA ^{32}P) is also released from the core

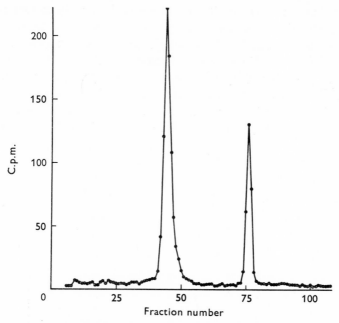

Fig. 6. Acrylamide gel electrophoresis of proteins of Sindbis virus. Purified Sindbis virus (Pfefferkorn & Hunter, 1963) was subjected to electrophoresis after SDS-mercaptoethanol treatment (Summers *et al.* 1965).

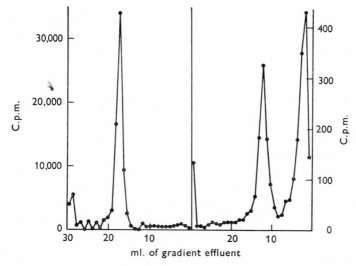

Fig. 7. Effect of Na deoxycholate on sedimentation of Sindbis virus: 'Core particles'. Purified Sindbis virus labelled with ¹⁴C amino acids was sedimented in 15–30 % sucrose gradients for 2 hr. (left panel). A small portion of the preparation was treated with 0·2 % Na deoxycholate before sedimentation, releasing the 'core' particle which sediments more slowly than the whole virus (right panel).

particle by this treatment, it was concluded that the DOC released
protein is the lipoprotein and efforts are under way to purify and
characterize this molecule. Ultimately it can be hoped that the synthesis
and entry of this protein into the membrane of the infected cells can be
examined. This should provide valuable information on the broad
general topic of how membranes are assembled as well as information
about the insertion of virus products into cellular membranes, a process
which is quite possibly fundamental to virus transformations of cells.

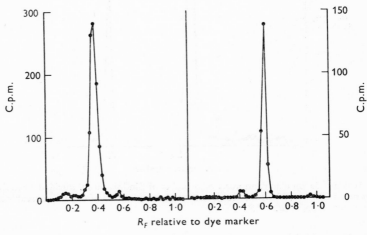

Fig. 8. Sindbis virus proteins. Protein labelled virus was treated with DOC as in Figure 7
and subjected to sedimentation analysis. The protein of the core particle and of the protein
released by DOC treatment were collected and analysed by acrylamide gel electrophoresis.
Left panel, DOC released protein; right panel, protein of core particle.

CONCLUSION

The study of virus specific functions has obviously been one of the key
research areas responsible for our present understanding of the biochemistry of gene expression. This has been especially true in bacteriophage research, but also studies on animal virus replication have
contributed information in this general area. There are many additional
opportunities open for the animal virologist. Indeed because somatic cell
genetics in animal cells cannot be utilized with the effectiveness that can
be achieved in bacterial genetics, the position can be taken that the most
effective means to study animal cell biology is to choose an appropriate
virus to introduce a controlled set of genes which perform or cause to be
performed the set of events one wishes to analyse. The obvious limitation to this approach is that we do not know a virus which will cause
any and every event of interest in animal cell biology. It is nevertheless

true that a vast array of interesting problems can be attacked, through the analysis of virus functions at the level of molecular interactions. One possible example of such a problem is the study of the synthesis and entry into organized structures of membrane proteins derived from viral genes.

REFERENCES

BALTIMORE, D. & FRANKLIN, R. M. (1962). Preliminary data on a virus-specific enzyme system responsible for the synthesis of viral RNA. *Biochem. biophys. Res. Commun.* **9**, 388.

BALTIMORE, D., GIRARD, M. & DARNELL, J. E. (1966). Aspects of the synthesis of poliovirus RNA and the formation of virus particles. *Virology*, **29**, 179.

BRENNER, S., JACOB, F. & MESELSON, M. (1961). An unstable intermediate carrying information from genes to ribosomes for protein synthesis. *Nature, Lond.* **190**, 576.

CHOPPIN, P. W. (1964). Multiplication of a myxovirus (SV5) with minimal cytopathic effects and without interference. *Virology*, **23**, 224.

DALES, S., EGGERS, H. J., TAMM, I. & PALADE, G. (1965). Electron microscopic study on the formation of poliovirus. *Virology*, **26**, 379.

DARNELL, J. E. (1962). Early events in poliovirus infection. *Cold Spring Harb. Symp. quant. Biol.* **27**, 149.

DARNELL, J. E., BALTIMORE, D., GIRARD, M., SUMMERS, D. F. & MAIZEL, J. V. (1967). The synthesis and translation of poliovirus RNA. In *Molecular Biology of Viruses*. Ed. J. S. Colter. New York: Academic Press.

EDGAR, R. S. & WOOD, W. B. (1966). Morphogenesis of bacteriophage T4 in extracts of mutant-infected cells. *Proc. natn. Acad. Sci. U.S.A.* **55**, 498.

FRANKLIN, R. M. & BALTIMORE, D. (1962). Patterns of macromolecular synthesis in normal and virus-infected mammalian cells. *Cold Spring Harb. Symp. quant. Biol.* **27**, 175.

GIRARD, M., BALTIMORE, D. & DARNELL, J. E. (1967). The poliovirus replication complex: site for synthesis of poliovirus RNA. *J. molec. Biol.* **24**, 59.

HERRIOTT, R. M. (1951). Nucleic-acid free T2 virus 'ghosts' with specific biological action. *J. Bact.* **61**, 752.

HOFFMAN-BERLING, H. & MAZÉ, R. (1964). Release of male-specific bacteriophages from surviving host bacteria. *Virology*, **22**, 305.

JOKLIK, W. K. (1966). The poxviruses. *Bact. Rev.* **30**, 33.

JUNGWIRTH, C. & JOKLIK, W. K. (1965). Studies on 'early' enzymes in HeLa cells infected with vaccinia virus. *Virology*, **27**, 80.

LURIA, S. E. & DARNELL, J. E., JUN. (1967). Review. In *General Virology*, chapter 14. New York: Wiley.

MAIZEL, J. V. (1963). Evidence for multiple components in the structural protein of type 1 poliovirus. *Biochem. biophys. Res. Commun.* **13**, 483.

MAIZEL, J. V. (1966). Mechanical fractionation of acrylamide gel electrophorograms: radioactive adenovirus proteins. *Science, N.Y.* **151**, 988.

MAIZEL, J. V., PHILLIPS, B. A. & SUMMERS, D. F. (1967). Composition of artificially produced and naturally occurring empty capsids of poliovirus type 1. *Virology*, **32**, 692.

MANDEL, B. (1967). The relationship between penetration and uncoating of poliovirus in HeLa cells. *Virology*, **31**, 702.

MCAUSLAN, B. R. (1963). The induction and repression of thymidine kinase in the poxvirus infected HeLa cell. *Virology*, **21**, 383.

PENMAN, S., BECKER, Y. & DARNELL, J. E. (1964). A cytoplasmic structure involved in the synthesis and assembly of poliovirus components. *J. molec. Biol.* **8**, 541.

PENMAN, S., SCHERRER, K., BECKER, Y. & DARNELL, J. E. (1963). Polyribosomes in normal and poliovirus infected HeLa cells and their relationship to messenger RNA. *Proc. natn. Acad. Sci. U.S.A.* **49**, 654.

PENMAN, S. & SUMMERS, D. F. (1965). Effects on host cell metabolism following synchronous infection with poliovirus. *Virology*, **27**, 614.

PFEFFERKORN, E. R. & HUNTER, H. S. (1963). The source of the ribonucleic acid and phospholipid of Sindbis virus. *Virology*, **20**, 446.

REICH, E., FRANKLIN, R. M., SHATKIN, A. J. & TATUM, E. L. (1961). Effect of actinomycin D on cellular nucleic acid synthesis and virus production. *Science, N.Y.* **134**, 556.

RUBIN, H. (1962). Response of cell and organism to infection with avian tumor viruses. *Bact. Rev.* **26**, 1.

STOKER, M. (1964). Regulation of growth and orientation in hamster cells transformed by polyoma virus. *Virology*, **24**, 165.

SUMMERS, D. F. & MAIZEL, J. V. (1967). Disaggregation of HeLa cell polysomes after infection with poliovirus. *Virology*, **31**, 550.

SUMMERS, D. F., MAIZEL, J. V. & DARNELL, J. E. (1965). Evidence for virus-specific noncapsid proteins in poliovirus-infected HeLa cells. *Proc. natn. Acad. Sci. U.S.A.* **54**, 505.

SUMMERS, D. F., MAIZEL, J. V. & DARNELL, J. E. (1967). The decrease in size and synthetic activity of poliovirus polysomes late in the infectious cycle. *Virology*, **31**, 427.

WILLEMS, M. & PENMAN, S. (1966). The mechanism of host cell protein synthesis inhibition by poliovirus. *Virology*, **30**, 355.

ZIMMERMAN, E. F., HEETER, M. & DARNELL, J. E. (1963). RNA synthesis in poliovirus-infected cells. *Virology*, **19**, 400.

THE STRUCTURE AND FUNCTION
OF VIRAL NUCLEIC ACID

L. V. CRAWFORD

Medical Research Council Experimental Virus Research Unit,
Institute of Virology, Glasgow, Scotland

INTRODUCTION

The nucleic acids of viruses offer a unique opportunity to correlate structure and function by virtue of their infectivity. In the most favourable situation where the extracted nucleic acid, free from virus protein, is infectious the effect of different treatments upon the nucleic acid can be determined directly. This is of course basically an unnatural situation and in many cases it has not been possible to obtain active nucleic acid preparations, but even where infectious nucleic acid cannot be obtained it is still possible to follow the effect of treating the nucleic acid contained in the virus particle both by biological and physical criteria.

Infectivity is not the exclusive property of one type of viral nucleic acid; single- and double-stranded RNA and DNA are all infectious in particular cases. The structure of the infective nucleic acid molecule also may be linear or circular. In the face of this variety we may inquire what basic features are essential for the biological activity of a virus nucleic acid. Infectivity may be determined simply by introducing the nucleic acid into susceptible cells and looking for progeny virus. In some cases the level of activity obtained by such a direct assay is low and assay with 'helper' is used. In this case the recipient cell is first infected with 'helper' virus and then exposed to the virus nucleic acid. The requirements for infectivity in the absence of any helper are clearly the most exacting. The virus nucleic acid outside the cell has to get through a number of barriers, depending on the system involved, before reaching the site within the host cell at which replication occurs.

The nucleases on and near the cell surface constitute the first barrier and these may account to a large extent for the low specific infectivity of viral nucleic acids in many systems. This is in one sense a purely artificial difficulty, arising from the fact that the virus nucleic acid is deprived of its protective protein coat. In some systems surface nucleases may however be important in normal infection with virus particles, for instance in the systems where host controlled modification of the nucleic acid is a pre-requisite for successful infection.

The cell wall and cell membrane present the next barrier to penetration and are particularly important in bacterial systems. Some of the differences between the requirements for infectivity in animal virus systems and in bacterial virus systems may arise from the difficulty of penetrating the bacterial cell wall. For example, this might explain the fact that double-stranded RNA is infectious in animal cells but analogous double-stranded RNA is not infectious in bacterial cells.

Even after penetration into the cell the virus nucleic acid has still to find its way to a site of replication without inactivation by the nucleases present. In a bacterial cell this may be relatively easy, but may be less so in animal or plant cells. Here the cells are larger and the interior is divided up by many internal membranes.

Nucleic acid transcription

Having reached the site of replication the viral nucleic acid is then dependent upon the pre-existing enzymes of the cell for the first stages of its replication. Firstly, for the synthesis of a complementary strand on incoming single-stranded DNA and, secondly, for the transcription of the double-stranded DNA or RNA into messenger RNA. The only exception to this requirement for transcription would be single-stranded RNA which is able to act as messenger directly. For the synthesis of a complementary DNA strand and for transcription into messenger RNA the viral nucleic acid must possess the appropriate attachment sites corresponding to the affinities of the host-cell polymerases.

Messenger RNA formed by transcription of virus DNA and virus RNA acting directly as messenger appear to have rather different properties. RNA extracted from viruses containing single-stranded RNA acts as a very efficient template for *in vitro* amino acid incorporation, possibly because it can be read by mature ribosomes. On the other hand the messenger RNA of DNA viruses appears to be similar to that of the host cell, in that it acts rather inefficiently as template *in vitro*. This may be due to some requirement for the attachment of an immature ribosome for the initiation of reading. This difference might be responsible for the different abilities of heated cells to support the growth of RNA and DNA viruses reported by Gharpure (1965). Any treatment which interrupted the supply of ribosomal RNA or of nascent ribosomes would affect the reading of DNA coded messenger but not of virus RNA acting directly as messenger.

Nucleic acid replication

The viral nucleic acid must also be replicated and again the requirements for RNA and DNA viruses are rather different. Double-stranded virus DNA may be replicated by host enzymes, or at least by virus coded enzymes analogous to those of the host, since the overall processes of viral DNA replication and host DNA replication are similar. Viral DNA with unusual features, such as single-stranded DNA, must involve additional steps which will be discussed later. Viral RNA replication, on the other hand, is different from anything occurring on a substantial scale in the cell before infection. Accordingly, the enzymes involved would be expected to be virus coded, and this has been shown to be so in several cases.

The processes of transcription, translation, and replication, are dependent on the nucleotide sequence of the nucleic acid. In addition to this requirement for information in the nucleic acid sequence, the earlier stages of the process of infection may impose certain additional requirements for the physical structure of the nucleic acid. We may now consider in detail particular systems where the physical and biological properties of the virus nucleic acid have been examined in some detail.

RNA VIRUSES

Structurally the simplest viral nucleic acid is that of the small RNA containing viruses of plants, animals or bacteria. The virus particles may have either cubic or helical symmetry. The virus RNA appears to be a single linear polynucleotide chain comprising 3,000 to 6,000 nucleotides. In view of the fact that circularity is required for infectivity in many of the systems discussed subsequently it is reasonable to ask whether we can be certain these RNAs are really linear. In rod-shaped viruses such as tobacco mosaic virus (TMV) the structure of the virus makes it unlikely that the RNA in the virus particle is circular. Some of the bacteriophages containing circular DNA are of course rod-shaped but here the structure of the virus particle is quite different from that of TMV. This argument cannot in any case be applied to RNA containing viruses with cubic symmetry. However, particles have been made artificially from TMV protein reconstituted on turnip yellow mosaic RNA and these particles may be infectious (Matthews & Hardie, 1966). These particles are rod-shaped so that, if turnip yellow mosaic virus RNA is circular, as has been suggested from its physical properties (Strazielle, Benoit & Hirth, 1965) then this circularity may not be essential for infectivity. Chemical evidence also supports a linear structure for virus RNAs, both TMV and phage f2 RNA having a constant 3'-end. The RNA chain terminates: — —pyrimidine-adenine in both cases (Lee & Gilham, 1965; Sugiyama, 1965). The base at the opposite, 5'-end of phage f2 RNA is also adenine (Takanami, 1966). The fact that the bases at the ends of the RNA molecule are always the same is consistent with the idea that the molecule is linear, or less likely, that it is circular with some preferential point at which the ring opens, for example between two adenines.

The effect of removing nucleotides from the end of the RNA chain has been studied in the case of TMV (Steinschneider & Fraenkel-Conrat, 1966 a,b). Although there is some difficulty in correlating the bulk properties of the RNA with those of the infectious RNA molecules in systems where the particle:infectivity ratio is very high it seems that the infectivity of the RNA is very sensitive to alteration or removal of nucleotides from the 3'-end of the RNA. Change or removal of the terminal adenosine residue caused the loss of most of the infectivity, but 20 % did survive this treatment. The presence or absence of phosphate at the end of the chain had little effect.

The characteristic nucleotide sequence of the virus RNA, particularly at the ends of the molecule, must be most important both in replication and translation. It has been shown that the RNA synthetases of phages $Q\beta$ and MS 2 are highly specific in recognizing their homologous RNAs, and also that the enzymes are much more active on intact molecules than on fragments (Haruna & Spiegelman, 1965). This leads to the idea that the ends of the RNA molecule must be unusual in structure and that both ends, possibly joined by hydrogen bonds to form a double-stranded region, are required for the synthetase to act. If this is correct then the RNA of these phages is functionally circular at this stage in replication, even though it is physically linear under other conditions. The same may also be true for the other virus RNAs where the RNA of the virus particle is a linear single strand.

The replication of RNA viruses involves a double-stranded replicative intermediate. In animal virus systems this double-stranded RNA is infectious but in bacterial virus systems it does not appear to be infectious even to spheroplasts until denatured (Ammann, Delius & Hofschneider, 1964). This difference might be due simply to the difficulty of penetrating the bacterial cell wall but a more probable explanation is mentioned later. The fact that the RNA has a double-stranded structure would not in itself seem to provide a sufficient reason, since both single-stranded and double-stranded DNA are infectious in other bacterial systems. In some animal virus systems, for example poliovirus, the specific infectivity of double-stranded RNA is actually higher than that of single-stranded RNA (Koch, Quintrell & Bishop, 1966). This may be due, at least in part, to the greater resistance of the double-stranded RNA to the ribonucleases found in cells and media.

The initiation of infection by the replicative form of a single-stranded RNA virus is in fact very different from initiation by the single-stranded form, in spite of the fact that double-stranded RNA is a natural inter-mediate in the replication of single-stranded RNA. When the incoming RNA is single-stranded it has the opportunity of first acting as messenger and thereby producing the virus coded polymerase(s) required for subsequent steps. The double-stranded RNA is not capable of doing this and must depend on the action of enzymes already present in the cell before infection. To obtain single-stranded RNA from the incoming double-stranded material there are two alternatives. Either the comple-mentary strand can be stripped off or a single-stranded copy can be made using the double-stranded RNA as template. The second alternative is perhaps easier to envisage, being analogous to the normal transcription of double-stranded DNA into messenger RNA. The critical difference between animal and bacterial cells may be that animal cells contain an enzyme capable of transcribing a double-stranded RNA template into messenger RNA. Alternatively, the enzyme might be a destructive one capable of destroying one strand of double-stranded RNA selectively.

It is interesting to compare the process of infection initiated by replicative form RNA, with that of reovirus where the RNA of the virus particle itself is double-stranded (Gomatos & Tamm, 1963a). Quantita-tive comparisons between the various systems are difficult to make since the conditions used were so different. Experiments with poliovirus RNA have shown that the infectivity of the double-stranded RNA was much more sensitive to pretreatment of the cells with actinomycin D than was the infectivity of single-stranded viral RNA (Koch, Quintrell & Bishop, 1967). Actinomycin D at a concentration of 0·4 μg./ml. depressed the relative infectivity of the double-stranded RNA (as compared to that of single-stranded RNA) to one-tenth of the control (no drug). In the reovirus system where the cells were infected with intact virus, a concentration of 1 μg./ml. was required to get a comparable effect (Shatkin, 1965; Kudo & Graham, 1965). However, the cells to be infected with poliovirus RNA were exposed to the drug for several hours before infection, whereas in the reovirus system the drug was added at the same time or after the virus. Also the effect on poliovirus RNA infectivity was determined by scoring the number of infectious centres produced by infection of the cells pretreated with the drug. The quantity measured with reovirus was the yield of virus in a single cycle of infection. In the absence of any direct comparison it will be assumed for the sake of the argument that all these differences compensate for each other. If the two systems show comparable sensitivity to actinomycin D this may indicate that the early stages of infection in these two cases are similar and involve an enzyme which is moderately sensitive to the drug, although not so sensitive as the cellular DNA primed RNA polymerase.

Several plant viruses also appear to contain double-stranded RNA, showing that this type of nucleic acid is not a unique property of animal viruses (Gomatos & Tamm, 1963 b). However, no bacterial viruses with double-stranded RNA have been reported and it might be predicted from the argument above that none will be found.

DNA VIRUSES

Bacteriophage φX 174

The situation with DNA viruses is more complex than with the RNA viruses as there is a greater variety of nucleic acid structure to be considered. The DNA of φX 174 in the virus particle is in the form of a single-stranded ring and the circularity of the molecule appears to be essential for its infectivity (Fiers & Sinsheimer, 1962). This immediately raises the question which recurs many times subsequently—is circularity *per se* essential or does the opening of the ring almost always destroy some essential function by splitting the corresponding cistron into two parts? It is still not possible to give a definite answer to this question.

Replication of φX 174 DNA involves a double-stranded replicative form (RF) and it is possible to compare the relative infectivity of the RF and denatured forms derived from it with that of the single-stranded virus DNA. In general the infectivity of the double-stranded form is lower than that of the single-stranded DNA, even though both are assayed in the same spheroplast system where differential effects of the bacterial wall should be minimal. The supercoiled RF DNA, comprising two intact circular polynucleotide strands, has one-twentieth the specific infectivity of the single-stranded virus DNA. However, denaturation of the RF increases its specific infectivity 25-fold, making this material more infectious than the single-stranded DNA (Burton & Sinsheimer 1965; Sinsheimer, Lawrence & Nagler, 1965). This highly infectious denatured DNA nevertheless retains the high resistance to inactivation by ultraviolet light which is characteristic of double-stranded DNA. This suggests that it is not the presence of the complementary strand which depresses the infectivity of the double-stranded DNA, but its physical structure. The isolated complementary strand of φX 174 indeed appears to be infectious by itself (Rüst & Sinsheimer, 1967). Since this system is one of the few in which the infectivity of single-stranded, double-stranded, and denatured double-stranded DNA can be compared, these results are particularly important. Unfortunately the infectivity of any DNA preparation is affected by the conditions of assay and different forms of the DNA may be affected to different extents. This has been shown to be the case with φX 174. The infectivity of φX 174 RF is more strongly inhibited by the presence of *Escherichia coli* DNA than that of single-stranded φX 174 DNA (Burton & Sinsheimer, 1965; Jaenisch, Hofschneider & Preuss, 1966). In the latter paper the infectivity of RF component II (unsupercoiled circular molecules) was found to be increased very little by denaturation, zero to threefold. The effect of denaturation on supercoiled molecules was not determined, although the infectivity of both supercoiled and unsupercoiled molecules was depressed by *E. coli* DNA. The bacterial DNA may come from broken spheroplasts in the assay or be present in the phage DNA preparation used. The larger the amount of bacterial DNA present, the more the infectivity of the RF will be depressed, and the greater will be the increase of infectivity on denaturation. It is possible therefore that the specific infectivity of φX 174 RF and that of single-stranded φX 174 DNA would be similar in the absence of *E. coli* DNA.

All the various forms of φX 174 DNA rely to a large extent on the action of host enzymes to carry out the initial stages of the infectious process, as discussed in detail by Sinsheimer in this Symposium. Thus since a host enzyme, probably one normally concerned with DNA repair, is involved in adding a complementary strand to the incoming DNA the same system can apparently function with an incoming complementary strand. The next stage, the circularization of the newly formed complementary strand, is also likely to be done by a host enzyme of the type described by Gellert (1967) which is active on λ DNA. Yet a third host enzyme is responsible for the replication of the RF once formed (Denhardt, Dressler & Hathaway, 1967) and a fourth one, DNA primed RNA polymerase, is responsible for transcribing the virus messenger RNA from the RF. This dependence of the virus on the host is reflected in the striking variation in the relative infectivity of various forms of the virus DNA when assayed on irradiated bacteria or bacteria with lesions in particular enzymes of DNA metabolism. One of the bacterial enzymes has been shown to be essential for virus replication but not to be essential for the growth of the bacteria (Denhardt *et al.* 1967). This is the enzyme which replicates RF, giving rise to progeny RF.

Bacteriophage λ

Further evidence for the importance of the host in dictating the require-ments for infectivity of virus DNA comes from studies with λ DNA. Two systems are available for the assay of λ DNA infectivity. The recipient bacteria are either preinfected with 'helper' phage or treated with lysozyme and versene to remove part of the cell wall and convert them to spheroplasts (Kaiser & Hogness, 1960; Meyer, Mackal, Tao & Evans, 1961; Young & Sinsheimer, 1967). In the first system, which has been used most extensively, the 'helper' phage is suitably marked genetically to allow it to be differentiated from the phage from which the DNA was derived. Although the infectious entity involved here appears to be the whole DNA molecule, the system is in some respects similar to 'marker rescue'. The incoming DNA may rely to some extent on the 'helper' phage for performance of essential functions. Infectivity in the 'helper' system is therefore somewhat different from that in the sphero-plast system.

Lambda DNA in the virus particle is probably a linear double-stranded molecule in which the 5'-ends of the two strands project as short, single-stranded regions about twenty nucleotides long. The two ends have complementary sequences and can join reversibly by hydrogen bonds to make the molecule circular. These cohesive ends are very intimately involved with the infectivity of the DNA in the 'helper' assay (Kaiser & Inman, 1965). Alteration of the ends of the DNA molecule has a striking effect on its infectivity, 'repair' of the single-stranded regions by addition of nucleotides to the 3'-end of the other strand causing the loss of infectivity. This effect can be reversed by removing the added nucleotides by exonuclease treatment (Strack & Kaiser, 1965). Another interesting point which comes out of these and other studies on the nature of the cohesive ends of lambda DNA is that the presence or absence of phosphate on the terminal nucleotide is not important for infectivity (Wu & Kaiser, 1967). Apparently cells contain phosphatases and kinases capable of converting the terminal nucleotide to the required state. A similar result was also obtained with TMV RNA, as mentioned previously.

The separated single strands of λ DNA may also have some infectivity, again in the 'helper' assay (Brody et al. 1964; Young & Sinsheimer, 1964). It is not known whether one or both of the strands is infectious, but the presence of 'helper' in the assay, combined with the ability of the host to repair DNA, make it likely that both strands would carry infectivity. Since both the complementary strands of the DNA are present in the same samples it is also possible that the infectivity observed is the result of annealing of the strands to re-form double-stranded molecules. More definite evidence for the infectivity of single-stranded DNA came from studies on the denaturation of circular λ DNA. Alkaline denaturation of unsupercoiled circular molecules of DNA produces one circular single strand and one linear single strand, as described below. Lambda DNA sedimenting as circular single strands in alkaline sucrose gradients was found to be highly infectious in the spheroplast assay (Kiger, Young & Sinsheimer, 1967). In fact the specific infectivity of this DNA, although variable, was greater than that of native DNA from phage particles. Most of this infectivity had the high sensitivity to u.v. light characteristic of single-stranded DNA, so that the possibility that the infectivity was due to the formation of double-stranded DNA by annealing during the assay can definitely be ruled out here (E. T. Young & J. A. Kiger, personal communication).

As in the case of ϕX 174, λ DNA undergoes conversion to physically different forms during replication and again during maturation. A circular molecule results from the joining of the cohesive ends of the DNA by hydrogen bonds, leaving a discontinuity in each of the strands. These discontinuities are then healed by alkali stable bonds, giving a supercoiled circular molecule. This operation occurs after λ infection and extracts of sensitive, lysogenic, or induced lysogenic bacteria can perform the healing operation *in vitro* (Gellert, 1967). The enzyme concerned seems to be present in uninfected bacteria, rather than being induced by phage infection. Molecules in which only one of the discontinuities has been healed, or supercoiled molecules in which one single-strand scission has occurred, are circular but unsupercoiled. On denaturation therefore the two strands can separate, giving one circular and one linear single strand. The supercoiled form of λ DNA is not infectious in the 'helper' assay, but is infectious in the 'spheroplast' assay (Bode & Kaiser, 1965; Young & Sinsheimer, 1967). This under-lines the importance not only of the structure of the virus DNA but also of the type and state of the recipient host. Although it has not been shown conclusively that the *supercoiled* circular form of λ DNA is an obligatory intermediate in the replication of the virus, it is very probable that it is. Genetic evidence suggests that some circular form must be involved.

Recent work has indicated that the cistrons in different regions of λ DNA are read in different directions and therefore on opposite strands of the DNA. This might not have been expected in view of the finding that one strand of λ DNA appeared to be dispensable for infectivity and one indispensable (Fox & Meselson, 1963). Damage to the virus DNA inside the virus particle, as a result of the action of visible light on bromouracil incorporated into one or other of the strands of the DNA, inactivated only half the virus at a high rate. Possibly damage to the region(s) of the DNA concerned with some early function(s) is lethal whereas damage elsewhere can be repaired later.

Polyoma virus

Although polyoma virus is an animal virus and contains double-stranded DNA it is functionally similar in many respects to ϕX 174. The situation is complicated by the fact that the virus has at least two biological activities and these are normally assayed in different host cells. The lytic activity of the virus is displayed in mouse cells and corresponds to infectivity. The transforming activity of the virus is normally assayed in hamster cells. The relationship between lytic and transforming activity is not clear, involving as it does the comparison of different cell types as well as different activities. Transformation in this context means the alteration of the properties of the recipient cell so that it can grow and divide under conditions where the untransformed cell remains static. In contrast to the lytic cycle where the cell is killed and virus progeny produced, transformation results in cell survival and no virus progeny. There is therefore no *a priori* reason to expect that the two activities will depend on exactly the same properties of the virus DNA. Since no progeny virus is produced in transformation the requirements for this process may well be less exacting, and some functions involved in the production of progeny virus, for instance the formation of coat protein, may be dispensable. Infection with polyoma virus also causes the stimulation of cellular DNA synthesis in static cells but it is not clear whether this activity is merely an early stage of the other two activities mentioned or whether it is independent of them. We shall return to this point later.

The DNA extracted from polyoma virus is unusual in that it comprises three physical components—I, II and III (Weil & Vinograd, 1963). Components I and II are virus DNA, whereas component III appears to contain mouse DNA, derived from the cells in which the virus is propagated (Winocour, 1967). This variety of biological properties of the virus and variety of DNA components make it essential to correlate the physical and biological properties of the separated DNA species.

The intact DNA of the virus, component I, comprises two poly-nucleotide strands, each separately continuous and circular. Under the usual conditions of examination in aqueous solution the DNA is supercoiled. This is apparently because, when the molecule was made, the number of turns in the basic Watson–Crick double helix was less than normal for double-stranded DNA of this size. The deficient turns are then supplied by the twisting of the molecule as a whole (Vinograd *et al.* 1965). The sense of the supercoiling and the number of supercoiling turns have been determined by a variety of methods based on the examination of the denaturation behaviour of the DNA and the effect of intercalating drugs on the physical properties of the DNA (Crawford &

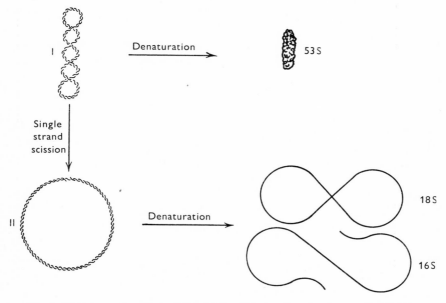

Fig. 1. Diagrammatic representation of the denaturation of polyoma virus DNA.

Black, 1964; Lebowitz, Watson & Vinograd, 1966; Crawford & Waring, 1967). Estimates for the number of supercoiling turns obtained by different methods are in reasonable agreement, ranging from about 12 to 15–20 turns per molecule of polyoma DNA.

Component II is derived from component I by the introduction of one or more single-strand scissions into the molecule, allowing the super-coiling turns to be released. In addition to these naturally occurring forms of the DNA there are other forms obtained by denaturation. Collapse of component I under denaturing conditions gives an extremely compact form (53 S) in which both strands of the DNA are still present, since their circularity prevents them from separating from each other. Denaturation of component II gives rise to one circular strand (18 S) and one linear single-strand (16S) if only one single-strand scission was initially present in the double-stranded molecule (Fig. 1).

The infectivity of different forms of the DNA in mouse cells was examined by Dulbecco & Vogt (1963). In the native DNA both components I and II were infectious (Fig. 2). Components II and III were not separated sufficiently to allow a decision on the activity of component III. With alkali denatured DNA the intact collapsed molecule

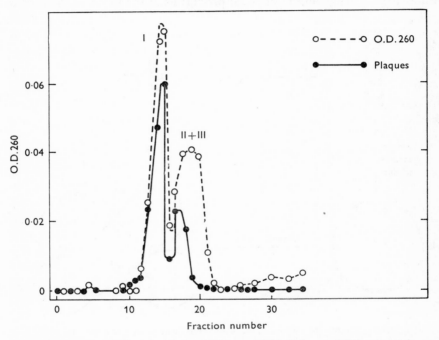

Fig. 2. Band sedimentation of polyoma DNA at pH 7·5. Polyoma DNA extracted once with phenol and sedimented at 35 K rev./min. in CsCl of density 1·50 for 4 hr. (modified from Dulbecco & Vogt, 1963).

(53S) was highly infectious and there was also activity in the slower running peak which comprised the separated single strands of the DNA (Fig. 3). The positions of the peaks of infectivity and optical density were consistent with the circular single strands (18S) being infectious and the linear strands (16S) being non-infectious. As in the case of λ DNA it is possible that the infectivity in the 18S region of the gradient was due to annealing of the DNA, since both of the complementary strands were present in the same fractions. This point can only be settled unequivocally by testing the activity of each of the single strands separately or by determining the u.v. sensitivity of the infectivity.

Since component III in the native DNA comprises linear fragments of mouse DNA there are clearly two types of linear single strands in the denatured DNA, mouse DNA derived from component III and viral DNA derived from the strands of component II in which there was a scission. The situation is otherwise analogous to that described for

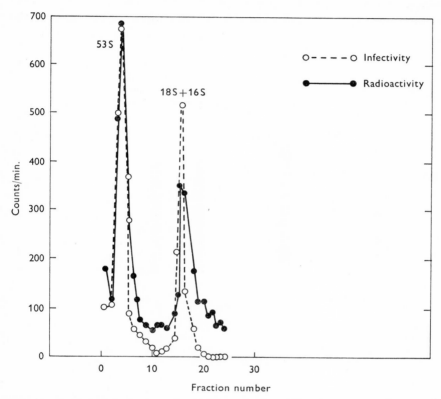

Fig. 3. Band sedimentation of polyoma DNA at pH 12·5. ³H-thymidine labelled polyoma DNA component I was brought to pH 12·5 with 0·1 M phosphate buffer and sedimented at 35 K rev./min. for 2 hr. in a preformed gradient of CsCl, mean density 1·5, containing phosphate buffer (0·01 M, pH 12·5). Both radioactivity and infectivity were determined in all fractions (modified from Dulbecco & Vogt, 1963).

ϕX 174 and can be summarized as follows. The essential feature for infectivity is possession of one intact circular strand. In the light of what is already known about the ability of cells to add a complementary strand to single-stranded DNA and to heal single-stranded nicks, thus making linear strands circular, it seems likely that either strand of the DNA could be capable of initiating infection. The events occurring in infection with single-stranded polyoma DNA would therefore be almost identical to those occurring normally in ϕX 174 infection, and presumed to occur in infection of animal cells with viruses containing single-stranded DNA.

Since transformation involves changes in cellular functions it might have been reasonable to think that component III was the active molecule in transformation. However, this was shown not to be the case (Crawford, Dulbecco, Fried, Montagnier & Stoker, 1964). Actually the reason for doing these experiments was connected with the possibility of

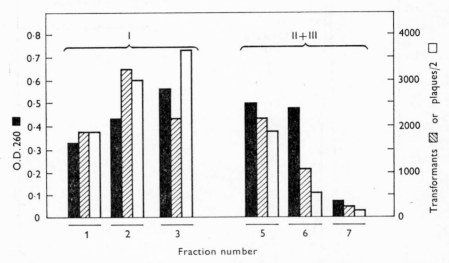

Fig. 4. Band sedimentation of polyoma DNA at pH 7·5. Polyoma DNA was layered on a linear gradient of sucrose (20–5 % in 0·1 M NaCl, 0·01 M EDTA) and centrifuged for 7·5 hr. at 35 K rev./min. The transforming and plaque forming activities and optical density of each of the fractions was determined (modified from Crawford *et al.* 1964).

defectiveness, either structural or functional, being required for trans-formation. The nature of component III was then not understood and it was thought that this was the most likely candidate for a defective form of the virus DNA. In fact the experiments showed that the requirements for lytic and transforming activities were similar and that component I possessed both activities (Fig. 4).

Although the nature of component III is now known its functions, if any, are not. The fact that virus particles do contain host DNA makes the situation analogous to transduction. There is no clear evidence that any genetic material is transferred from one cell to another by polyoma virus, but it might be argued that the ability of polyoma virus to stimu-late host DNA synthesis is due to such transfer. This seems unlikely but it should be possible to decide soon whether or not it is so. It is difficult to prove that component III is completely lacking in transforming and lytic activity, especially since it may include a few linear molecules of virus DNA. In certain circumstances these linear molecules of virus DNA might be infectious and this would have a very important bearing on the minimum requirements for infectivity. If each of the polynucleotide strands of component I were held in the form of a ring by some special 'linker', then it might be possible to open the ring at this point, without interrupting the polynucleotide strand. The nature of the hypothetical 'linker' is unspecified but it could be a polypeptide or simply a bond different from the other internucleotide bonds in the molecule. There does not now seem to be any evidence for the presence of such a 'linker'. Data which previously seemed to support the idea of a 'linker' can now be better explained in other ways. For example, the conversion of component I to component II by reducing agents is now thought to result from introduction of single-strand scissions into the DNA, rather than from any action on disulphide bonds in a hypothetical 'linker' (J. Vinograd, personal communication). Random opening of the ring would be expected to produce one molecule in 5000 infectious, assuming that there was only one point at which the ring could open and still retain infectivity, out of the 5000 internucleotide bonds in each strand of polyoma virus DNA. The data available at present do not exclude this possibility and one can therefore not distinguish between circularity *per se* being required for infectivity here and alternative explanations.

DISCUSSION

The basic essential for infectivity in all cell systems must be the possession of a complete genome. A few exceptional functions seem to be dispensable in some large viruses, but the pressure of selection seems to have freed viruses of all but the last traces of non-essential material. Almost any alteration or removal of nucleotides from viral nucleic acids seems to result in loss of infectivity. Thus the minimum essential requirement would be for an intact linear polynucleotide comprising all the information of the virus. TMV RNA may be an example of an infectious RNA of this type. Molecules which are physically linear may still contain interacting sequences at either end which can, by uniting reversibly, make the molecule functionally circular. Evidence has been presented that this is the case with MS 2 and $Q\beta$ RNAs when they react with their homologous replicases. Some sort of reversible circularization might also explain some of the unusual physical properties of turnip

yellow mosaic virus RNA. It has not yet been possible to study the
replication of the plant virus RNAs by their homologous replicases. If
this were done a requirement for functional circularity might be found
there also. It is easy to see why circularity might be useful in a system of
nucleic acid replication and transcription, both in allowing the rejection
of incomplete molecules of RNA and in avoiding the necessity for
attachment and detachment of the polymerases from the nucleic acid
template.

The presence of ribonucleases in all biological systems and the
sensitivity of single-stranded RNA to these enzymes might suggest that
viruses containing double-stranded RNA might possess considerable
selective advantage. In fact very few viruses containing double-stranded
RNA are known and these few are restricted to animals and plants. The
advantage of ribonuclease resistance of the RNA must be balanced by
the complications introduced into replication by the possession of a
double-stranded structure. The fact that single-stranded viral RNA can
function directly as messenger makes it independent of host cell
polymerases. Double-stranded RNA does not have this advantage and
seems to have to rely on host enzymes for transcription of the viral
RNA to give messenger RNA.

DNA viruses are like their hosts in that they carry their information
in DNA. RNA viruses are the exception to the rule that organisms carry
their genetic information in DNA. Physically single-stranded DNA from
viruses such as ϕX 174 is similar in some respects to single-stranded
RNA. There is, however, no evidence that single-stranded DNA acts
directly as messenger under natural conditions. Instead the incoming
viral DNA is copied to give rise to a complementary strand. This strand,
while in the double-stranded replicative form, is copied to give rise to
the viral messenger RNA. The messenger RNA thus has the same sense
as the viral DNA strand. Here the circularity of the viral DNA is
retained in the circular replicative form, in which both the strands are
circular. The replication of the replicative form gives rise to more
circular molecules, although the semi-conservative replication of a circular molecule must involve the opening of at least one of the circular
strands at the beginning of replication and the closure of the strands
of the daughter molecules at the end of replication. The same process
must occur in the production of progeny single-stranded viral DNA.
All the molecules involved in all these processes have at least one
circular strand so that circularity is something which is retained.

In the case of λ DNA physical circularity is not a permanent feature
of the molecule but is rather a configuration adopted at a certain stage

in replication. Again some advantages can be suggested for possession of a circular configuration during replication and transcription. The fact that the virus DNA inside the virus particle is linear and only potentially circular may reflect the requirements for entry into the bacterial cell by injection. The viruses which are known to have super-coiled, circular, double-stranded DNA in the virus particle are all animal viruses. Viruses appear to penetrate into animal cells as particles, rather than undergoing separation of the nucleic acid from its protein coat at the cell surface. The stage of injection is therefore avoided, together with all the structural requirements it imposes. The configuration of polyoma virus DNA in the intact virus can therefore be similar to that of the replicative forms of other viruses, such as ϕX 174 and λ. Polyoma virus is then an instance where the virus particle actually contains the replicative form of the virus DNA.

ACKNOWLEDGEMENTS

I am indebted to Professor R. Dulbecco for permission to use the Figures 2 and 3 in this article and to the National Academy of Sciences for permission to use Figures 2, 3 and 4.

REFERENCES

AMMANN, J., DELIUS, H. & HOFSCHNEIDER, P. H. (1964). Isolation and properties of an intact phage specific replicative form of RNA phage M12. *J. molec. Biol.* **10**, 557.

BODE, V. & KAISER, A. D. (1965). Changes in the structure and activity of λ DNA in a superinfected immune bacterium. *J. molec. Biol.* **14**, 399.

BRODY, E., COLEMAN, L., MACKAL, R. P., WERNINGHAUS, B. & EVANS, E. A., JUN. (1964). Properties of infectious deoxyribonucleic acid from T1 and λ bacteriophage. *J. biol. Chem.* **239**, 285.

BURTON, A. & SINSHEIMER, R. L. (1965). The process of infection with bacteriophage ϕX174. VII. Ultracentrifugal analysis of the replicative form. *J. molec. Biol.* **14**, 327.

CRAWFORD, L. V. & BLACK, P. H. (1964). The nucleic acid of Simian Virus 40. *Virology*, **24**, 388.

CRAWFORD, L., DULBECCO, R., FRIED, M., MONTAGNIER, L. & STOKER, M. (1964). Cell transformation by different forms of polyoma virus DNA. *Proc. natn. Acad. Sci. U.S.A.* **52**, 148.

CRAWFORD, L. V. & WARING, M. (1967). Supercoiling of polyoma virus DNA measured by its interaction with ethidium bromide. *J. molec. Biol.* **25**, 23.

DENHARDT, D. T., DRESSLER, D. H. & HATHAWAY, A. (1967). The abortive replication of ϕX174 DNA in a recombination-deficient mutant of *Escherichia coli. Proc. natn. Acad. Sci. U.S.A.* **57**, 813.

DULBECCO, R. & VOGT, M. (1963). Evidence for a ring structure of polyoma virus DNA. *Proc. natn. Acad. Sci. U.S.A.* **50**, 236.

FIERS, W. & SINSHEIMER, R. L. (1962). The structure of the DNA of bacteriophage ϕX174. III. Ultracentrifugal evidence for a ring structure. *J. molec. Biol.* **5**, 424.

Fox, E. & Meselson, M. (1963). Unequal photosensitivity of the two strands of DNA in bacteriophage λ. *J. molec. Biol.* **7**, 583.

Gellert, M. (1967). Formation of covalent circles of lambda DNA by *E. coli* extracts. *Proc. natn. Acad. Sci. U.S.A.* **57**, 148.

Gharpure, M. A. (1965). A heat-sensitive cellular function required for the replication of DNA viruses but not RNA viruses *Virology*, **27**, 308.

Gomatos, P. & Tamm, I. (1963a). The secondary structure of Reovirus RNA. *Proc. natn. Acad. Sci. U.S.A.* **49**, 707.

Gomatos, P. & Tamm, I. (1963b). Animal and plant viruses with double-helical RNA. *Proc. natn. Acad. Sci. U.S.A.* **50**, 878.

Haruna, I. & Spiegelman, S. (1965). Recognition of size and sequence by an RNA replicase. *Proc. natn. Acad. Sci. U.S.A.* **54**, 1189.

Jaenisch, R., Hofschneider, P. H. & Preuss, A. (1966). Über infektiöse Substrukturen aus *Escherichia coli* Bakteriophagen. VIII. On the tertiary structure and biological properties of ϕX174 replicative form. *J. molec. Biol.* **21**, 501.

Kaiser, A. D. & Hogness, D. S. (1960). The transformation of *Escherichia coli* with deoxyribonucleic acid isolated from bacteriophage λ dg. *J. molec. Biol.* **2**, 392.

Kaiser, A. D. & Inman, R. B. (1965). Cohesion and the biological activity of bacteriophage λ DNA. *J. molec. Biol.* **13**, 78.

Kiger, J. A., jun., Young, E. T. & Sinsheimer, R. L. (1967). The infectivity of single-stranded rings of phage lambda DNA. *J. molec. Biol.* (in the Press).

Koch, G., Quintrell, N. & Bishop, J. M. (1966). An agar cell-suspension plaque assay for isolated viral RNA. *Biochem. biophys. Res. Commun.* **24**, 304.

Koch, G., Quintrell, N. & Bishop, J. M. (1967). Differential effect of actinomycin D on the infectivity of single- and double-stranded poliovirus RNA. *Virology*, **31**, 388.

Kudo, H. & Graham, A. F. (1965). Synthesis of Reovirus ribonucleic acid in L cells. *J. Bact.* **90**, 936.

Lebowitz, J., Watson, R. & Vinograd, J. (1966). Determination of the number of twists in polyoma DNA. *Second International Biophysics Congress of the International Organization for Pure and Applied Biophysics, Vienna*, 1966.

Lee, J. C. & Gilham, P. T. (1965). Determination of terminal sequences in viral and ribosomal ribonucleic acids. *J. Am. chem. Soc.* **87**, 4000.

Matthews, R. E. F. & Hardie, J. D. (1966). Reconstitution of RNA from spherical viruses with tobacco mosaic virus protein. *Virology*, **28**, 165.

Meyer, F., Mackal, R. P., Tao, M. & Evans, E. A. (1961). Infectious deoxyribonucleic acid from λ bacteriophage. *J. biol. Chem.* **236**, 1141.

Rüst, P. & Sinsheimer, R. L. (1967). The process of infection with bacteriophage ϕX174. XI. Infectivity of the complementary strand of the replicative form. *J. molec. Biol.* **23**, 545.

Shatkin, A. J. (1965). Actinomycin and the differential synthesis of Reovirus and L cell RNA. *Biochem. biophys. Res. Commun.* **19**, 506.

Sinsheimer, R. L., Lawrence, M. & Nagler, C. (1965). The process of infection with bacteriophage ϕX174. VIII. Centrifugal analysis in alkaline media of the RF DNA at various stages of infection. *J. molec. Biol.* **14**, 348.

Steinschneider, A. & Fraenkel-Conrat, H. (1966a). Studies of nucleotide sequences in tobacco mosaic virus ribonucleic acid. III. Periodate oxidation and semicarbazone formation. *Biochemistry*, **5**, 2729.

Steinschneider, A. & Fraenkel-Conrat, H. (1966b). Studies of nucleotide sequences in tobacco mosaic virus ribonucleic acid. IV. Use of aniline in stepwise degradation. *Biochemistry*, **5**, 2735.

Strack, H. B. & Kaiser, A. D. (1965). On the structure of the ends of lambda DNA. *J. molec. Biol.* **12**, 36.

STRAZIELLE, C., BENOIT, H. & HIRTH, L. (1965). Particularités structurales de l'acide ribonucléique extrait du virus de la mosaique jaune du navet. II. *J. molec. Biol.* **13**, 735.

SUGIYAMA, T. (1965). 5′-Linked end group of RNA from bacteriophage MS2. *J. molec. Biol.* **11**, 856.

TAKANAMI, M. (1966). The 5′-termini of *E. coli* ribosomal RNA and f2 bacteriophage RNA. *Cold Spring Harb. Symp. quant. Biol.* **31**, 611.

VINOGRAD, J., LEBOWITZ, J., RADLOFF, R., WATSON, R. & LAIPIS, P. (1965). The twisted circular form of polyoma viral DNA. *Proc. natn. Acad. Sci. U.S.A.* **53**, 1104.

WEIL, R. & VINOGRAD, J. (1963). The cyclic helix and cyclic coil forms of polyoma viral DNA. *Proc. natn. Acad. Sci. U.S.A.* **50**, 730.

WINOCOUR, E. (1967). In the apparent homology between DNA from polyoma virus and normal mouse synthetic RNA. *Virology*, **31**, 15.

WU, R. & KAISER, A. D. (1967). Mapping the 5′-terminal nucleotides of the DNA of bacteriophage λ and related phages. *Proc. natn. Acad. Sci. U.S.A.* **57**, 170.

YOUNG, E. T. & SINSHEIMER, R. L. (1964). Novel intra-cellular forms of lambda DNA. *J. molec. Biol.* **10**, 562.

YOUNG, E. T. & SINSHEIMER, R. L. (1967). Vegetative λ DNA. I. Infectivity in a spheroplast assay. *J. molec. Biol.* (in the Press).

THE DISAGGREGATION AND ASSEMBLY OF SIMPLE VIRUSES

R. LEBERMAN

Medical Research Council Laboratory of Molecular Biology, Hills Road, Cambridge

INTRODUCTION

In classical organic chemistry of natural products a compound whose structure was to be determined was first analysed and then synthesized. The synthetic and natural compounds were then compared and shown to be identical by physical and chemical tests. In an analogous way one aspect of the study of the structural organization of a simple virus consists of breaking down the infective particle, isolating and characterizing its component parts and then attempting to reassemble these into the original material. A truly reconstituted virus particle should then satisfy two criteria, it should be infectious and it should be structurally identical to the native virus, in that it has the same morphology and stability. This does not mean that only successful reconstitution experiments can give structural information; well-defined intermediate or variants can also give information on how a virus particle is constructed. This article will be confined to the problems encountered with simple or minimal viruses, where the virus particles have a regular morphology and have only protein and nucleic acid as their main chemical components. The large complex bacteriophages will not be discussed although their structural components are probably examples of self-assembly systems.

In order to carry out such studies on a simple virus one must be able to separate and characterize the two major structural components, protein and nucleic acid, in an undenatured condition. To define this condition is probably simpler for the nucleic acid component than for the protein. Thus the criteria that the isolated nucleic acid is infectious, and/or that its molecular weight corresponds to the nucleic acid content of the virus particle, are relatively easily investigated in most cases. On the other hand, the definition of 'native' viral protein is more difficult and the criteria used for TMV protein (cf. Caspar (1963) and Anderer (1963) in their reviews of TMV) are essentially that the protein will polymerize into rods of the same diameter as the virus particle and will also recombine with viral RNA to form infectious rods indistinguishable from the original virus particles. In other words, nativeness of the protein is defined by ability to reassemble into the protein coat with or without viral nucleic acid. Conditions for performing these operations for TMV protein are known, but with other viruses they have to be determined, so that the experimenter does not know whether or not he has native protein until he finds the conditions for reassembly. In addition to the problem of nativeness, the possibility exists that there are more than one species of structural protein in some viruses and that the proportions and perhaps even the order of addition of the protein species added to the viral nucleic acid may be important for reassembly. The usual aim in obtaining native virus protein is therefore to isolate the protein in a soluble form where its molecular weight corresponds to the chemical protein subunit or some small polymer of it.

Once soluble protein and high molecular weight or infectious nucleic acid have been isolated these are mixed together—usually with the protein:nucleic acid ratio greater than the ratio found in the original virus—under conditions of pH, ionic strength, etc., where it is hoped reassembly into the virus particle will occur. The term 'self-assembly' was first introduced and the concept discussed in detail by Caspar & Klug (1962). They describe the process as one similar to crystallization where the protein subunits crystallize together with the nucleic acid under suitable solvent conditions to form virus particles. Under these conditions this is the lowest energy state of the two components. As with crystallization there may be a nucleation step, e.g. the binding of a critical number of subunits to a specific point on the nucleic acid, and once this occurs the protein subunits rapidly crystallize into the protein coat. Other (morphopoietic) factors may be required for the assembly of virus particles, but at the time of writing the evidence for these is indirect (Kellenberger, 1966).

The simple viruses consist of two general morphological types; the rod shaped, which may be rigid or flexous and the isometric or spherical virus. All examples of the latter so far examined have been shown to have icosahedral symmetry. Examples of disaggregation and reassembly experiments on viruses from these morphological groups will be described, and it is hoped that the examples will illustrate all aspects of this type of investigation.

ISOLATION OF VIRUS PROTEINS AND NUCLEIC ACIDS

Virus proteins

Methods for the isolation of virus protein have been recently reviewed by Knight (1963). Most of the methods originate from studies on TMV and often involve the denaturation of protein followed by a renaturation procedure, and the choice of any method might well depend on how easily the virus protein under study renatures. Some of the methods have the advantage that nucleic acid is isolated by the same procedure.

The most general methods appear to be: (a) mild alkaline treatment, where presumably the nucleic acid is hydrolysed; (b) cold 66 % acetic acid in which the protein is soluble and the nucleic acid insoluble; (c) dissociation of the virus with detergent, followed by fractional precipitation to separate protein and nucleic acid. Some alternative methods are described later in the sections on cowpea chlorotic mottle virus and turnip crinkle virus.

There is evidence that some simple viruses contain more than one kind of structural protein. The first indication that poliovirus may contain at least two structural proteins was reported by Maizel (1963) on the results of disc electrophoresis of poliovirus protein. Electrophoresis of the protein in 8 M urea showed the presence of two major and two minor bands. In a further paper (Summers, Maizel & Darnell, 1965) reference is made to poliovirus having three structural proteins, but all evidence published thus far is based only on acrylamide gel electrophoresis studies with no characterization of the protein bands. A much more definite result was obtained by Rueckert & Duesberg (1966) for mouse encephalitis virus. They isolated by disc-electrophoresis two major components from mouse encephalitis virus protein and showed by amino acid analysis and tryptic peptide mapping that there are at least two, and probably three, different major structural proteins. At this stage nothing is known about the location and function of the different proteins compared with, say, adenovirus (see D. H. Watson, p. 216, this Symposium).

Virus nucleic acids

Isolation procedures for virus nucleic acids have also been reviewed by Knight (1963). The most commonly used method is degradation of the virus with phenol so that two phases are formed. Denatured protein is soluble in the phenolic phase and the nucleic acid in the aqueous phase. Improved RNA preparation can be achieved by using bentonite to remove contaminating nucleases (Fraenkel-Conrat, Singer & Tsugita, 1961). An alternative method which may have wide applicability is described in the section on turnip crinkle virus.

ROD-SHAPED VIRUSES

Tobacco mosaic virus (TMV)

The disaggregation and repolymerization of the protein and *in vitro* reassembly of this virus is well documented, but even for this virus the detailed interactions between the viral nucleic acid and the protein are not yet understood, and there are at least two theories on the nature of the assembly process (Caspar, 1963; Lauffer, 1966).

The virus can be dissociated into its nucleic acid and protein by a number of reagents (Anderer, 1963), the choice of reagent depending on whether the desired product is protein or nucleic acid. The usual methods of isolating TMV protein are to dissociate the virus, either in cold 66 % acetic acid (Fraenkel-Conrat, 1957), or under mild alkaline conditions pH 9·5–10·5 in either buffered or unbuffered solution—the protein obtained by the latter method is termed A-protein ('alkalischer' protein, Schramm, Schumacher & Zillig, 1955). Newmark & Myers (1957) suggested that alkanolamine buffers might facilitate the production of A-protein; in practice this is found to be so and the effect may be due to the interaction of the alkanolamine with TMV–RNA. The soluble protein eventually obtained by any method is now generally called A-protein and except at extreme conditions of pH or dilution is in a state of limited aggregation. The lowest stable aggregate has a sedimentation constant of 4–4·6S and is apparently the trimer of the ultimate chemical subunit of molecular weight 17,400.

The methods favoured for producing high molecular weight infectious RNA are treating the virus either with phenol (Gierer & Schramm, 1956) or with phenol and bentonite (Fraenkel-Conrat *et al.* 1961). The nucleic acid obtained by these methods is infectious and has a molecular weight of about 2×10^6 corresponding to the total RNA content of the virus particle.

Polymerization of TMV protein without RNA

Caspar (1963) has discussed in considerable detail the polymerization of TMV protein. His review covers most of the experimental and theoretical aspects and also presents a theory for the stable intermediate states of the self-assembly process. Subsequently Lauffer and his co-workers have published a series of papers on TMV-protein polymerization, one of which (Lauffer, 1966) presents a model of the system differing from Caspar's model on the question of the specificity of inter-subunit bonds.

That TMV protein forms a number of well-defined aggregates is now firmly established. The composition and structure of these aggregates depend on the conditions under which they are produced and the stability of some aggregates is not yet known. However, there are at least four definable stable states of aggregation of TMV protein, the predominance of any one depending on the polymerization conditions.

The soluble TMV A-protein, with a sedimentation constant between 4–4·6S, was predicted (Caspar, 1963) and shown to be a trimer of the chemical subunit (Banerjee & Lauffer, 1966). As with all the polymers of TMV protein, it would be difficult to define all the conditions under which it is stable without a phase diagram which would cover protein concentration, pH, ionic strength and temperature. In general terms, it is stable at low protein and salt concentrations, pH in the range 10·5 to 6·5 and temperature in the region of 0–5°. At very low protein concentrations and high pH it dissociates into monomer and is also monomeric in 66 % acetic acid. Raising the temperature and salt concentration produces polymerization into higher aggregates and indicates that the bonding between protein subunits is mainly 'hydrophobic'.

The next higher stable aggregate is the two-turn disc, which has about the same diameter as the virus and sedimentation constant of 20–26S. This consists of two rings, each containing 17 protein subunits (Finch, Leberman, Chang Yu-Shang & Klug, 1966). The preliminary X-ray diffraction studies cannot rule out the possibility that there may be dyad axes in the plane of the disc. If this is proved to be the case it would mean that the two rings of a disc are packed head to head (or tail to tail). Such a result would profoundly affect any discussion on the two models of TMV polymerization (Caspar, 1963; Lauffer, 1966).

The third stable aggregate of TMV protein is the rod-like structure that is a stack of the two-turn discs. These stacked-disc rods (Pl. 1, fig. 1) can be obtained in a great variety of lengths, depending on polymerization conditions, from short polymers consisting of a few discs to rods many times the length of the virus. It is not known, and probably is difficult to investigate, whether any particular lengths of stacked-disc rod are more stable than others. This is because the stacking process is a linear crystallization of the two-turn disc particles, and long rods may grow at the expense of short rods; in fact, it is often observed that old preparations of polymerized TMV protein consist of extraordinarily long stacked-disc rods.

The last, and in many ways the most interesting, polymer of TMV protein to be considered is the helical rod. This polymer, in which the protein subunits are packed in the same way as in the virus (Franklin, 1955), is the one that is widely believed to be the usual end product of polymerizing TMV protein. It is, in fact, obtained only under carefully controlled conditions and appears to be unstable outside these conditions. Observations in this laboratory (Leberman & von Sengbusch, unpublished results) have borne out Caspar's (1963) speculation that the helical repolymerized protein can be obtained only at low ionic strength and with careful adjustment of the pH to about 5 (Pl. 1, fig. 1). Conditions for producing the helical polymer of TMV protein are not, at the present, more closely defined than this. Dissolution of these helical rods occurs very easily and they are apparently much less stable than the stacked-disc rods.

In the two variant rod-shaped polymers of TMV protein, the stacked-disc and the helical, the bonds between subunits are clearly not identical, but are likely to be very similar. The precise degree of similarity will have to await further investigation. However, it is clear that within a ring of the stacked-disc structure the lateral bonds between subunits are very similar to those in the helical structure, and the remaining question relates to the bonds between rings of subunits. The stacked-disc structure is, apparently, the more stable structure under a wide variety of polymerizing conditions, but more work is required to define more exactly conditions for obtaining either this or the helical polymer.

Reconstitution of TMV

At this point in time, when workers in a number of laboratories are undertaking the apparently more difficult task of reconstituting spherical viruses, it may prove worthwhile to look more carefully at what has been achieved in the reconstitution of TMV. An examination of the experimental conditions for the combination of TMV protein with various nucleic acids may provide some guiding principles for the reconstitution of other viruses. The *in vitro* reconstitution of infective TMV particles, first reported in 1955 by Fraenkel-Conrat & Williams, is the prime example of self-assembly of a biological system, and since that time the Berkeley group have described improved methods for isolating the viral protein and nucleic acid, and for obtaining higher yields of reconstituted virus.

Fraenkel-Conrat & Williams (1955) carried out their reconstitution experiments by reacting a 1 % protein solution with one-tenth its weight of TMV–RNA at pH 6 in 0·03 M acetate at 3° for at least 24 hr.

Subsequently this method was improved by working at lower protein concentrations and higher temperatures and pH. Fraenkel-Conrat & Singer (1959) described a set of conditions for the reconstitution of TMV which could produce 30–80 % of the activity of the original virus. These conditions are 0·1 % protein and 0·005 % RNA in 0·1 M pyrophosphate pH 7 at room temperature for 16 hr. In a later paper (Fraenkel-Conrat & Singer, 1964), these conditions were slightly modified to incubation at 30° for 6 hr. at 0·1 M pyrophosphate pH 7·25.

The attribution to pyrophosphate of some property which facilitates the reconstitution process (Fraenkel-Conrat & Singer, 1959) must be treated with a certain amount of caution. This also applies to the apparent inhibition by Tris (Fraenkel-Conrat & Singer, 1964). Studies on the efficiency of reconstitution in pyrophosphate, phosphate, Tris and various combinations of the three, were performed on the basis of the effect of the *concentrations* of the buffers. Since at pH 7·25 the ionic strength of equimolar solutions of pyrophosphate (prepared according to Fraenkel-Conrat & Singer, 1959) phosphate and Tris would be approximately in the ratio 6:2:0·8, it is difficult to distinguish between specific buffer ion and ionic strength effects. An example to illustrate this point is that Fraenkel-Conrat & Singer (1964) found that in 0·1 M phosphate at pH 7·1 reconstitution is only 55 % as efficient as reconstitution in 0·1 M pyrophosphate at pH 7·25; on the other hand, reconstitution in 0·05 M pyrophosphate at pH 7·3, where the ionic strength is more comparable to the 0·1 M phosphate, is only 11 % as efficient. From these observations one could argue that phosphate was the better reconstitution medium. Similar arguments could be applied to the data for media containing Tris, where no reconstitution was observed, but in which the highest ionic strength solution was 0·1 compared with about 0·6 for 0·1 M pyrophosphate.

More recently Matthews (1966) has reported another, lower pH optimum for the reconstitution of TMV which has a higher optimum temperature. At pH 4·8 and at 60° in 0·25 M phosphate, higher yields of reconstituted rods could be obtained than in 0·25M phosphate at pH 6·7 and at 30°, but the pH 4·8 product was less infective and not as resistant to ribonuclease as the pH 6·7 product.

Although a set of conditions for reconstituting TMV can be accurately described, which of the variables are most important has not, as yet, been determined. Factors such as temperature, pH, and ionic strength, obviously affect the course of the reaction as they do in the polymerization of TMV protein, and merit further study. It is by no means certain that the 'optimum' conditions for reconstituting the virus will be very different from the conditions for producing helical polymerized protein, but what is clear is that at high ionic strength interaction of the nucleic acid with the protein produces a helical structure where the protein alone might be expected to form stacked-disc rods.

'Reconstituted' hybrids of strains of TMV

Fraenkel-Conrat & Singer (1957) described the reconstitution of virus rods with various combinations of nucleic acids and proteins from masked, yellow aucuba, Holmes ribgrass and common strains of TMV. All reconstitutions gave infective particles, and symptoms produced in each case corresponded to those produced by the strain supplying the nucleic acid. It was also observed that the particles produced by the interaction of Holmes ribgrass RNA with TMV protein were as infective as those produced by TMV–RNA and TMV protein, whereas the original Holmes ribgrass virus had only about 5 % of the infectivity of TMV.

Holoubek (1962) reconstituted TMV protein with RNA from three chemical mutants of TMV and the Holmes ribgrass strain of the virus. All formed infectious rods but the infectivities of the mixed reconstituted viruses were greater than of the strain providing the encaspulated nucleic acid. This was interpreted as being due to TMV protein being more efficient in adsorbing the virus to the sites of infection in the host plant than the mutant or Holmes ribgrass protein.

'Reconstitution' with foreign RNA

In their early studies on the reconstitution of TMV Fraenkel-Conrat & Williams (1955) found that TMV protein did not combine with TYMV–RNA under their reconstitution conditions. This is in contrast to the results of Matthews (1966) who found he could combine infectious TYMV–RNA with TMV–protein either at pH 4·8 or 6·7 in 0·25M phosphate. This 'reconstituted' material had some TYMV infectivity on Chinese cabbage, which could be removed by ribonuclease treatment. A possible interpretation of these observations is that the TYMV–RNA is inefficiently or incompletely coated with TMV protein. An alternative explanation, similar to that advanced by Holoubek, is that TMV protein can coat infectious TYMV–RNA but that TMV protein cannot bind to the infectible sites in Chinese cabbage.

In addition to different results obtained with TYMV–RNA by Matthews and Fraenkel-Conrat & Williams, other conflicting results on binding 'foreign' RNA have been obtained. Hart & Smith (1956), using conditions similar to the original Fraenkel-Conrat & Williams reconstitution, found that yeast RNA could be coated with TMV protein to produce sedimentable particles. Under conditions of higher pH, ionic strength and temperature, Fraenkel-Conrat & Singer (1964) did not obtain any significant amount of 'reconstitution' with either wheat germ or ascites cell RNA.

'Reconstitution' with synthetic polynucleotides

Hart & Smith (1956) reacted TMV protein with the homopolyers poly-A, poly-C, poly-U, poly-I and copolymers poly-AU and poly-AGUC. They found that under their conditions of 0·03 M acetate pH 6 at 5° all these synthetic polymers formed sedimentable rods with the protein. The main differences from their conditions and those used by Fraenkel-Conrat & Williams (1955) were lower protein concentrations (0·2 %) and a higher protein:nucleic acid ratio (2:1).

Fraenkel-Conrat & Singer (1964) in a more detailed study, under different conditions, found that out of a large number of different synthetic polymers only those with a high A content, poly-A, and poly-I, combined with TMV protein to any extent. Reconstitution with poly-A was examined in the greatest detail, and in 0·1 M pyrophosphate at 30° the reaction has a pH optimum of about pH 6·6 compared with pH 7·25 for TMV–RNA. It was also observed that the combination of poly-A and TMV protein proceeds nearly as well in 0·1 M Tris at pH 6·7 as with 0·1 M pyrophosphate. An explanation of the differences between the Hart & Smith and the Fraenkel-Conrat & Singer results may be that under the Hart & Smith conditions TMV-protein itself would polymerize and in doing so non-specifically entrap any RNA (Caspar, 1963). This explanation is hardly valid since the Hart & Smith conditions were very close to the original conditions used by Fraenkel-Conrat & Williams, and in fact their protein concentration was a factor of five times lower. It is possible that both sets of workers were correct in their observations since they were performing their experiments under entirely different conditions. Examination of the data provided by Fraenkel-Conrat & Singer shows that under their optimum conditions for reconstituting TMV–RNA with TMV–protein very little poly-A combines with TMV–protein but at lower pH values the reaction proceeds favourably. By the same token one might argue that under other conditions poly-C, poly-U and poly-G (or any other natural or synthetic polynucleotide) can combine with TMV protein.

It is difficult to reach any conclusions on the specificity of the reaction between synthetic poly-nucleotides and TMV protein which may throw some light on the specific reaction between TMV–RNA and TMV–

protein. Under a limited range of conditions there may be an apparent specificity, but unless a wide range of conditions are examined there are dangers in being too categorical. Thus, although poly-C or synthetic polymers with a high C content do not combine with TMV protein in 0·1 M pyrophosphate at 30° and pH values around 7, TYMV–RNA has been reported to form stable nucleoprotein complexes under similar conditions (Matthews, 1966) even though it contains about 37 % cytidylic acid.

Bacteriophage fd

This virus is one of a group of filamentous DNA containing bacteriophages. It is much longer and thinner than TMV, about 8000 Å long and 50 Å diameter. The DNA appears to be single-stranded and circular with a molecular weight of $1·7 \times 10^6$.

Various methods for isolating fd-protein were investigated by Knippers & Hoffmann-Berling (1966a) and they concluded that only disaggregation of the virus with phenol produced nucleic acid-free protein. Since treatment with phenol is a standard method for isolating nucleic acids from viruses, this procedure can be used to isolate both the protein and DNA components.

The chemical molecular weight of fd-protein monomer is about 9000 and ultracentrifugal studies (Knippers & Hoffmann-Berling, 1966a) show that this can be obtained in 1 % SDS solution. In other solvents the protein aggregates and in 0·1 M pyrophosphate not very regular rod-shaped particles of similar diameter to the phage, but much shorter, have been obtained. The studies of Knippers & Hoffmann-Berling do show, however, that fd-protein is much more difficult to disaggregate than TMV–protein, the protein being fairly aggregated even in 8 M urea or 6 M guanidine-HCl. The existence of definite stable polymers of fd-protein has not yet been investigated fully.

Successful reconstitution experiments by Knippers & Hoffmann-Berling (1966b) were carried out by the following procedure. DNA solution was added to excess protein in 8 M urea in which the protein is in an oligomeric state. The reaction medium was then changed to 0·01 M pyrophosphate pH 7·0 at room temperature by stepwise dialysis, first with decreasing concentrations of urea and finally pyrophosphate buffers. The conditions they describe were adopted after studying the reconstitution process under various other conditions and in different buffers. No interaction between protein and nucleic acid occurred outside the pH range 6·0–8·1 and maximum interaction was found at pH 6·8. Interaction was found in other buffer systems but pyrophosphate appeared to produce more stable products. The reaction proceeded as well at 37° as at room temperature but not at all at 0°. No interaction was observed under conditions in which the protein would be monomeric.

The products obtained in these reconstitution studies were of two types. The main product consisted of irregular nucleoprotein particles containing about 17 % DNA, compared to 11·6 % for the intact phage. The particles were non-infectious for both bacteria and spheroplasts, whereas the DNA extracted from these complexes possessed the normal infectivity towards spheroplasts. In the electron microscope these complexes appeared to be irregular filamentous particles similar to the rod-shaped aggregates of fd-protein found by Knippers & Hoffmann-Berling (1966a). The minor product appeared to be fully reconstituted active phage in its infectivity, DNase resistance and inactivation by fd antiserum. The yield of reconstituted virus was, however, very low and varied with DNA preparation.

The possibility was investigated that different DNA preparations contained varying proportions of DNA molecules with some protein subunits still bound to them, and that these subunits could be nucleation centres for the crystallization of the protein coat around the DNA. The result showed that DNA molecules which recombined with fd-protein to produce active phage particles contained 5–10 % of the original protein coat. Of these DNA molecules only one in 10^7 produced active particles, possibly indicating that the residual protein has to be at a specific point on the DNA molecule to initiate the crystallization of the protein around the nucleic acid or that this coating process is seldom completed.

With foreign single-stranded DNA irregular complexes were formed with fd-protein, but none were formed with native *Escherichia coli* DNA. The irregular complexes formed with ϕX174 DNA were non-infectious, although ϕX174 DNA can be coated *in vivo* with fd-protein to produce infectious particles (Knippers & Hoffmann-Berling, 1966c).

SPHERICAL VIRUSES

ϕX174

The first experimental evidence for the reconstitution of a spherical virus comes from the work of Takai (1966) on the bacteriophage ϕX174. This virus is a small DNA containing bacteriophage, particle weight $6·2 \times 10^6$ avogrammes, in which the nucleic acid molecule is in the form of a closed ring (Fiers & Sinsheimer, 1962).

Phage DNA and protein were isolated by phenol and acetic acid

degradations respectively. When the isolated components were incubated together at 37° in 5×10^{-3}M Tris pH 7·5 in the presence of Ca^{2+}
for 10 min., most of the viral DNA sedimented with added *E. coli*
C cells, indicating that nucleoprotein complexes were being formed
which adsorbed to the bacteria. The reaction was strongly dependent on
the concentration of Ca^{2+} which could be effectively replaced by
Zn^{2+}, Cd^{2+} or Mn^{2+}. Deoxyribonuclease-treated viral DNA did not
sediment with *E. coli* C and preincubation of the nucleic acid with Ca^{2+}
only provided a partial protection. However, once the DNA-protein
complex had been formed in the presence of the cells it was DNase
resistant.

In the absence of *E. coli* C, DNA-protein complexes were formed but
were heterogeneous and their size depended on the relative concentrations of DNA and protein in the reaction mixture. Thus, when a twofold
excess of protein was used and the reaction incubated at 37° for 12 hr. in
the presence of 5×10^{-3}M Ca^{2+}, 50 % of the complexes formed had
sedimentation constants greater than 200S, whereas the virus has a
constant of 110S. On the other hand, using a twofold excess of DNA
under the same conditions most (94 %) of the complexes had sedimentation constants less than 30S.

From the point of view of reconstitution, the most significant result
mentioned by Takai is that of a preliminary experiment in which
ϕX174 DNA prepared in the presence of 10^{-3}M Mg^{2+} (i.e. twice the
usual concentration) was incubated with an equivalent amount (in
terms of phage composition) of ϕX174 protein for 12 hr. in 10^{-3}M Ca^{2+}.
The complex formed was found to sediment in a sucrose gradient at the
same rate as intact ϕX174. If Ca^{2+} was omitted no such complex was
formed.

These last-mentioned observations provide evidence that it should
be possible to reconstitute a spherical virus, and also that the viral
nucleic acid may need to have a definite tertiary structure in order to
complex with the virus protein. The formation of both virus and RNA
structure appears to depend on interaction with Mg and Ca ions, whose
roles do not appear to be interchangeable. The way in which addition of
E. coli C cells catalyses the formation of nucleoprotein complexes is not
understood, a possible explanation being that a group of protein subunits are adsorbed to the cell wall and are oriented by it so that co-
aggregation with the DNA is favoured (cf. Hutchinson, Edgell &
Sinsheimer, 1967).

MS 2 and fr

These two viruses will be discussed together since they are both members of a class of small, spherical, RNA-containing bacteriophages, which have a particle weight of about 3.5×10^6 avogrammes and contain about 30 % RNA. The RNA codes for only three proteins: a RNA polymerase, the coat protein, and a maturation protein which is required to produce complete, active phage.

Sugiyama, Hebert & Hartman (1967) reported the formation of two types of nucleoprotein complexes between isolated MS2–RNA and MS2–protein depending on relative concentrations of the two components. Complex I appears to be composed of a molecule of RNA and a small number of protein subunits, whereas Complex II is a spherical particle with the approximate dimensions of the intact virus. Spherical particles, whose properties are remarkably similar to Complex II, were also obtained by Hohn (1967) in a study on the self-assembly of fr.

Viral nucleic acids and protein were isolated by phenol and acetic acid methods respectively. These were then mixed together under somewhat different conditions for the two viruses. For MS2, a protein:RNA mixture in the ratio of about 1:1 incubated at 37° for 4 hr. in 0·1 M Tris at pH 7 led to the encapsulation of some of the RNA into a complex with a sedimentation constant of 69S compared with 81S for the virus. A similar complex with a sedimentation constant of about 70S was obtained on incubating a 3:1 mixture of fr-protein:fr-RNA for 24 hr. at 4° in Tris at pH 7·2 and ionic strength about 0·15. In the electron microscope these complexes appear to be very similar to the original virus particles. Both reassembled particles contained less RNA than intact virus and were biologically inactive. In these properties they resemble the defective particles of amber mutants of bacteriophage f2 grown in non-permissive hosts (Lodish, Horiuchi & Zinder, 1965). Hohn (1967) has suggested that the defectiveness of both the *in vivo* and *in vitro* assembled particles is due to the absence of the maturation factor.

Hohn (1967) also reported that he could obtain protein shells of fr, containing less than 1 % RNA, by freezing and thawing the virus. He could not reassemble these shells from the fr-protein isolated from the virus with acetic acid and the dissociated protein by itself showed little tendency to aggregate even at relatively high ionic strengths. The preliminary results show that spherical nucleoprotein particles with similar morphology to intact virus particles can be reassembled from the RNA and protein of these bacteriophages. Undoubtedly, since they provide the best systems for studying the assembly of virus particles, more results will be forthcoming on the requirements and conditions for reassembling intact, active virus.

Cowpea chlorotic mottle virus

The most detailed study so far reported on the self-assembly and reconstitution of spherical virus is that of cowpea chlorotic mottle virus (CCMV). This RNA-containing plant virus is one of a group of physically related viruses (others being broad bean mottle virus, BBMV, and brome grass mosaic virus, BGMV) and has a particle weight of about 4.6×10^6 avogrammes, contains about 24% RNA and has a protein shell composed of 180 chemical subunits.

Bancroft, Hills & Markham (1967) have described in detail the *in vitro* assembly of a variety of organized structures from the disaggregated virus. The virus, which is stable at pH values between 3 and 6, swells above pH 7·0 in the absence of Mg^{2+} and becomes susceptible to nuclease attack. The products of ribonuclease and T1 degradation can then polymerize into a number of morphologically distinct structures, the relative amounts produced depending on the degradation procedure. Besides the virus (250Å dry diameter) and the swollen virus six types of reaggregation products were found. These are ellipsoids, small spheres 160Å diameter, double-shelled spheres of outer diameter 250Å, double-shelled spheres 340Å diameter, narrow tubes 160Å diameter, and wide tubes 250Å. From the ratio of the ultraviolet absorbancies at 260 and 280 mμ for the various products it would appear that they all contain some, but probably different amounts of, degraded RNA, with none containing as much as the original virus. The smaller double-shelled spheres and the wide tubes, some, and perhaps all, of which contain narrow tubes, are only produced in quantity with T1 digestion.

The 180 chemical subunits of CCMV are arranged on the surface of the particle into 32 morphological units of 20 hexamers and 12 pentamers at the vertices of an icosahedron. The structures proposed by Bancroft *et al.* (1967) for all the reaggregation products of the virus are composed of these morphological units in various arrangements. Thus, the small particle is thought to be made up of 12 pentamers at the corners of an icosahedron (T = 1 lattice, Caspar & Klug, 1962) and the outer shell of the large double-shelled sphere is thought, on the grounds of its size, to have a structure similar to human wart virus (T = 7, see Klug & Finch, 1965). The structures of two tubular variants were analysed by optical diffraction of electron micrographs and appear to be helical arrangements of hexamers, 'rounded off' at the ends with pentamers.

The virus can also be degraded at pH 7·15 in 1 M NaCl. If it is then dialysed back to 0·05 M phosphate pH 6·5, a nucleoprotein with the same sedimentation constant as the swollen virus is obtained in about a 30 % yield, and by dialysis against 0·1 M acetate pH 5 a faster sedimenting form is obtained which has the same appearance as virus in the electron microscope.

Similar experiments with brome grass mosaic virus produced ellipsoidal and small spherical particles by the nuclease digestion procedure. With the NaCl degradation similar results to those with CCMV were obtained.

All of the above degradations and reassembly experiments were performed without the separation of the protein and nucleic acid components. In a subsequent study Bancroft & Hiebert (1967) described the apparent reconstitution of infectious virus from the isolated CCMV–protein and CCMV–RNA. Viral RNA was prepared by phenol treatment of the virus, and protein by degrading the virus in 1 M NaCl at pH 7·4 and removing the nucleic acid by centrifugation. Virus protein and RNA were mixed in approximately the same proportions as they appear in the virus and dialysed against a low ionic strength (approximately 0·02) Tris buffer pH 7·4, containing 5×10^{-3} M Mg^{2+}, for 2 hr. at 4°. The reaction mixture was then fractionated by centrifugation in a sucrose density-gradient in 0·1 M acetate pH 5. The majority of the nucleoprotein formed sedimented at the same rate as intact virus, was infectious, and resistant to snake venom phosphodiesterase. The serological properties, ultraviolet spectrum, density and appearance in the electron microscope of the reassembled material are all reported similar to those of the virus. All these observations indicate that a true reconstitution of a spherical virus, or at least a nucleoprotein very similar to a virus, has been achieved by these workers.

Similar reconstitutions can be carried out for BGMV and BBMV (J. B. Bancroft, private communication). Various hybridization experiments have also been performed with CCMV, BGMV and BBMV; thus CCMV–RNA has been encapsulated with BGMV and BBMV protein, and BGMV–RNA with CCMV and BBMV. All give infectious particles resistant to snake venom phosphodiesterase, but the infectivities are not as high as the products of homologous reconstitution experiments. With ribonucleic acids from f2, MS2, TMV and plant ribosomes it is possible to obtain organized spherical particles by interaction with CCMV–protein and in the case of TMV, the particles are infectious.

Turnip crinkle virus (TCV)

Turnip crinkle virus is a plant virus containing about 20 % RNA. The virus is a spherical particle 300 Å diameter and structurally resembles the more familiar tomato bushy stunt virus. The particle weight is of the order of 9×10^6 avogrammes, but this is, at present, being redetermined. This virus has been the subject of structural investigations by physical and chemical methods in this laboratory and some general observations have been described (Klug, Finch, Leberman & Longley, 1966).

TCV was one of the first viruses which on degradation produced what might be a protein core (Haselkorn, Hills, Markham & Rees, 1961). These initial observations were borne out in our early studies on the degradation of the virus in 0·1 M ethanolamine pH 10·5 and 0° when one of the products, which was predominantly protein, was a small spherical particle about 180 Å diameter. Subsequent systematic investigations still do not allow us to decide whether this small particle is a core present in the native virus or only a reaggregation product of the degraded virus.

Degradation studies have been made on the virus with different pH values, ionic strengths and buffers. The picture that we now have is that the virus degrades more easily the higher the ionic strength and pH and the degradation is independent of the nature of the buffer. Providing the ionic strength is sufficiently high, the virus can be degraded completely to soluble protein and nucleic acid at pH 8 and above. There is some variation from preparation to preparation in ease of degradation and to counteract this our standard procedure is to degrade the virus at 0° at pH 9 in 0·1 M Tris and 0·5 M KCl. After 1 hr. the reaction mixture is neutralized. If it is examined in the analytical ultracentrifuge (Pl. 1, fig. 2) it is generally found to have two components with sedimentation constants of about 4S and 30S, compared with the virus of 126S.

The 4 S material is the virus protein and has a molecular weight of about 50,000 (S. C. Harrison, R. Leberman & A. Klug, unpublished results). This is about three times the lowest molecular weight for the chemical subunit obtained by amino acid composition of the protein (Symons, Rees, Short & Markham, 1963). The nature of this protein is at present under investigation—there is, however, no indication at the moment that there is more than one species of protein. The amount of 30S material varies with the preparation and is apparently the intact RNA molecules.

If the neutralized degradation mixture is dialysed overnight against 0·1 M NaCl at 4° the solution becomes opalescent and a representative sedimentation pattern is shown in Plate 1, fig. 2. The degraded material reaggregates to give a number of faster sedimenting species, the fastest having a sedimentation constant corresponding approximately to intact virus and the slowest being about 45S. There are never any aggregates sedimenting between the 4S protein and the 45S component. Fractionation of the reaggregated material by sucrose-gradient centrifugation and examination of the fractions in the electron microscope shows that the fastest sedimenting material is similar in appearance to virus with some indication of a regular surface structure but with a centre either less dense or more accessible to negative stain (Pl. 2, fig. 2). The 45S material corresponds to the small particles found earlier by ethanolamine degradation. These particles have the same surface structure as the intact virus particles in that both types of particles show dimer clustering (Pl. 2, fig. 1).

Treatment of the disaggregated material with RNase generally results in the removal of some of the faster sedimenting components with a corresponding increase in the 4S protein and small particle peaks on dialysis. In some instances more small particles are obtained than could be present if they were virus cores. This demonstrates that the small particles can be formed as reaggregation products, composed of protein from the outer shell and probably small pieces of RNA. However, these observations do not rule out the possibility that the small particles also pre-exist in intact virus particles.

In order to carry out reconstitution studies with isolated TCV–protein and TCV–RNA, use was made of degradation results already described. TCV–protein can be isolated from the degraded virus in $0.5\,M$ KCl by removing the nucleic acid by centrifugation (cf. CCMV). Traces of non-sedimentable RNA can then be eliminated by ammonium sulphate precipitation. The 30S high molecular weight (ca. 2×10^6) RNA in $0.5\,M$ KCl is in a compact form and can be easily separated from the protein and broken down nucleic acid by gel filtration on Sephadex G-200. The nucleic acid ($E260/280 = 2.2$) comes out at the void volume and is well separated from the protein fractions (Fig. 1).

Our initial experiments show that virus-like particles can be produced by mixing protein with nucleic acid in the ratio of 4:1 and reducing the ionic strength of the solution to practically zero. The particles produced sediment in a sucrose gradient in about the same place as the intact virus and, although spherical, are heterogeneous in size with no clear surface structure (Pl. 2, fig. 2). Further experiments on the course of the reconstitution are under way to discover if there is a requirement for a metal ion such as Mg^{2+} or Ca^{2+} and the effects of pH and ionic strength.

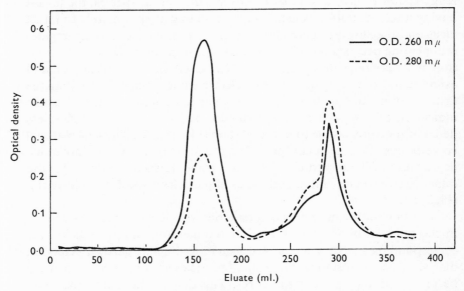

Fig. 1. Separation of TCV–RNA and TCV–protein on Sephadex G-200. 10 ml. sample of neutralized degradation mixture (containing *ca.* 80 mg. virus) applied to column of G-200 (3·2 cm. × 40 cm.) equilibrated with 0·5 M KCl. Eluant 0·5 M KCl, flow rate 15 ml./hr.

DISCUSSION

The theory of self-assembly of viruses set out by Caspar & Klug (1962) describes the process as being akin to crystallization in which the two components of a simple virus will spontaneously combine to form virus particles since this is their lowest energy state. At the time the theory was described, only the helical rod-shaped virus TMV had been reconstituted; now ample experimental evidence has been produced to verify the theory for spherical viruses. Once initiated, the self-assembly process for a helical virus is fairly obvious, additional subunits clicking into identical positions where they make the maximum number of bonds with their neighbours. The self-assembly of spherical viruses is, however, by no means obvious since a *closed* spherical shell is to be built out of identical subunits. The way this may be achieved is set out in the quasi-equivalence theory of Caspar and Klug in which all chemical subunits, using the same bonds, are situated in approximately equivalent environments in an icosahedral surface lattice. A consequence of this theory is that the chemical subunits may cluster into groups in order to maximize the number of bonds, these clusters or morphological units giving a particular virus its characteristic appearance in the electron microscope.

In solution, intermediate aggregates may exist which are different from the structure units, and the virus may be assembled from these 'building units' (Klug & Caspar, 1960). These intermediate aggregates need not be the same as the morphological units but may be related to a grouping of units found in the complete protein shell.

The dimensions of a rod-shaped virus particle are determined by the length of its nucleic acid and the size and shape and hence the number of protein subunits required to coat the nucleic acid. Similarly, the diameter of a spherical virus is determined by these properties, but with a spherical virus the secondary and tertiary structure of the nucleic acid may also be important. Evidence in favour of this for some viruses can be derived from the reconstitution studies on ϕX174 where there appears to be a requirement for Mg^{2+} and Ca^{2+} ions, and from the degradation and reconstitution studies on CCMV, where Mg^{2+} appears to play an important role. On the other hand, the shape of a virus is determined primarily by the bonds that can be made by the protein subunits and their shape. This is exemplified by the CCMV studies. Nuclease digestion of the expanded virus apparently leaves the morphological units intact and these can then aggregate into various organized structures which have different forms from the virus, but in which the local bonding patterns between the morphological units in these variants forms are similar to those in the virus. More strikingly, perhaps, TMV–RNA can be encapsulated by the protein of the spherical CCMV to give an infectious spherical particle.

The tubular variants of spherical viruses were considered by Caspar and Klug to be mistakes in the assembly of the virus and are frequently found naturally in viruses of the polyoma and papilloma groups. With the CCMV variants the 'mistakes' appear to be due to the absence of intact viral nucleic acid, required to impose restrictions on their structure. Clearly very little DNA is present in the tubular variants found in preparations of polyoma and papilloma viruses, and in the polyheads of T4 bacteriophage; whether there are significant amounts has yet to be determined. With CCMV, however, the results indicate that there is still some nucleotide material associated with the morphological units which aggregate into tubes. In this context the results of Kaper & Halperin (1965) on TYMV are of interest since they found that about 30 equal-sized pieces of RNA were produced in the virus by mild alkaline degradation, and this virus, like CCMV, has 32 morphological units.

The problems associated with carrying out the self-assembly of a simple virus *in vitro* are now apparently those of isolating the nucleic acid and protein components and then determining the best conditions for the assembly reaction to proceed. Indications of the conditions for reassembly can be given by degradation studies. As we have seen, TMV is disaggregated under conditions which would tend to weaken hydrophobic bonds and can be reconstituted under conditions which strengthen them. With CCMV and TCV, however, high ionic strength of the degradation procedures leads to the breakage of salt links, presumably between the nucleic acid and the protein, and reaggregation occurs at low ionic strength. This may indicate that in some viruses the bonds between subunits are not predominantly hydrophobic but that the nucleic acid may serve to cement the subunits together. Other viruses may have a mixture of both types of bonds, thus in TYMV, where the nucleic acid is not just a ball inside a protein shell but is intimately associated with and interlaced between the protein units (Klug, Longley & Leberman, 1966), a nucleic acid-free protein shell of the same structure as the virus occurs naturally, or can be produced artificially by degrading the nucleic acid *in situ* (Kaper, 1960). Whether or not it will be possible to reassemble such a protein shell without the assistance of nucleic acid is still an open question (cf. phage fr).

The mechanism by which the crystallization of the protein subunits around the nucleic acid is initiated is not understood. With TMV it is not known whether the reassembly of the virus starts at one particular end of the RNA molecule, although there is evidence that the virus itself is polar with subunits being removed preferentially from one end on alkali (Harrington & Schachman, 1956), urea (Buzzell, 1962) and detergent degradation (Hart, 1955; May & Knight, 1965). The results of May & Knight show that protein subunits are stripped from the 3′-hydroxyl end of the RNA by sodium dodecyl sulphate. The results with fd can be interpreted as indicating a nucleation point on the DNA ring; if this is so, how this locus is identified by the protein or whether there is a special protein at this point is yet to be determined. The results obtained with spherical viruses provide no evidence for nucleation centres on their nucleic acids but like fd a small number of protein units may still be attached to their nucleic acids and have not been detected. Alternatively the nucleic acids of spherical viruses may be closed rings (not necessarily closed by covalent bonds)—this is so for ϕX174 and there is some evidence that TYMV–RNA (Strazielle, Benoit & Hirth, 1965) and TCV–RNA (R. Leberman, A. Klug & P. Heap, unpublished results) are circular. The topological advantages of a closed ring of nucleic acid being packed inside an icosahedral shell of protein units have been discussed by Klug & Finch (1960).

The details of the interactions between a virus nucleic acid and protein may have to await the results of X-ray diffraction studies. Until that time structural information can be obtained by the type of investigations described above. There are still many questions unanswered, such as whether or not some of the larger spherical viruses have cores, and what is the function of the different structural proteins that some viruses contain. Until now the disaggregation and reassembly of only a handful of viruses have been studied; the next few years will undoubtedly see a greater activity in this field which may then fill in the chemical details of virus construction.

ACKNOWLEDGEMENTS

I wish to thank Professor J. B. Bancroft for permission to refer to his results prior to publication and Dr A. Klug for valuable discussions during the preparation of this manuscript.

REFERENCES

ANDERER, F. A. (1963). Recent studies on the structure of tobacco mosaic virus. *Adv. Protein Chem.* **18**, 1.

BANCROFT, J. B. & HIEBERT, E. (1967). Formation of an infectious nucleoprotein from protein and nucleic acid from a small spherical virus. *Virology*, **32**, 354.

BANCROFT, J. B., HILLS, G. J. & MARKHAM, R. (1967). A study of the self-assembly process in a small spherical virus. *Virology*, **31**, 354.

BANERJEE, K. & LAUFFER, M. A. (1966). Polymerization-depolymerization of tobacco mosaic virus protein. VI. Osmotic pressure studies of early stages of polymerization. *Biochemistry*, **5**, 1957.

BUZZELL, A. (1962). Action of urea on tobacco mosaic virus. *J. Am. chem. Soc.* **82**, 1636.

CASPAR, D. L. D. (1963). Assembly and stability of the tobacco mosaic virus particle. *Adv. Protein Chem.* **18**, 37.

CASPAR, D. L. D. & KLUG, A. (1962). Physical principles in the construction of regular viruses. *Cold Spring Harb. Symp. quant. Biol.* **27**, 1.

FIERS, W. & SINSHEIMER, R. L. (1962). The structure of the DNA of bacteriophage ϕX174. III. Ultracentrifugal evidence for a ring structure. *J. molec. Biol.* **5**, 424.

FINCH, J. T., LEBERMAN, R., CHANG YU-SHANG & KLUG, A. (1966). Rotational symmetry of the two-turn disk aggregate of tobacco mosaic virus protein. *Nature, Lond.* **212**, 349.

FRAENKEL-CONRAT, H. (1957). Degradation of tobacco mosaic virus with acetic acid. *Virology*, **4**, 1.

FRAENKEL-CONRAT, H. & SINGER, B. (1957). Virus reconstitution. II. Combination of protein and nucleic acid from different strains. *Biochim. biophys. Acta*, **24**, 540.

FRAENKEL-CONRAT, H. & SINGER, B. (1959). Reconstitution of tobacco mosaic virus. III. Improved methods and the use of mixed nucleic acids. *Biochim. biophys. Acta*, **33**, 359.

FRAENKEL-CONRAT, H. & SINGER, B. (1964). Reconstitution of tobacco mosaic virus. IV. Inhibition by enzymes and other proteins, and the use of polynucleotides. *Virology*, **23**, 354.

FRAENKEL-CONRAT, H., SINGER, B. & TSUGITA, A. (1961). Purification of viral RNA by means of bentonite. *Virology*, **14**, 54.

FRAENKEL-CONRAT, H. & WILLIAMS, R. C. (1955). Reconstitution of active tobacco mosaic virus from its inactive protein and nucleic acid components. *Proc. natn. Acad. Sci. U.S.A.* **41**, 690.

FRANKLIN, R. E. (1955). Structural resemblance between Schramm's repolymerised A-protein and tobacco mosaic virus. *Biochim. biophys. Acta*, **18**, 313.

GIERER, A. & SCHRAMM, G. (1956). Infectivity of ribonucleic acid from tobacco mosaic virus. *Nature, Lond.* **177**, 702.

HARRINGTON, W. F. & SCHACHMAN, H. K. (1956). Studies on the alkaline degradation of tobacco mosaic virus. I. Ultracentrifugal analysis. *Archs Biochem. Biophys.* **65**, 278.

HART, R. G. (1955). Electron microscopic evidence for the localisation of ribonucleic acid in the particles of tobacco mosaic virus. *Proc. natn. Acad. Sci. U.S.A.* **41**, 261.

HART, R. G. & SMITH, J. D. (1956). Interactions of ribonucleotide polymers with tobacco mosaic virus protein to form virus-like particles. *Nature, Lond.* **178**, 739.

HASELKORN, R., HILLS, G. J., MARKHAM, R. & REES, M. W. (1961). The structure of turnip crinkle virus. *Abst. 1st Int. Congr. Biophys. Stockholm*, p. 293.

HOHN, T. (1967). Self-assembly of defective particles of the bacteriophage fr. *Europn J. Biochem.* **2**, 152.

HOLOUBEK, V. (1962). Mixed reconstitution between protein from common TMV and RNA from different TMV strains. *Virology*, **18**, 401.

HUTCHISON, C. A., EDGELL, M. H. & SINSHEIMER, R. L. (1967). The process of infection with bacteriophage ϕX174. XII. Phenotypic mixing between electrophoretic mutants of ϕX174. *J. molec. Biol.* **23**, 553.

KAPER, J. M. (1960). Preparation and characterization of artificial top component from turnip yellow mosaic virus. *J. molec. Biol.* **2**, 425.

KAPER, J. M. & HALPERIN, J. E. (1965). Alkaline degradation of turnip yellow mosaic virus. II. *In situ* breakage of the ribonucleic acid. *Biochemistry*, **4**, 2434.

KELLENBERGER, E. (1966). Control mechanisms in bacteriophage morphopoiesis. *Ciba Foundation Symposium* on *Principles of Biomolecular Organisation*, p. 192.

KLUG, A. & CASPAR, D. L. D. (1960). The structure of small viruses. *Adv. Virus Res.* **7**, 225.

KLUG, A. & FINCH, J. T. (1960). The symmetries of the protein and nucleic acid in turnip yellow mosaic virus: X-ray diffraction studies. *J. molec. Biol.* **2**, 201.

KLUG, A. & FINCH, J. T. (1965). Structure of viruses of the papilloma-polyoma type. I. Human wart virus. *J. molec. Biol.* **11**, 403.

KLUG, A., FINCH, J. T., LEBERMAN, R. & LONGLEY, W. (1966). Design and structure of regular virus particles. *Ciba Foundation Symposium* on *Principles of Biomolecular Organisation*, p. 158.

KLUG, A., LONGLEY, W. & LEBERMAN, R. (1966). Arrangement of protein subunits and the distribution of nucleic acid in turnip yellow mosaic virus. I. X-ray diffraction studies. *J. molec. Biol.* **15**, 315.

KNIGHT, C. A. (1963). *Chemistry of Viruses*. Vienna: Springer-Verlag.

KNIPPERS, R. & HOFFMANN-BERLING, H. (1966a). A coat protein from bacteriophage fd. I. Hydrodynamic measurements and biological characterisation. *J. molec. Biol.* **21**, 281.

KNIPPERS, R. & HOFFMANN-BERLING, H. (1966b). A coat protein from bacteriophage fd. II. Interaction of the protein with DNA *in vitro*. *J. molec. Biol.* **21**, 293.

KNIPPERS, R. & HOFFMANN-BERLING, H. (1966c). A coat protein from bacteriophage fd. III. Specificity of protein-DNA association *in vivo*. *J. molec. Biol.* **21**, 305.

LAUFFER, M. A. (1966). Polymerization-depolymerization of tobacco mosaic virus protein. VII. A model. *Biochemistry*, **5**, 2440.

PLATE I

Fig. 1

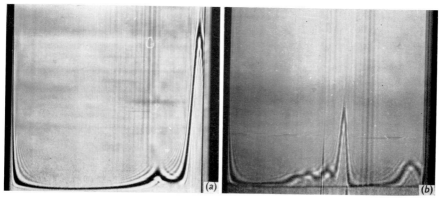

Fig. 2

PLATE 2

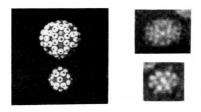

Fig. 1

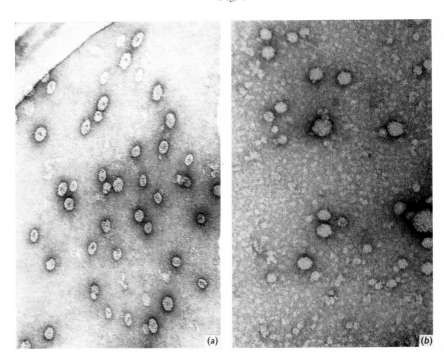

Fig. 2

LODISH, H. F., HORIUCHI, K. & ZINDER, N. D. (1965). Mutants of the bacteriophage f2. V. On the production of noninfectious phage particles. *Virology*, **27**, 139.

MAIZEL, J. V. (1963). Evidence for multiple components in the structural protein of type 1 poliovirus. *Biochem. biophys. Res. Commun.* **13**, 483.

MATTHEWS, R. E. F. (1966). Reconstitution of turnip yellow mosaic virus RNA with TMV protein subunits. *Virology*, **30**, 82.

MAY, D. S. & KNIGHT, C. A. (1965). Polar stripping of protein subunits from tobacco mosaic virus. *Virology*, **25**, 502.

NEWMARK, P. & MYERS, R. W. (1957). Degradation of tobacco mosaic virus by alkanolamines. *Fedn Proc.* **16**, 226.

RUECKERT, R. R. & DUESBERG, P. H. (1966). Non-identical peptide chains in mouse encephalitis virus. *J. molec. Biol.* **17**, 490.

SCHRAMM, G., SCHUMACHER, G. & ZILLIG, W. (1955). Uber die Stuktur des Tabakmosaikvirus. III. Der Zerfall in alkalischer Losung. *Z. Naturforsch.* **10** b, 481.

STRAZIELLE, C., BENOIT, H. & HIRTH, L. (1965). Particularités structurales de l'acide ribonucléique extrait du virus de la mosaique jaune du Navet. II. *J. molec. Biol.* **13**, 735.

SUGIYAMA, T., HEBERT, R. R. & HARTMAN, K. A. (1967). Ribonucleoprotein complexes formed between bacteriophage MS2 RNA and MS2 protein *in vitro*. *J. molec. Biol.* **25**, 455.

SUMMERS, D. F., MAIZEL, J. V. & DARNELL, J. E. (1965). Evidence for virus-specific non-capsid proteins in poliovirus-infected HeLa cells. *Proc. natn. Acad. Sci. U.S.A.* **54**, 505.

SYMONS, R. H., REES, M. W., SHORT, M. N. & MARKHAM, R. (1963). Relationship between the ribonucleic acid and protein of some plant viruses. *J. molec. Biol.* **6**, 1.

TAKAI, M. (1966). Complex-formation by protein and nucleic acid. I. Evidence for specific interaction between ϕX174 phage protein and ϕX174 phage deoxyribonucleic acid *in vitro*. *Biochim. biophys. Acta*, **119**, 20.

EXPLANATION OF PLATES

PLATE 1

Fig. 1. Electron micrographs. (*a*) tobacco mosaic virus. (*b*) Repolymerized TMV protein—stacked disc rod. (*c*) Repolymerized TMV protein—helical rod.

Fig. 2. Sedimentation patterns of degraded and 'reconstituted' TCV. (*a*) TCV degraded at pH 9, 0° at ionic strength 0·55, neutralized to pH 7. (*b*) Degraded TCV after dialysis against 0·1 M NaCl for 16 h. at 4°. Sedimentation is from right to left. Pictures taken 20 min. after attaining speed of 35,600 rev./min.

PLATE 2

Fig. 1. Electron micrographs of turnip crinkle virus and small particle compared with models viewed down a local two-fold axis. The large particle consists of 90 symmetrically disposed knobs and the small particle of 30. These knobs occupy precisely the position to be expected from the clustering of 180 and 60 subunits respectively into dimers.

Fig. 2. Electron micrographs of 'reconstituted' TCV. (*a*) Virus-like particles obtained without separating protein and nucleic acid. (*b*) Particles obtained by mixing protein and nucleic acid in 0·5 M pH 7 and desalting the mixture with Sephadex G. 25.

THE STRUCTURE OF ANIMAL VIRUSES IN RELATION TO THEIR BIOLOGICAL FUNCTIONS

D. H. WATSON*

Department of Virology, The University, Birmingham

INTRODUCTION

The visualization of virus particles was perhaps one of the first advantages claimed for the electron microscope by its early proponents in the 1930s and it is therefore not surprising to find viruses among the first objects to be examined in the new instrument (Kausche, Pfankuch & Ruska, 1939). The tobacco mosaic virus particle was indeed an almost invariable part of the programme in electron microscope demonstrations in the post-war era. Further, by 1941, Riedel & Ruska were attempting to make particle counts in the electron microscope and it seemed that electron microscopy was destined to play an important part in the quantitative biological studies on viruses which were developing rapidly at that time. The following 20 years were marked by great advances in both electron microscopy and virology, yet, except in the bacteriophage field, little information was gained on virus structure and even less on its relation to biological function. Such was still the position at the time of the last virus symposium of the Society, and Burnet (1960) was probably reflecting the general climate of virological opinion at that time when he said that 'the return from the great amount of money, time and effort put into the studies of viruses with the electron microscope has been rather small' and pointed out that electron microscopy of virus particles as then practised demanded elaborate purification and preparative technique and accordingly 'cannot be used as a practical tool for diagnosis or quantitation of a virus preparation'.

At that very time, however, Brenner & Horne (1959) described their simple new negative contrast technique in which particles were examined against an electron dense matrix of potassium phosphotungstate. The principle of negative contrast had of course been used in light microscopy for decades and the future scientific historian may well be amazed that negative contrast methods were not introduced into electron microscopy by analogy rather than as the result of an accident; Brenner & Horne,

* Member of the Medical Research Council Virus Research Group.

like Hall (1955) and Huxley (1957) before them, had originally noted
the negative contrast effect in preparations imperfectly washed after
attempted positive staining with phosphotungstate.

The negative contrast technique has two characteristics of outstanding
importance. First, it is relatively unaffected by impurities of small
molecular size which disappear in the background; secondly, it reveals
many intricate and highly characteristic structural features in virus
particles. This latter property means that the method is therefore also
unaffected by larger sized impurities which can be clearly differentiated
from virus particles. Accordingly, the way was then opened for the
subsequent exploitation of electron microscopy as a diagnostic and
quantitative tool in virology. The structural detail revealed by negative
contrast produced a wealth of information about virus architecture. This
has been extensively reviewed elsewhere (Horne & Wildy, 1964) and
such problems are not the concern of this paper, which will attempt
rather to relate the structures revealed on virus particles to biological
function.

It should be obvious that this kind of correlation demands a quantitative approach to electron microscopy, and particle counting techniques
are therefore of prime importance. Neglect of this point and adoption
of 'look and see' methods often lead to subjective impressions acting as
the foundation of false conclusions. Techniques for particle counting are
described elsewhere (Wildy & Watson, 1962); usually the virus count is
derived from a comparison of the numbers of virus particles and reference particles of known count. Use of negative contrast in counting
preparations allows in many cases the solution of the principal problem
of counting, succinctly expressed by Isaacs (1957) as being 'not how to
count but what to count'. One cautionary note should perhaps be
sounded: many viruses (including some of the RNA tumour viruses) do
not exhibit a very characteristic morphology in negative contrast;
accordingly, in such cases, identification of miscellaneous round blobs
as virus particles can be dangerously misleading.

Two main aspects of the topic will be considered: first, the relationship to function of differential particle morphology; and secondly, the
function of individual components of particles revealed by electron
microscopy.

The biological significance of bacteriophage structure has already
been extensively discussed by Bradley (1965) and elsewhere in this
symposium Bancroft considers the significance of differential particle
morphology in certain plant viruses. Accordingly this contribution will
be limited to consideration of animal viruses.

DIFFERENTIAL PARTICLE MORPHOLOGY
AND FUNCTION
'Full' and 'empty' particles

The existence of non-infective virus particles lacking nucleic acid was first demonstrated by Markham & Smith (1949) who isolated 'top component' particles of lower density than the infective particles by ultracentrifugation of preparations of turnip yellow mosaic virus. When Brenner & Horne (1959) showed that some particles of turnip yellow mosaic virus were completely penetrated by phosphotungstate to give a hollow ring effect (cf. Pl. 1, fig. 1), it seemed natural to assume that these represented the particles of 'top component'. From this it was but a short step to the hypothesis that *all* phosphotungstate penetrated particles were 'empties' and hopes ran high that electron microscopy could quickly differentiate 'fulls' and 'empties'. Unfortunately parallel biophysical, biochemical and biological studies have only been carried out in a few instances and such integrated studies have generally indicated the need for caution in interpretation. Thus Markham (1962) has shown that some 'bottom component' particles are apparently easily emptied in the preparative procedures and appear 'empty' in the microscope. Crawford, Crawford & Watson (1962) demonstrated that 'top component' particles of polyoma virus do appear 'empty' and 'bottom component' particles 'full' (Pl. 1, fig. 1), provided care is taken with the preparative procedure: in some instances, however, 'top component' particles appeared 'full' in the electron microscope.

Apparently 'empty' particles have been described for a number of other viruses. Thus herpes virus particles (Pl. 2) can appear with apparently 'empty' cores, although no 'top component' particles have so far been isolated in the ultracentrifuge. Here again external conditions can frequently affect the appearance of the particles. Thus, as will be shown later, enveloped herpes particles, agglutinated by antiserum to the host cell in which the virus was grown, frequently appear with 'empty' cores, although this is almost certainly artifactual.

'Top component' polio virus particles can be produced by heating the virus and the particles differ in antigenicity from the native particles. A convincing morphological correlation was provided by Hummeler, Anderson & Brown (1962) who demonstrated, in the electron microscope, agglutination of 'empty' particles by antiserum to heated virus, and of 'full' particles by antiserum to native virus. In this case then there is good correlation of morphology and biological function.

The correlation appears equally good with another enterovirus,

ECHO 12, where non-infective haemagglutinating particles of lower buoyant density than infective particles appear 'empty' in negative contrast.

It is relevant to consider further the significance of these 'empty' particles. In most animal virus preparations the number of physical particles greatly exceeds the number of infective particles and the presence of 'empty' particles obviously contributes to this excess. However, in polyoma virus with a particle/infectivity ratio of over 100 only about one half of the particles are defective in this way and 'bottom component' still has a particle/infectivity ratio of about 80. Certainly in preparations of some viruses there are other particles with cores partly penetrated by phosphotungstate, but even assuming that this appearance reflects a real defect in the particle, and is not artifactual, the number of apparently morphologically perfect particles greatly exceeds the number of infective particles. Morphological defects, as revealed by electron microscopy, plainly play only a small part in causing high particle/infectivity ratios.

There still remains the problem of how these particles originate. They may of course be the result of damage in preparative procedures such as disruption of the cell. However, sections of infected cells frequently show coreless particles although in one of the few instances when a thin section study was carried out in parallel with negative contrast experiments, there appeared to be little correlation between the numbers of apparently coreless particles seen in thin section, and of apparently 'empty' particles in negative contrast preparations of cell homogenates (Watson, Wildy & Russell, 1964). Halperen, Eggers & Tamm (1964) also conclude that coreless ECHO 12 particles are not produced by disruption of complete particles, since the proportion of coreless particles is the same in extracellularly released virus as in virus from disrupted cells. The general view is that 'empty' particles are assembled at the same time as virus maturation occurs, but that they do not represent a developmental form of complete virus particles. Thus, Halperen *et al.* (1964) found a constant low proportion of coreless particles throughout the growth cycle and conclude that both types of particle are formed at the same rate. By contrast Smith (1963) has claimed that there is a high proportion of 'empty' particles in herpes-infected cells at early stages of the growth cycle. Watson *et al.* (1964), on the other hand, found no evidence for this in their quantitative studies on the growth cycle of herpes virus. At the stage before the number of particles increases, that is where the particles counted represent residual inoculum virus, a higher proportion of the particles appears to be penetrated by phosphotungstate and this may account for Smith's findings.

In summary, then, 'empty' particles are only a minor contributory cause of high particle/infectivity ratios. They are not a developmental form of mature particles, although evidence is now mounting that they can be self assembling. The question of self assembly is dealt with by Leberman elsewhere in this symposium. Perhaps, however, the most startling feature of investigations of 'empty' particles by electron microscopy is the tacit assumption that phosphotungstate penetration inevitably indicates emptiness and vice versa, while the controlled investigations involving parallel studies by other techniques have by no means justified this assumption.

Enveloped and naked particles of herpes virus

Capsids of herpes group viruses are unusual in that some of them possess an outer coat or envelope (Wildy, Russell & Horne, 1960) as shown in Plate 2, fig. 1. Enveloped particles are only clearly identifiable where phosphotungstate has penetrated the envelope and delineated the characteristic structure of the capsid as in Plate 2, figs. (a) and (b). In some preparations of herpes virus there are appreciable numbers of particles of the approximate size of enveloped particles which are un-penetrated by phosphotungstate (Pl. 1, fig. 2). Sometimes a faint outline of the capsid can be seen but to be certain of identifying all enveloped particles in herpes virus preparations it is necessary to dry the grids and then re-wet with phosphotungstate. Such preparations show no un-penetrated particles and a correspondingly larger number of clearly identifiable enveloped particles.

Thin sections of infected cells show single ringed particles (presumably naked particles), some of which apparently acquire an outer covering of nuclear, cytoplasmic or cellular membrane (Morgan, Ellison, Rose & Moore, 1954; Epstein, 1962). A combination of electron microscopic and serological methods has confirmed the cellular nature of the herpes envelope (Watson & Wildy, 1963). Incubation of herpes virus suspensions with antiserum to the cell in which the virus was grown caused selective agglutination of enveloped particles (Pl. 3, fig. (a)). (It may be noted in passing that a much higher proportion of the enveloped particles were penetrated by the phosphotungstate after agglutination. Since there seems no reason why these particles should have lost their core, this observation serves to underline the doubtful validity of assuming that such penetration always indicates empty cores.) The agglutination was quite specific; enveloped particles grown in BHK 21 cells were agglutinated by antiserum to BHK 21 cells but not by antiserum to HeLa cells and *vice versa*.

Enveloped particles increase in number later in the growth cycle than do naked particles (Watson *et al.* 1964). The proportion of enveloped particles in any preparation depends on the time of harvesting and also on the cell type: HeLa and HEp 2 cells yield a higher proportion of enveloped particles than do BHK 21 cells. Extracellular virus contains a higher proportion of enveloped particles than intracellular virus and this finding is consistent with the idea that enveloped particles 'bud' out into surrounding medium as suggested by Epstein (1962). However, considerable numbers of naked particles do occur in the supernatant medium, presumably because of cell lysis.

The function of the envelope remains obscure. Enveloped particles were more readily adsorbed to cells than were naked ones (Holmes & Watson, 1963) but this may reflect only the size difference. Watson *et al.* (1964) found the number of infective particles significantly exceeded the number of enveloped particles in some preparations and concluded that probably both naked and enveloped particles were infective, although they did not exclude the possibility of the envelope conferring greater stability on the infective particle. This view was challenged by Smith (1964), who separated naked and enveloped particles on caesium chloride density gradients and found infectivity associated only with enveloped particles. However, the recovery of infectivity from caesium chloride gradients is very poor and the loss in infectivity might well be selective for naked particles. Spring & Roizman (1967) show herpes particles to be unstable in caesium chloride gradients and that prior treatment of the preparations with formalin results in the retention of greater amounts of DNA and complement-fixing antigen in the region of the virus band than are found in untreated preparations. They conclude that at the moment there is no basis for differentiating enveloped and naked particles with respect to biological activity.

As with the question of 'empty' particles it is clearly important to be certain that the differentiation of enveloped and naked particles is not biased by the experimental procedures. The best evidence on this point comes from parallel studies by ultramicrotomy which showed comparable proportions of enveloped particles by the two techniques. Plainly the question will only be finally answered when the two types of particle are separated with minimal loss of biological activity.

The envelope has been shown to possess host cell antigen. Whether it also possesses viral antigens has been open to some doubt. Antiserum to herpes virus (prepared in a heterologous cell system) agglutinated naked herpes particles (Pl. 3, fig. (b)) but apparently not enveloped particles. Clearly this is paradoxical since enveloped particles are thought to be infective and yet this antiserum neutralized most of the infectivity of the preparation. In any event other evidence does suggest that the envelope contains at the very least an antigen differing from uninfected host cell antigens. Thus Roizman & Roane (1961) showed that infected cells were susceptible to immune lysis by virus antibody and complement, and Watkins (1964) showed that sensitized sheep erythrocytes adhere to herpes infected cells, the effect being abolished by previous treatment with antiserum to the virus. Further Roizman & Spring (1967) show that antiserum made to virus grown in a heterologous cell system also produced immune lysis with complement. Furthermore, the alteration in immunological specificity of the cellular membrane was shown by them to be genetically determined since the reaction between cells infected with different mutants was greater with the homologous antiserum than any heterologous antisera.

The non-agglutination of enveloped particles may of course be entirely due to an insufficiently high concentration of the appropriate antibody. It may also be of importance to note that enveloped particles are formed from nuclear, cytoplasmic or cellular membranes, while Roizman's and Watkins's findings relate to the cellular membrane only. Further, the concentration of such antigenic sites on the cellular membrane would not need to be high to explain the immunological specificity of the cell membrane, while specific agglutination of enveloped particles might demand a higher concentration of specific antigen in the envelope. Nevertheless, the weight of the evidence does appear to be in favour of the envelope containing virus antigen, a fact that confers even greater fascination on the question of the function of the envelope.

Defective interfering particles of vesicular stomatitis virus

Vesicular stomatitis virus preparations consist mainly of bullet-shaped particles 1800 Å × 650 Å containing an internal striated component surrounded by an envelope with projecting spikes (Howatson & Whitmore, 1962). Also present are a number of smaller roughly spherical particles also possessing envelopes studded with spikes. Howatson & Whitmore believed these might be artifactual, but Hackett (1964) pointed out that no such particles could be produced by disruption of bullet-shaped particles. She found that high multiplicity infection and serial undiluted passage of the virus gave stocks in which the second spherical component predominated. Cooper & Bellett (1959) had reported that such methods of propagation resulted in a reduced yield of infective virus but that the stocks contained a 'transmissible com-ponent' (T) which interfered with the replication of infective virus. Accordingly, Hackett (1964) identified the smaller spherical particles in her preparation as the T component.

Confirmation of this identification has come from density gradient studies on vesicular stomatitis virus. Brown, Cartwright & Almeida (1966) showed separation of three complement-fixing components by centrifugation of vesicular stomatitis virus preparations. One of these corresponded to complete particles, while one of the others contained smaller spherical particles similar to those described by Hackett (1964). These observations were extended by Huang, Greenawalt & Wagner (1966) and Hackett, Schaffer & Madin (1967) who separated the smaller particles from complete particles by zonal centrifugation in sucrose gradients. They showed by radioactive labelling that RNA was in-corporated in the T particles, which, like the complete particles, could inhibit cellular RNA synthesis. Finally, Huang & Wagner (1966) showed that these purified T particles interfered with the growth of infective vesicular stomatitis virus. The capacity to interfere was completely destroyed by ultraviolet radiation, although such treatment had no effect on the ability of either complete or T particles to inhibit cellular RNA synthesis. It appears likely that viral RNA is necessary for the proper functioning of the interfering power of T particles and accord-ingly Huang et al. (1966) suggest that the T particles contain a portion of the viral genome. The replication of T particles requires simultaneous infection with complete particles although it is not yet certain whether infection with complete virus particles alone can result in production of T particles.

Incomplete and filamentous forms of influenza virus

As with vesicular stomatitis virus, undiluted serial passage of influenza virus at high multiplicity gives lower yields of infective virus. The yield of haemagglutinin remains unchanged and so the resultant 'incomplete' virus is characterized by lower infectivity/haemagglutination ratios (10^2–10^3 instead of 10^6 for standard virus). Preparations of incomplete virus contained many irregularly shaped particles of heterogeneous size. By contrast, standard virus particles were much more uniform in size and shape (Barry, Waterson & Horne, 1962; Moore, Davies, Levine & Englert, 1962).

Filamentous forms of influenza were first described by Mosley & Wyckoff (1946). After earlier claims that they were not infective, Ada, Perry & Abbott (1958) showed the reverse to be true. Freshly isolated strains have been shown to contain a higher proportion of filaments (Choppin, Murphy & Tamm, 1960), although passage in eggs apparently results in reversion to spherical forms. Burnet & Lind (1957), who were able to maintain a filamentous strain very largely in this form by limit dilution passage, concluded that the conversion was mutational and Kilbourne (1963) regards the filamentous or spherical form as a genetic trait controlled by the virus nucleic acid.

Blough (1963) has shown that increased proportions of filamentous virus particles are produced when influenza virus is grown in the presence of vitamin A alcohol. He concludes that since vitamin A alters the cell membrane, the production of filaments is related to the surface state of the cell. Choppin (1963) has pointed out that this does not necessarily mean that normal production of filaments is not a genetically controlled process: the surface state of the cell may well be determined by interaction with the virus genome, although it can of course be affected by environmental conditions such as the presence of vitamin A.

Some artifactual differences

Running through the preceding sections has been the theme that one has to be certain that differences revealed by electron microscopy are real. It has been pointed out several times that certain differences may be artifactual. In concluding the part of this paper dealing with morphological heterogeneity of virus preparations it is perhaps salutary to mention some differences that being artifactual have no biological significance. In at least one case biological significance was tentatively assigned to the observed difference.

The first case to be described is that of the 'capsulated' and 'mulberry' forms of vaccinia virus which are shown in Plate 5, fig. 1. These were first described by Nagington & Horne (1962) and were later investigated by Westwood et al. (1964). The capsulated form is larger than the mulberry form and the simplest explanation of the difference was that the capsulated form was the mature form and represented the mulberry form inside an outer membrane or envelope. However, if virus was pelleted from preparations in which both forms were shown by negative contrast, sections of the pellet showed all particles to be of the same apparently 'mature' form. Further, the proportion of the two types observed in negative contrast was dependent on the preparative procedure used. In particular if the grids were dried in preparation and then re-wet the proportion of capsulated forms increased. It therefore appears that the two forms are identical and that the capsulated form results from phosphotungstate penetration of the mulberry form with consequent partial separation of the outer layer (hence the slightly larger size of the capsulated form). The penetration means that the threaded structure of the outer layer is no longer revealed. The whole process adds further weight to the earlier stricture that all phospho-tungstate penetrated particles should not be presumed to be 'empty'.

Difficulties in penetration of phosphotungstate leading to uncertain identification of enveloped particles of herpes virus have been described in a preceding section. It would seem that enveloped particles into which phosphotungstate has penetrated have definitely been damaged, but it is not safe to assume that all such penetrated particles exist in such a damaged form in the original sample before preparation of the grid. As with the mulberry forms of vaccinia re-wetting results in penetration of all the particles and it is very possible that the particles which appear penetrated without intentional re-wetting may have been accidently re-wet in preparation. This seems all the more likely since Harris & Westwood (1964) showed that preparation of a grid in the normal way followed by storage, resulted in most of the particles of vaccinia virus being penetrated, while grids stored under a vacuum showed far lower proportions of penetrated particles. Thus re-wetting has probably taken place by adsorption of moisture by the dried phosphotungstate from the atmosphere.

These findings mean that great care has to be exercised in interpreting differential penetration of particles by phosphotungstate. In general, unless there is supporting evidence from sectioning studies or differences in sedimentation properties, it is perhaps best to assume all such differences to be artifactual.

THE FUNCTION OF INDIVIDUAL COMPONENTS OF THE VIRION

Antigenic components of adenovirus

Adenovirus was the first virus which was demonstrated to have a capsid with icosahedral symmetry by the negative contrast method (Horne, Brenner, Waterson & Wildy, 1959) and it is perhaps appropriate that it is the first such virus for which there is detailed knowledge of the biological activity of the components revealed by electron microscopy. This knowledge has led to a nomenclature for the soluble antigens of adenovirus based on morphology in place of the several earlier nomenclatures which differed between each group of workers (Ginsberg, Pereira, Valentine & Wilcox, 1966). We shall follow this recommended nomenclature in this paper.

In parallel with these studies it was shown that the adenovirus particle possessed an even more intricate structure than the earlier work of Horne *et al.* (1959) had suggested. Valentine & Pereira (1965) showed that a fibre 200 × 20 Å was attached to each of the 12 capsomeres lying on the fivefold axes of symmetry at the vertices of the icosahedron. To this fibre is attached a 40 Å knob (Pl. 4, fig. (*a*)).

Cells infected with adenoviruses contain three soluble antigens. These can be separated by various chromatographic or electrophoretic techniques. The first is the so-called group specific complement-fixing antigen which as its name implies is shared by most adenovirus serotypes of human and animal origin. This antigen can also be obtained by disruption of purified virus particles by alkaline dialysis. The antigen separated from infected cells or from purified particles is shown to be identical morphologically with the 240 capsomeres on the particle in positions other than on the vertices (Wilcox, Ginsberg & Anderson, 1963). Since these capsomeres have six neighbours they are called 'hexons' and the group specific antigen is described as the 'hexon antigen' (Pl. 4, fig. (*b*)).

The second soluble antigen is the so-called early cytopathic factor or toxin which causes detachment of infected cells from the glass. One of the components produced by trypsin digestion of this antigen is identical to the third soluble antigen, the type specific complement-fixing antigen. It seems that these two antigens share two determinants, the toxin antigen possessing an additional determinant. Once again these antigens can be identified in disrupted virus particles and the early cytopathic factor is shown to correspond to the assembly of capsomere + fibre + knob (Valentine & Pereira, 1965) as shown in Plate 4, fig. (*c*). Since these capsomeres have five neighbours they are called 'pentons' and the toxin is referred to as 'penton antigen'. The third soluble antigen is found to correspond to the fibre + knob which presumably explains why it shares antigenic determinants with the cytopathic factor. The component of penton antigen which is not shared with the fibre + knob antigen—presumably the capsomere (or penton base) itself—is the one responsible for cell detachment and direct haemagglutination and has a subgroup specificity. The two components of fibre + knob antigen are responsible for indirect haemagglutination which requires the presence of heterotypic adenovirus antiserum.

The fibre + knob appendage is very reminiscent of the bacteriophage tail and this resemblance led Valentine & Pereira (1965) to speculate that the fibre might, like the bacteriophage tail, play some role in adsorption to the cell. Further, the fibre + knob has at least one type specific component and it appeared likely, since neutralization was type specific, that this would be the component which reacts with neutralizing antibody. However, Wilcox & Ginsberg (1963) showed that antibody to hexon antigen was neutralizing, although this neutralization was type specific. In fact it now appears that this antigen possesses two determinants, one type specific and one group specific: thus if antiserum to the hexon antigen of adenovirus type 5 is absorbed with type 4 hexon antigen, the antiserum loses the power to react in complement-fixation tests with type 4 hexon but not with type 5 hexon antigen. Absorption of the antiserum with homologous antigen abolishes all complement-fixing abilities. Pereira (private communication) has confirmed that antisera to hexon antigen neutralize infectivity but finds that antisera to penton antigen display virtually no neutralizing activity. The role of the fibre and knob in infection would thus not appear to be correlated to adsorption.

Norrby (1964, 1966) has made similar investigations on the structure of antigens of adenovirus type 3. In particular he has investigated the structure of the haemagglutinins. These are of two types: 'direct' which will agglutinate green monkey erythrocytes, and indirect, which will agglutinate only in the presence of a heterotypic adenovirus antiserum. Norrby found the direct haemagglutinin was an aggregate of penton antigens, while the indirect haemagglutinin consisted of free penton antigens. The result is in contrast to that obtained with adenovirus type 5, where free penton antigen was found to be capable of causing direct haemagglutination.

The results of these studies on the biological activity of the capsomeres have some bearing on ideas of virus architecture in general. Horne & Wildy (1961) suggested that hexons consisted of six structure units, the penton bases consisting of five such units, and they imagined the virus to be assembled from pre-formed capsomeres. Caspar & Klug (1962) depicted viruses as being built directly from structure units. The appearance of the penton bases and hexons in electron microscopy would then be the result of the location of structure units at particular sites on the shell. The results have certainly shown that capsomeres exist as such in infected cells, but of course this could still be the result of aggregation of excess structure units and the particle itself could still be directly constructed from structure units. Implicit in both approaches is the idea that the structure units are identical, although this is not a fundamental requirement of either theory. The differing antigenic properties of the pentons and hexons would seem to suggest that they have no common structure unit but it is of course known that the way protein units aggregate influences their antigenic structure. Thus glutamic dehydrogenase exists in several forms and polymers differ in antigenicity from each other and from monomer (Talal, Tomkins, Mushinski & Yielding, 1964). Further, Maizel (1966) showed that adenovirus particles contained 10 different species of protein molecule by acrylamide electrophoresis of purified adenovirus particles disrupted by sodium dodecyl sulphate. This would suggest that each of the component antigens of the particle would be separable into several molecules by sodium dodecyl sulphate and indeed Pereira & Martin (private communication) have shown this to be so. However, as yet there is no information about the antigenic properties of these fragments.

One final problem arises in connexion with the fibre and knob structures revealed on the electron micrographs of Valentine & Pereira (1965). During growth of adenovirus in infected cells large aggregates of 'crystals' of virus particles are seen in the nuclei (Morgan, Howe, Rose & Moore, 1956). The particles in such aggregates are very closely packed and the disposition of the fibres and knobs would appear to be a problem here since the centre-to-centre distance of the particles in the aggregates seems to be no greater than the capsid diameter without projections. Similar comments can be made about the electron microscopic studies of Dales (1962) on the early stages of infection of HeLa cells with adenovirus. In both thin sections and whole mount preparations of infected cells by negative contrast the capsids are shown in close apposition with the cellular membrane. The fibre + knob structures are only seen on preparations of highly purified virus. They may of course be obscured by the presence of contaminating protein in less pure preparations but it seems at least possible that like the fibres of bacteriophage particles they may normally be folded down near the particle and only become extended as the result of some specific process in infection or by some external effect in manipulation.

The components of influenza virus

Like other myxoviruses, influenza virus particles are composed of a helical nucleoprotein component packed in an outer envelope covered with projecting spikes (Horne, Waterson, Wildy & Farnham, 1960). Influenza particles are able to adsorb to erythrocytes of many species (causing haemagglutination) and, further, to elute from them again owing to the action of the enzyme neuraminidase. The haemagglutinin and neuraminidase must reside in the outer coat of the virus but the relationship of each to one another and to the characteristic spikes on the surface of the particle is still far from clear and there have been many conflicting reports dealing with these points.

Hoyle (1952) showed that ether treatment of influenza virus resulted in the separation of the envelope from the nucleoprotein fraction of the particle and this method has formed the basis of much subsequent work on the relation of the structure of influenza virus to its function, although Laver (1963) has pointed out that the method is not very effective. Unfortunately other methods which may give more efficient separation at the molecular level may also destroy biological activity. This dilemma is a frequent obstacle in attempts to 'dissect' viruses into individual components whose activities can be characterized. Following on earlier studies on the morphology of the haemagglutinating component released by ether splitting, Hoyle, Horne & Waterson (1961) studied it by negative contrast and showed the presence of small 'rosettes' which resembled aggregates of the external spikes of the complete virus particle. It would seem very likely that the haemagglutinin forms part of these spikes. Blough (1964) and Reginster (1966) have both claimed to find haemagglutinating particles which do not possess spikes but, as Fazekas de St Groth (1964) has pointed out, only a few spikes on an otherwise 'bald' particle would be quite sufficient to cause agglutination.

The location of the neuraminidase is more obscure. Hoyle *et al.* (1961) claimed that trypsin treatment which destroyed the ability of influenza virus to elute from red cells also resulted in the loss of spikes from the particles. They concluded that the neuraminidase was released from the particles and must accordingly reside in the spikes. However, Noll, Aoyagi & Orlando (1962) found that although this treatment did indeed release neuraminidase the spikes remained on the particle. They further found that the haemagglutinin remained associated with the particle. Further evidence of the distinct nature of the haemagglutinin and neuraminidase was obtained by Laver (1963, 1964) who disrupted the virus with sodium dodecyl sulphate. Electrophoresis allowed separation of the neuraminidase from a protein which in the case of one strain of influenza virus caused haemagglutination. These experiments suggest strongly that the two activities reside in different molecules since, unlike trypsin digestion, detergent treatment does not break covalent bonds.

Blough (1967) has recently used a ferritin-tagging procedure in an attempt to locate the neuraminidase molecules. Ferritin is attached to fetuin, a glycoprotein which can act as a substrate for neuraminidase. Influenza infected cells are exposed to the labelled fetuin after glutaraldehyde treatment which serves the dual purpose of fixing the material for ultramicrotomy and of selectively inhibiting the haemagglutinin without affecting the neuraminidase activity. After embedding and thin sectioning, the sites of neuraminidase activity are indicated by the characteristic ferritin particles. Unfortunately the method only reveals the site as being generally on the exterior of the particle and infected cells.

It has been accepted that the influenza virus coat contained host cell protein since Smith, Belyavin & Sheffield (1953) were unable to rid virus preparations of host cell protein by repeated adsorption and elution from red cells. Ananthanarayan (1954) disputed this, claiming that the purified virus still contained fragments of chorioallantoic membrane as impurity but not as an integral part of the particle. Duc-Nguyen, Rose & Morgan (1966), using ferritin tagged antibodies to host cell and virus, demonstrated a progressive conversion from host to virus specificity of the cell membrane and concluded that host cell protein was not an integral part of the particle, although even the light 'labelling' of the virus particles with antibody to host cell would seem to indicate that it is not possible to reach such a definite conclusion from these results. Studies by Laver & Webster (1966) seem to bear out the conclusions of Smith *et al.* (1953); they found inhibition of haemagglutination by influenza viruses by antibody to the host cell in which the virus was grown.

Apart from the haemagglutinating antigen, influenza particles contain a complement-fixing antigen identical to the soluble antigen found in cells infected with the virus (Schäfer, 1957). This activity corresponds to the internal nucleoprotein component released from the outer envelope by ether treatment. This appears to consist of structure units 30 Å in diameter possibly arranged in a helical formation which is 90 Å wide (Hoyle *et al.* 1961). The diameter of the nucleoprotein is a valuable taxonomic aid in classification of myxoviruses (Waterson, 1962); the influenza group of particles all have nucleoprotein of 90 Å diameter; that of the paramyxovirus group is 180 Å wide.

'Subviral particles' and 'inner capsids'

This section is concerned with evidence for ordered protein structure or inner capsids within the normal capsid and the relationship of these to the subviral particles found in some virus preparations.

Horne (1962) reported that in preparations of infective canine hepatitis virus (a virus of the adeno-group) there were apparently empty particles whose diameter was smaller than that of the complete virus particle. A structural component of approximately the same size could be distinguished in some complete particles of the virus.

A similar internal component was reported for reovirus by Mayor, Jamison, Jordan & Mitchell (1965) who were also able to find free inner components. These were usually phosphotungstate penetrated and on caesium chloride gradients were found together with 'empty' particles of normal size. They could be obtained by trypsin digestion of complete virus particles.

Klug, Finch, Leberman & Longley (1965) have also reported the presence of small inner capsids of papilloma viruses which once again were observed inside complete particles and also lying free. They observed a similar component in preparations of turnip crinkle virus and were able to produce the internal component by alkaline degradation of complete particles. They suggested that the virus nucleic acid is packed between the two protein shells. The main point of interest is that there is a serological relationship between protein of the two shells.

The main hazard in identifying such inner capsids is the presence of a contaminating virus. It is noteworthy that the 'adeno-associated virus' (AAV) found in several adenovirus preparations (Atchison, Casto & Hammon, 1965; Hoggan, Blacklow & Rowe, 1966) was originally claimed to be a component of adenovirus itself (Mayor, Jamison, Jordan & Melnick, 1965). These particles are 200 Å in diameter and can be separated from adenovirus by density gradient centrifugation or ultrafiltration. Mayor, Jamison, Jordan & Melnick (1965) claimed to find more small particles after fluorocarbon treatment of adenovirus preparations and concluded that they had been produced by disruption of complete adenovirus particles. Unfortunately particle counts were not made to determine whether or not the increase in the number of small particles was accompanied by a parallel decrease in complete adenovirus particles. In fact the small particles were found to be antigenically distinct from adenovirus (Atchison *et al.* 1965; Hoggan *et al.* 1966) and contaminated stocks of adenovirus could be freed from small particles by passing them in the presence of antiserum to the associated virus particles. Further Hoggan *et al.* (1966) were unable to disrupt purified preparations of adenovirus particles to yield the small particles. It can only be assumed that the fluorocarbon treatment used by Mayor, Jamison, Jordan & Melnick (1965) released small particles from larger pieces of cell debris thus apparently producing them in higher numbers.

The author has had a similar experience in a study of trypsin-treated polyoma virus. A number of small particles were apparent in the preparation (Pl. 5, fig. 2). However, particle counting before and after trypsin treatment revealed that the number of normal size polyoma particles had not decreased, and accordingly the increased numbers of smaller particles can only be due to disaggregation or release of particles from cellular debris. These particles could still of course represent inner capsids of polyoma virus even although they could not be obtained by disruption of polyoma virus particles. The observation does, however, underline the conclusion drawn from the results with the adeno-associated virus, that it is not justifiable to assume that all small particles found in virus preparations are 'inner capsids', nor that an increase in the number of such particles seen after treatment of a preparation is due to disruption of larger particles unless supporting quantitative data are available.

Easterbrook (1966) has isolated cores from vaccinia virus by using non-ionic detergents. The cores look very similar to those seen in sectioned particles. They also resemble the immature forms found in homogenates of infected cells. Unfortunately, no studies have been made on the biological activity of such cores. Joklik (1964) has suggested that in infection of cells with pox viruses, the outer coat of the particle is digested away non-specifically to reveal the core, which then induces synthesis of an uncoating protein. Studies with purified cores might therefore be of interest in further elucidation of these processes.

Takehara & Schwerdt (1967) have recently reported the isolation of infective 'subviral particles' from preparations of myxoma and fibroma viruses. These are separable from complete virus particles by ultra-filtration and are neutralizable by virus antiserum. Takehara & Schwerdt conclude they cannot be as large as cores from the pore size of ultrafilter used. However, relation of ultrafiltration data to morphological size is always difficult and it is quite possible that their particles are indeed cores. Unfortunately, in this case where biological data are available no electron microscopy was done. Admittedly the number of infective subviral particles was very low and it is possible that there would be insufficient particles to detect in the electron microscope. However, it is very likely that the particle/infectivity ratio of such components is very high and so the preparations may contain a sufficiently high concentration of particles for electron microscopy.

CONCLUSION

In this paper some examples have been given of the conclusions that can be drawn about the relationship between structure and function of the virion. Stress has been laid on possible misinterpretations that can arise through over-enthusiastic acceptance of the old adage 'seeing is believing'. It would seem appropriate to conclude, as did Morgan & Rose (1959) in the last symposium on viruses, by stating that 'electron microscopy by itself, disassociated from allied or complementary techniques, can raise new questions, but it cannot provide the answers'.

ACKNOWLEDGEMENTS

I am indebted to Professor P. Wildy who first interested me in the relationship of structure and function, for much advice and encouragement. I am grateful to Dr H. G. Pereira for permission to quote unpublished results, to Dr B. Roizman for sending me details of his results in advance and to other colleagues named in the text for permission to reproduce photographs. Thanks are due to Mr R. Dowler for preparation of some of the photographs for publication.

REFERENCES

ADA, G. L., PERRY, B. T. & ABBOT, A. (1958). Biological and physical properties of the Ryan strain of filamentous influenza virus. *J. gen. Microbiol.* **19**, 23.

ANANTHANARAYAN, R. (1954). The fabric of virus elementary bodies. *Br. J. exp. Path.* **35**, 381.

ATCHISON, R. W., CASTO, B. C. & HAMMON, W. McD. (1965). Adenovirus-associated defective virus particles. *Science*, **149**, 754.

BARRY, R. D., WATERSON, A. P. & HORNE, R. W. (1962). Incomplete forms of influenza virus. *Z. Naturf.* **17b**, 749.

BLOUGH, H. A. (1963). The effect of vitamin A alcohol on the morphology of myxoviruses. I. The production and comparison of artificially produced filamentous virus. *Virology*, **19**, 349.

BLOUGH, H. A. (1964). Role of the surface state in the development of myxoviruses. In *Cellular Biology of Myxovirus Infections. Ciba Foundation Symposium.* Ed. G. E. W. Wolstenholme and J. Knight. London: J. and A. Churchill.

BLOUGH, H. A. (1967). Viral neuraminidase: a cytochemical binding method using a glycoprotein coupled to ferritin. *Virology*, **31**, 514.

BRADLEY, D. E. (1965). The morphology and physiology of bacteriophages as revealed by the electron microscope. *Jl R. microsc. Soc.* **84**, 257.

BRENNER, S. & HORNE, R. W. (1959). A negative staining method for high resolution electron microscopy of viruses. *Biochim. biophys. Acta*, **34**, 103.

BROWN, F., CARTWRIGHT, B. & ALMEIDA, J. D. (1966). The antigens of vesicular stomatitis virus. I. Separation and immunogenicity of three complement-fixing components. *J. Immun.* **96**, 537.

BURNET, F. M. (1960). *Principles of Animal Virology*, 2nd ed. New York: Academic Press.

BURNET, F. M. & LIND, P. E. (1957). Studies on filamentary forms of influenza virus with special reference to the use of dark-ground microscopy. *Arch. Virusforsch.* **7**, 413.

CASPAR, D. L. D. & KLUG, A. (1962). Physical principles in the construction of regular viruses. *Cold Spring Harb. Symp. quant. Biol.* **27**, 1.

CHOPPIN, P. W. (1963). Multiplication of two kinds of influenza A2 virus particles in monkey kidney cells. *Virology*, **21**, 342.

CHOPPIN, P. W., MURPHY, J. S. & TAMM, I. (1960). Studies of two kinds of virus particles which comprise influenza A2 virus strains. III. Morphological characteristics: independence of morphological and functional traits. *J. exp. Med.* **112**, 765.

COOPER, P. D. & BELLETT, A. J. D. (1959). A transmissible interfering component of vesicular stomatitis virus preparations. *J. gen. Microbiol.* **21**, 485.

CRAWFORD, L. V., CRAWFORD, E. M. & WATSON, D. H. (1962). The physical characteristics of polyoma virus. I. Two types of particle. *Virology*, **18**, 170.

DALES, S. (1962). An electron microscopic study of the early association between two mammalian viruses and their hosts. *J. Cell Biol.* **13**, 303.

DUC-NGUYEN, H., ROSE, H. M. & MORGAN, C. (1966). An electron microscopic study of changes at the surface of influenza-infected cells as revealed by ferritin conjugated antibodies. *Virology*, **28**, 404.

EASTERBROOK, K. B. (1966). Controlled degradation of vaccinia virions *in vitro*: an electron microscopic study. *J. Ultrastruct. Res.* **14**, 484.

EPSTEIN, M. A. (1962). Observations on the release of herpes simplex virus from infected HeLa cells. *J. Cell Biol.* **12**, 589.

FAZEKAS DE ST GROTH, S. (1964). The antibody response. In *Cellular Biology of Myxovirus Infections. Ciba Foundation Symposium.* Ed. G. E. W. Wolstenholme and J. Knight. London: J. and A. Churchill.

GINSBERG, H. S., PEREIRA, H. G., VALENTINE, R. C. & WILCOX, W. C. (1966). A proposed terminology for the adenovirus antigens and virion morphological subunits. *Virology*, **28**, 782.

HACKETT, A. J. (1964). A possible morphologic basis for the autointerference phenomenon in vesicular stomatitis virus. *Virology*, **24**, 51.

HACKETT, A. J., SCHAFFER, F. L. & MADIN, S. H. (1967). The separation of infectious and autointerfering particles in vesicular stomatitis virus preparations. *Virology*, **31** 114.

HALL, C. E. (1955). Electron densitometry of stained virus particles. *J. biophys. biochem. Cytol.* **1**, 1.

HALPEREN, S., EGGERS, H. J. & TAMM, I. (1964). Complete and coreless hemagglutinating particles produced in ECHO 12 virus-infected cells. *Virology*, **23**, 81.

HARRIS, W. J. & WESTWOOD, J. C. N. (1964). Phosphotungstate staining of vaccinia virus. *J. gen. Microbiol.* **34**, 491.

HOGGAN, M. D., BLACKLOW, N. R. & ROWE, W. P. (1966). Studies of small DNA viruses found in various adenovirus preparations: physical, biological and immunological characteristics. *Proc. natn. Acad. Sci. U.S.A.* **55**, 1467.

HOLMES, I. H. & WATSON, D. H. (1963). An electron microscope study of the attachment and penetration of herpes virus in BHK 21 cells. *Virology*, **21**, 112.

HORNE, R. W. (1962). The comparative structure of adenoviruses. *Ann. N.Y. Acad. Sci.* **101**, 475.

HORNE, R. W., BRENNER, S., WATERSON, A. P. & WILDY, P. (1959). The icosahedral form of an adenovirus. *J. molec. Biol.* **1**, 84.

HORNE, R. W., WATERSON, A. P., WILDY, P. & FARNHAM, A. E. (1960). The structure and composition of the myxoviruses. I. Electron microscope studies of the structure of myxovirus particles by negative staining techniques. *Virology*, **11**, 79.

HORNE, R. W. & WILDY, P. (1961). Symmetry in virus architecture. *Virology*, **15**, 348.

HORNE, R. W. & WILDY, P. (1964). Virus structure revealed by negative staining. *Adv. Virus Res.* **10**, 102.

HOWATSON, A. F. & WHITMORE, G. F. (1962). The development and structure of vesicular stomatitis virus. *Virology*, **16**, 466.

HOYLE, L. (1952). Structure of the influenza virus. The relation between biological activity and chemical structure of virus fractions. *J. Hyg., Camb.* **50**, 229.

HOYLE, L., HORNE, R. W. & WATERSON, A. P. (1961). The structure and composition of the myxoviruses. II. Components released from the influenza virus particle by ether. *Virology*, **13**, 448.

HUANG, A. S., GREENAWALT, J. W. & WAGNER, R. R. (1966). Defective T particles of vesicular stomatitis virus. I. Preparation, morphology and some biologic properties. *Virology*, **30**, 161.

226   D. H. WATSON

HUANG, A. S. & WAGNER, R. R. (1966). Defective T particles of vesicular stomatitis virus. II. Biologic role in homologous interference. *Virology*, **30**, 173.

HUMMELER, K., ANDERSON, T. F. & BROWN, R. A. (1962). Identification of poliovirus particles of different antigenicity as seen in the electron microscope. *Virology*, **16**, 84.

HUXLEY, H. E. (1957). Some observations on the structure of tobacco mosaic virus. In *Proc. 1st European Conf. Electron Microscopy*. Stockholm: Almquist and Wiksell.

ISAACS, A. (1957). Particle counts and infectivity titrations for animal viruses. *Adv. Virus Res.* **4**, 111.

JOKLIK, W. K. (1964). The intracellular uncoating of pox virus DNA. II. The molecular basis of the uncoating process. *J. molec. Biol.* **8**, 277.

KAUSCHE, G. A., PFANKUCH, E. & RUSKA, H. (1939). Die sichtbarmung von pflanzlichen virus im Übermikroskop. *Naturwissenschaften*, **27**, 292.

KILBOURNE, E. D. (1963). Influenza virus genetics. *Prog. med. Virol.* **5**, 79.

KLUG, A., FINCH, J. T., LEBERMAN, R. & LONGLEY, W. (1965). Design and structure of regular virus particles. In *Principles of Biomolecular Organisation. Ciba Foundation Symposium*. Ed. G. E. W. Wolstenholme and M. O'Connor. London: J. and A. Churchill.

LAVER, W. G. (1963). The structure of influenza viruses. 3. Disruption of the virus particle and separation of neuraminidase activity. *Virology*, **20**, 251.

LAVER, W. G. (1964). Structural studies on the protein subunits from three strains of influenza virus. *J. molec. Biol.* **9**. 109.

LAVER, W. G. & WEBSTER, R. G. (1966). The structure of influenza viruses. IV. Chemical studies of the host antigen. *Virology*, **30**, 104.

MAIZEL, J. V. (1966). Acrylamide-gel electrophorograms by mechanical fractionation: radioactive adenovirus proteins. *Science, N.Y.* **151**, 988.

MARKHAM, R. (1962). The analytical centrifuge as a tool for the investigation of plant viruses. *Adv. Virus Res.* **9**, 241.

MARKHAM, R. & SMITH, K. M. (1949). Studies on the virus of turnip yellow mosaic. *Parasitology*, **39**, 330.

MAYOR, H. D., JAMISON, R. M., JORDAN, L. E. & MELNICK, J. L. (1965). Structure and composition of a small particle prepared from a simian adenovirus. *J. Bact.* **90**, 235.

MAYOR, H. D., JAMISON, R. M., JORDAN, L. E. & MITCHELL, M. W. (1965). Reoviruses. II. Structure and composition of the virion. *J. Bact.* **89**, 1548.

MOORE, D. H., DAVIES, M. C., LEVINE, J. & ENGLERT, M. E. (1962). Correlation of structure with infectivity of influenza virus. *Virology*, **17**, 470.

MORGAN, C., ELLISON, S. A., ROSE, H. M. & MOORE, D. H. (1954). Structure and development of viruses as observed in the electron microscope. I. Herpes simplex virus. *J. exp. Med.* **100**, 195.

MORGAN, C., HOWE, C., ROSE, H. M. & MOORE, D. H. (1956). Structure and development of viruses observed in the electron microscope. IV. Viruses of the RI–APC group. *J. biophys. biochem. Cytol.* **2**, 351.

MORGAN, C. & ROSE, H. M. (1959). Electron microscopic observations on adenoviruses and viruses of the influenza group. In *Virus Growth and Variation. Soc. gen. Microbiol. IXth Symp.* Ed. A. Isaacs and B. W. Lacey. Cambridge University Press.

MOSLEY, V. M. & WYCKOFF, R. W. G. (1946). Electron micrography of the virus of influenza. *Nature, Lond.* **157**, 263.

NAGINGTON, J. & HORNE, R. W. (1962). Morphological studies of orf and vaccinia viruses. *Virology*, **16**, 248.

NOLL, H., AOYAGI, T. & ORLANDO, J. (1962). The structural relationship of sialidase to the influenza virus surface. *Virology*, **18**, 154.

NORRBY, E. (1964). The relationship between the soluble antigens and the virion of adenovirus type 3. I. Morphological characteristics. *Virology*, **28**, 236.

NORRBY, E. (1966). The relationship between the soluble antigens and the virion of adenovirus type 3. II. Identification and characterization of an incomplete hemagglutinin. *Virology*, **30**, 608.

REGINSTER, M. (1966). Release of influenza virus neuraminidase by Caseinase C. of *Streptomyces albus* G. *J. gen. Microbiol.* **42**, 323.

RIEDEL, G. & RUSKA, H. (1941). Übermikroskopische Bestimmung der Teilchenzahl eines Sols über dessen aerodispersen Zustand. *Kolloidzeitschrift*, **96**, 86.

ROIZMAN, B. & ROANE, P. R. (1961). Studies on the determinant antigens of viable cells. I. A method and its application in tissue culture studies for enumeration of killed cells, based on the failure of virus multiplication following injury by cytotoxic antibody and complement. *J. Immun.* **87**, 714.

ROIZMAN, B. & SPRING, S. D. (1967). Alteration in immunologic specificity of cells infected with cytolytic viruses. *Proc. natn. Acad. Sci. U.S.A.* (in the Press).

SCHÄFER, W. (1957). Units isolated after splitting fowl plague virus. In *The Nature of Viruses. Ciba Foundation Symposium*. Ed. G. E. W. Wolstenholme and E. C. P. Millar. London: J. and A. Churchill.

SMITH, K. O. (1963). Physical and biological observations on herpes virus. *J. Bact.* **86**, 999.

SMITH, K. O. (1964). Relationship between the envelope and infectivity of herpes virus. *Proc. Soc. exp. biol. Med.* **115**, 814.

SMITH, W., BELYAVIN, G. & SHEFFIELD, F. W. (1953). A host protein component of influenza viruses. *Nature, Lond.* **172**, 669.

SPRING, S. D. & ROIZMAN, B. (1967). Studies of herpes simplex virus products in productive and abortive infection. I. Stabilization with formaldehyde and preliminary analyses by isopycnic centrifugation in CsCl. *J. Virol.* **1**, 294.

TAKEHARA, M. & SCHWERDT, C. L. (1967). Infective subviral particles from cell cultures infected with myxoma and fibroma viruses. *Virology*, **31**, 163.

TALAL, N., TOMKINS, G. M., MUSHINSKI, J. F. & YIELDING, K. L. (1964). Immuno-chemical evidence for multiple molecular forms of crystalline glutamic dehydro-genase. *J. molec. Biol.* **8**, 46.

VALENTINE, R. C. & PEREIRA, H. G. (1965). Antigens and structure of the adeno-virus. *J. molec. Biol.* **13**, 13.

WATERSON, A. P. (1962). Two kinds of myxovirus. *Nature, Lond.* **193**, 1163.

WATKINS, J. F. (1964). Adsorption of sensitized sheep erythrocytes to HeLa cells infected with herpes simplex virus. *Nature, Lond.* **202**, 1364.

WATSON, D. H. & WILDY, P. (1963). Some serological properties of herpes virus particles studied with the electron microscope. *Virology*, **21**, 100.

WATSON, D. H., WILDY, P. & RUSSELL, W. C. (1964). Quantitative electron micro-scope studies on the growth of herpes virus using the techniques of negative staining and ultramicrotomy. *Virology*, **24**, 523.

WESTWOOD, J. C. N., HARRIS, W. J., ZWARTOUW, H. T., TITMUSS, D. H. J. & APPLEYARD, G. (1964). Studies on the structure of vaccinia virus. *J. gen. Micro-biol.* **34**, 67.

WILCOX, W. C. & GINSBERG, H. S. (1963). Production of specific neutralizing antibody with soluble antigens of type 5 adenovirus. *Proc. Soc. exp. biol. Med.* **114**, 37.

WILCOX, W. C., GINSBERG, H. S. & ANDERSON, T. F. (1963). Structure of type 5 adenovirus. II. Fine structure of virus subunits. Morphological relationship of structural subunits to virus-specific soluble antigens from infected cells. *J. exp. Med.* **118**, 307.

Wildy, P., Russell, W. C. & Horne, R. W. (1960). The morphology of herpes virus. *Virology*, **12**, 204.

Wildy, P. & Watson, D. H. (1962). Electron microscopic studies on the architecture of animal viruses. *Cold Spring Harb. Symp. quant. Biol.* **27**, 25.

EXPLANATION OF PLATES

Plate 1

Fig. 1. (*a*) 'Full' and (*b*) 'empty' particles of polyoma virus in negative contrast. The 'empty' particles are penetrated by phosphotungstate and have electron dense cores. (Reproduced with permission from Fraenkel-Conrat, *Molecular Basis of Virology*, Reinhold Publishing Corporation, New York, 1968, and with permission of Dr E. A. C. Follett.)

Fig. 2. Enveloped particles of herpes simplex virus in which the capsids are poorly delineated because the phosphotungstate has barely penetrated the envelope. Identification of these particles is accordingly hazardous. (From Watson & Wildy (1963) by permission of Academic Press Inc.)

Plate 2

Different appearances of herpes simplex particles in negative contrast. (*a*) Enveloped particle, (*b*) enveloped particle with phosphotungstate penetrated core, (*c*) naked particle, (*d*) naked particle with penetrated core. (From Watson, Russell & Wildy (1963), *Virology*, **19**, 250, by permission of Academic Press, Inc.)

Plate 3

(*a*) Enveloped particles of herpes simplex virus grown in BHK 21 cells agglutinated after incubation with antiserum to BHK 21 cells. (From Watson & Wildy (1963) by permission of Academic Press Inc.) (*b*) Naked particles of herpes simplex virus grown in BHK 21 cells agglutinated after incubation with antiserum to infected RK 13 cells. Fibres of antibody molecules can be distinguished between the particles.

Plate 4

(*a*) Particle of adenovirus type 5 showing fibre + knob appendage of pentons. (*b*) Free hexons isolated from adenovirus infected cells. These are identical with the 240 capsomeres on the complete particle which have six neighbours. (*c*) Isolated pentons isolated from adenovirus infected cells. These consist of a large knob (penton base) with an attached fibre and smaller knob. This component is identical with the 12 capsomeres on the vertices of the complete particle. (From Valentine & Pereira (1965), by permission of the authors and Academic Press Inc.)

Plate 5

Fig. 1. The two appearances of vaccinia virus particle seen in negative contrast preparations: the mulberry (M) and capsulated (C) forms. The C form is produced by phosphotungstate penetration of the outer layer of the M form. (Kindly provided by Dr W. J. Harris. Crown copyright reserved. Reproduced with the permission of the Controller, H.M.S.O.)

Fig. 2. Small particles (S) seen in trypsin-treated preparation of polyoma virus (P). Particle counting shows no decrease in the number of complete polyoma particles after trypsin treatment so the small particles are not breakdown products of complete polyoma particles.

PLATE I

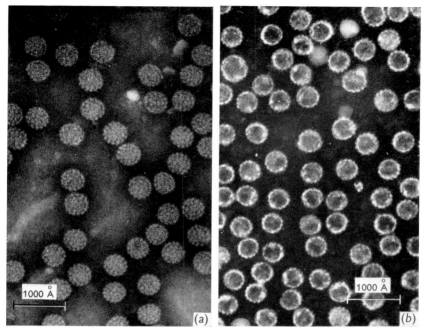

Fig. 1

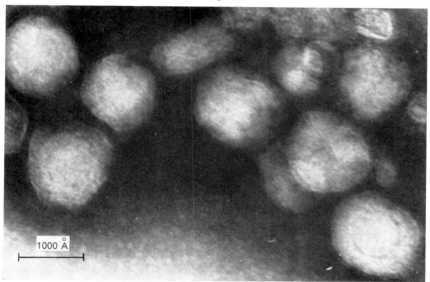

Fig 2

PLATE 2

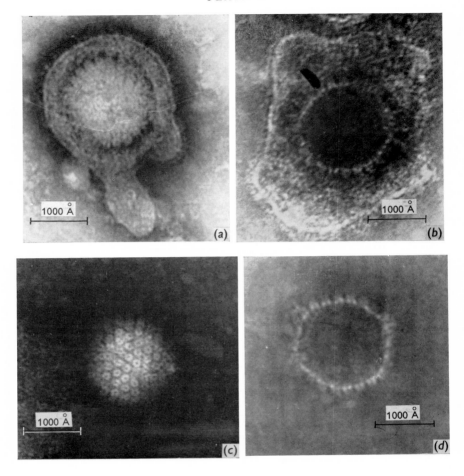

PLATE 3

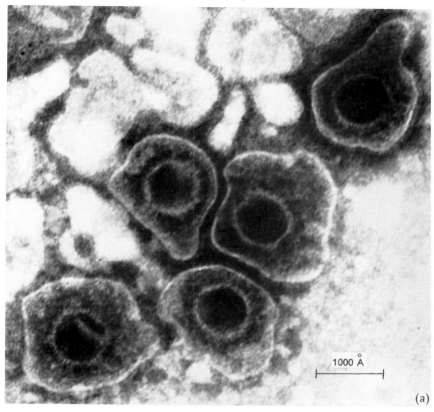

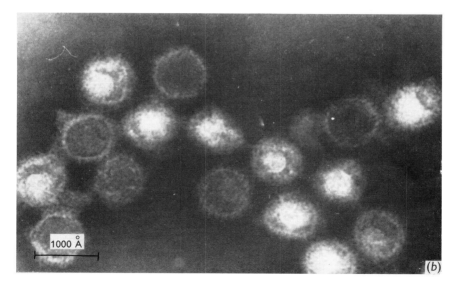

PLATE 4

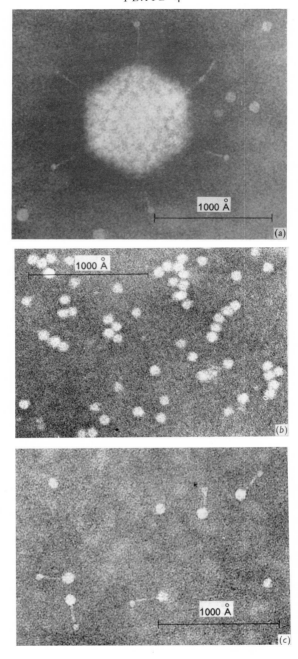

PLATE 5

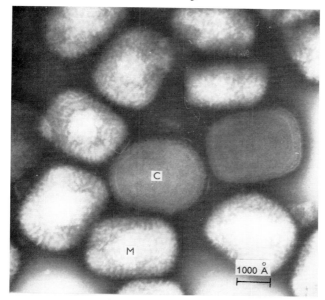

Fig. 1

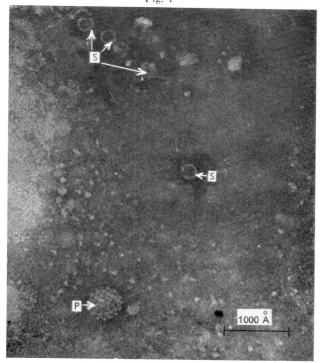

Fig. 2

PLANT VIRUSES: DEFECTIVENESS AND DEPENDENCE

J. B. BANCROFT

*Department of Botany and Plant Pathology, Purdue University,
Lafayette, Indiana, U.S.A.*

INTRODUCTION

It has been known for a number of years that infections caused by some plant viruses may result in the production of two or more nucleoprotein components. Most of the virus systems so far examined share the characteristic that only one of the components can potentially initiate an infection and can interact biologically with components defective in this regard. Suitable examples can be divided into three categories each having different characteristics.

The first and largest category contains viruses which generally yield two serologically similar nucleoprotein components one of which is partially deficient in RNA. Mixtures of partially deficient particles with those having a full complement of RNA are more infectious than would be expected on a basis of the infectivities of unmixed samples. The bean pod mottle-cowpea mosaic viruses of the squash mosaic family as well as alfalfa mosaic virus behave in this way, as do certain other viruses.

The second category contains the tobacco rattle virus which may produce serologically similar nucleoprotein rods of two lengths. The rods interact in such a way to suggest that the long rods carry the information necessary for multiplication and the short ones may carry that necessary for the production of coat protein required by both lengths of rod.

The third category contains the tobacco necrosis-satellite virus complex in which multiplication of the serologically distinct satellite virus is dependent on the multiplication of the tobacco necrosis virus.

This article is mainly limited to considerations of the characteristics of these plant virus systems, outlined in Figure 1, where interactions are biologically specific and can be related to some sort of reasonably well-defined physical deficiency.

VIRUSES PRIMARILY OF THE SQUASH MOSAIC FAMILY
Description of the components and their RNA

A large family of spherical plant viruses exists which may be called the squash mosaic family after its first clearly defined member (Rice *et al.* 1955). Viruses in the group are transmitted by beetles, have three sedimenting components and an adenlyic:cytidylic acid ratio of about 2:1. Although the biological and serological relationships among strains in this family are fairly complex, the various members of the group are quite alike physically. Thus, when squash mosaic virus, or bean pod mottle virus (BPMV) or the cowpea mosaic viruses (Agrawal, 1964) are mentioned, it will minimize confusion if they are regarded as being generally similar (Haselkorn, 1966). The squash mosaic family is distinct from the tobacco ringspot family which is also characterized by viruses which produce three sedimenting components upon infection.

Plants infected with any of the viruses of the squash mosaic family contain, in addition to a component containing a complete complement of RNA, a related nucleoprotein partially deficient in RNA as well as protein shells completely devoid of RNA. These components sediment at different rates, the fastest usually being called bottom (B), the next fastest middle (M) and the slowest, which is nucleic acid free, top (T) component. With BPMV, for example, the B, M, T components have sedimentation coefficients of 112S, 91S and 54S, respectively. For the viruses tested, the components of each system are serologically alike (Bancroft, 1962; Bruening & Agrawal, 1967), have the same amino acid composition (Mazzone, Incardona & Kaesberg, 1962), are electrophoretically similar—there can be a complication here (Bancroft, 1962) but it does not affect the main point—and have nucleic acids in M and B components the base ratios of which are similar but not necessarily identical (Semancik & Bancroft, 1964). The most detailed study of the physical properties of a virus in this group was made by Mazzone *et al.* (1962) of squash mosaic virus. They found that the diffusion coefficients of the T, M and B components were about the same and molecular weights of 4·5, 6·1 and 6·9 × 10^6 molecular weight units, respectively, were reported. Molecular weights of 1·6 and 2·4 × 10^6 were calculated for the RNA inside the M and B components. Sedimentation properties of phenol-extracted B and M-RNA from both BPMV and a cowpea mosaic virus indicate that M-RNA is a single large molecule roughly 50–60 % the size of B-RNA (Semancik & Bancroft, 1964; Bruening & Agrawal, 1967).

Infectivity and activation

It is necessary to compare the behaviour of a physical population with an infectious one in as many ways as possible to determine that what is believed to be virus is, in fact, infectious. The two nucleoprotein components of BPMV or cowpea mosaic virus are separated by only a few mm. after rate-zonal centrifugation (Brakke, 1960) in sucrose or centrifugation to equilibrium in CsCl. If density-gradient tubes are punctured at suitable levels and the samples withdrawn, the specific infectivity of M component is always at least 20–30 % of B component. Evidence suggesting that a nucleoprotein containing a nucleic acid molecule lacking about one-third to one-half of the total RNA functions normally should be treated with more scepticism than it often has received. If the simple expedient of droplet fractionating the complete population from the centrifuge tube is used, a fairly precise comparison between the nucleoprotein and infectivity distributions may be achieved and has led to the suggestion that BPMV M particles are not, by themselves, infective, whereas B particles probably are (Bancroft, 1962). Also, infection with preparations believed to contain only B particles gives rise to the three sedimenting components. Bruening & Agrawal (1967) made the same suggestion for cowpea mosaic virus B component after stressing the difficulty of eliminating the possibility of admixtures of particle types resulting in an activation effect as described next.

A detailed examination of the infectivity distribution in terms of a possible interaction between M and B particles of BPMV and cowpea mosaic virus showed that mixtures had specific infectivities that were higher, often by a factor of 8, than expected from the specific infectivities of unmixed samples (Wood & Bancroft, 1965; Bruening & Agrawal, 1967). This interaction required M-RNA since: (1) nucleic acid-free top component did not activate B particles (Wood & Bancroft, 1965); and (2) M-RNA when added to B-RNA caused the production of 10–20 times more lesions than produced by B-RNA alone (Bruening & Agrawal, 1967). Further, the interaction was specific in that M component from a cowpea mosaic virus which has a base ratio similar to that of BPMV, did not activate BPMV B component, whereas it did activate its homologous B component. Similarly, if the M component of BPMV was added to cowpea mosaic virus B component, no activation occurred, whereas BPMV M component activated its homologous B component (Wood & Bancroft, 1965). This type of interaction in which RNAs apparently must be closely related to one another in a definite way, is much more marked than, and is distinct from, non-specific interaction effects in which the added RNA may be unrelated to the viral RNA. For example, phenol extracts of the single-component brome mosaic virus contain infectious RNA molecules with a molecular weight of 1×10^6 and non-infectious ones of 7×10^5 and 3×10^5 (Bockstahler & Kaesberg, 1965). If the latter are added to infectious RNA, about twice as many lesions as expected are found. However, the same effect can be found if yeast or bacteriophage R17 RNA are added to the infectious RNA. The enhancement of tobacco mosaic virus infectivity with short particles derived from the virus may be of the same non-specific type (Hulett & Loring, 1965).

Speculation about infectivity activation

The question arises as to why the addition of M to B particles enhances infectivity. Since the same effect found with the nucleoprotein is also found when RNA is used, an explanation based on B particles having uncoating difficulties in the absence of M particles can be dismissed. Other suggestions have been made to account for the effect.

It has been proposed that there is an infectious class of B particles which contains unbroken phosphodiester chains and a non-infectious derivative class which does not. This assumption is certainly a reasonable one since the specific infectivity of purified BPMV or its RNA decreases as the length of infection time increases (Gillaspie & Bancroft, 1965a). Actually, all spherical plant viruses so far tested behave in this way and it has been clearly demonstrated for broad bean mottle virus (Kodama & Bancroft, 1964) that the loss of infectivity is due to some ageing process resulting in the rupture of RNA at specific points inside virus particles still in the plant. A similar effect in which the breaks may be hidden by heat-labile secondary forces has been demonstrated for turnip yellow mosaic virus (Haselkorn, 1962). It is suggested that if debilitated B component particles of BPMV or cowpea mosaic virus are associated with their corresponding M particles at the time of infection, the M-RNA might be able to compensate for the function lost by the damaged RNA permitting some kind of interaction resulting in the production of a lesion (Wood & Bancroft, 1965).

If the above theory that damaged particles can be reclaimed is correct, then a convenient explanation for the effect would be complementation or recombination. However, lesions which are believed to arise from interaction between M and B particles contain the same sort of virus, and RNA, as found from those believed to arise from single particles. This would tend to rule out a simple complementation scheme unless some sort of repair mechanism is involved. A reasonable approach to distinguish between complementation and recombination would be to use mutants that interact and that have different markers. Limited experiments with such mutants suggest that B particles control the progeny type at least in terms of lesion type and M:B component ratio (H. A. Wood, unpublished results).

Another hypothesis (Bruening & Agrawal, 1967) is based on the assumption that no damaged RNA molecules are involved and that most infections are initiated by mixtures of B and M particles containing two functionally distinct types of RNA of different sizes which somehow

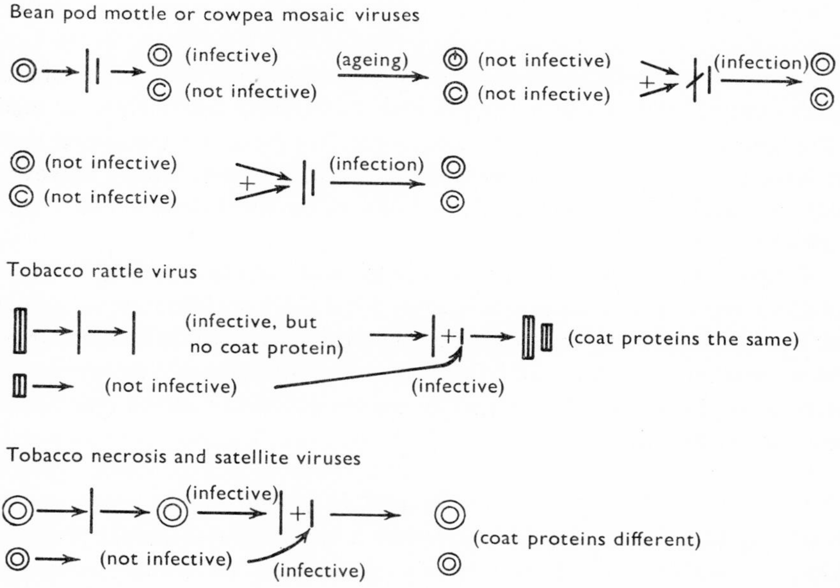

Fig. 1. Outlines of the three principal plant-virus systems showing various types of dependence. With the bean pod mottle or cowpea mosaic virus system, two schemes are shown. In the upper diagram, it is assumed a bottom component particle alone can infect to give rise to two sizes of RNA which are encapsulated by coat protein resulting in the formation of bottom and middle component nucleoprotein particles. The RNA of the bottom component particle may be cleaved during ageing and particles containing such RNA are shown to be helped by middle component particles. In the lower diagram, it has been assumed that the particles cannot infect by themselves, but can when mixed, cleaved RNA not being involved in the interaction. Tobacco rattle virus long particles are shown infecting by themselves to produce RNA which does not elicit the production of coat protein. If short particles are inoculated by themselves, they cannot replicate, but can in the presence of long particles or the RNA resulting from infection by long particles, causing the production of coat protein able to enclose RNA of both lengths. Tobacco necrosis virus is shown to multiply by itself resulting in the production of its own coat protein. Satellite virus is unable to multiply by itself but can in the presence of tobacco necrosis virus, the coat protein of the satellite virus being distinct from that of tobacco necrosis virus. Uncoated RNA is shown as a vertical line in all cases.

interact to greatly increase the efficiency of infection. Their argument is based on Kleczkowski's (1950) analysis that susceptible regions on an assay plant vary in the virus dosage requirements necessary for infection. They postulate that M particles may be particularly helpful in this regard. Their idea is supported in a negative sense by the observation that heating B-RNA did not disclose the existence of a significant number of hidden breaks (Bruening & Agrawal, 1967). H. A. Wood (unpublished results) has found, however, that most of the B-RNA dissociates upon dialysis against EDTA and water, suggesting that phosphodiester linkages are not responsible for the apparent integrity of much B-RNA.

Yet another possibility is that some normal B particles do infect and multiply but do not produce readily visible lesions, as is the case with some local infections caused by tobacco mosaic virus (Helms & McIntyre, 1962), but will in the presence of M particles. In this instance, the effect would not be on infection, *per se*, but rather on the appearance of infection. Assays on systemically responding hosts show, however, that the activation effect is not confined to local lesion hosts (unpublished observations).

Not enough is known about the initiation of infection by plant viruses to assess the importance of the specific M-RNA requirements necessary for activation in terms of susceptible site theory of multiplication or on the appearance of symptoms. It must be emphasized that all the interpretations given are inferential. The activation effect has been described but not explained.

Other viruses showing infectivity activation

Work on two viruses called the tobacco streak and necrotic ringspot viruses, unrelated to those already mentioned or to one another, indicates that additional factors may be involved in an activation or enhancement effect. Tobacco streak virus has two nucleoprotein components which sediment at 72 and 96S. The 72S component is not infective by itself, but its presence can cause the production of three times more lesions than normally found if added to purified but unfractionated virus (Fulton, 1967). Interestingly, the activation effect is host dependent—emphasizing some apparently important but unknown role of a host factor in studies of this type with this virus. The host effect need not be related to the ability of the non-infectious nucleoprotein to multiply in the assay plant in the presence of the infective one, as is shown by studies on necrotic ringspot virus. Infectivities of preparations of necrotic ringspot virus which contain a non-infectious nucleoprotein sedimenting at 72S (Fulton, 1959) and an infectious 95–100S nucleoprotein (S. Tolin, unpublished results) could not be enhanced by the addition of unrelated viruses, but could by the addition of strains related to the necrotic ringspot virus, but which were not detectably infectious on the assay host used for infectivity measurements (Fulton, 1962).

In short, bringing together data from all the systems, it appears that: (1) infectivity activation between related nucleoprotein particles containing different amounts of RNA may be a fairly widespread effect since it has usually been found when looked for; an apparent exception recently reported by van Kammen (1967) probably resulted from incomplete separation of components (cf. Bruening & Agrawal, 1967); (2) M-RNA is required; (3) the effect is specific; (4) the effect can be host mediated; (5) the deficient nucleoprotein may not have to multiply for enhancement to occur.

Origin of virus-related nucleoproteins partially deficient in RNA

A number of experiments designed to explain the origin or metabolic significance of virus-related nucleoproteins partially deficient in RNA have been made with several different multicomponent plant viruses. These observations are briefly discussed with the realization that different systems may behave differently.

Incomplete nucleoprotein particles, such as those comprising a middle component, could be preparative artifacts. Generally, however, there is no evidence that parts of a normal complement of RNA can leak from plant viruses under the normally mild conditions used for purification. The ratio of the three components of BPMV (1:7:7, for T, M, B components, respectively) remains about the same whether the virus be examined in the analytical centrifuge in crude sap immediately after it is expressed from tissue, or in the purified state with or without treatment with pancreatic ribonuclease. There has been one claim, since questioned (Schneider & Diener, 1966), that leakage may occur from purified tobacco ringspot virus kept in low ionic strength buffer (Stace-Smith, Reichmann & Wright, 1965). RNA can be induced to leak from turnip yellow mosaic virus (Kaper, 1960) and southern bean mosaic virus (unpublished results), but the conditions required are not normally those used in virus purification.

If components partially deficient in RNA are not preparative artifacts, then they must be formed in the plant. Three principal ideas about their origin have been considered. The first, which will be called the precursor theory, is that RNA-deficient components are precursors to those containing a full complement of RNA. This theory was suggested by

Matthews (1960) for the series of minor nucleoproteins associated with turnip yellow mosaic virus (Markham & Smith, 1949). The specific radioactivities of the RNA of the minor components isolated in CsCl decreased with an increase in nucleic acid content suggesting that they represented steps in the formation of virus. A subsequent detailed examination of the labelling data revealed other possibilities (Matthews, Bolton & Thompson, 1963).

The same idea has also been used to explain the existence of M component of tobacco ringspot virus (Diener & Schneider, 1966) which is more like BPMV than is turnip yellow mosaic virus. B component of tobacco ringspot virus was found to contain, in addition to RNA which sedimented at 32S and which was presumably infectious, a class of 24S RNA which sedimented at the same rate as M-RNA. The molecular weight of the 32S RNA was estimated to be about twice that of the 24S RNA, leading to the suggestion that RNA is made in two pieces and that single RNA pieces are encapsulated by coat protein to make M component but two pieces must be joined to form B component. An alternative and perhaps simpler interpretation for the presence of two sizes of RNA in B particles is that some RNA molecules break at a specific place while still being retained in the protein shell. The precursor theory has been rejected as an explanation for the appearance of M and B components of BPMV because both components take up ^{32}P at the same rate after a short labelling period (Semancik & Bancroft, 1964).

Lack of evidence for the precursor theory has led to other suggestions for the origin of components partially deficient in RNA. The degrade and coat hypothesis proposed for broad bean mottle virus (Aronson & Bancroft, 1962), suggests that only the complete genome of virus is initially made in plants, but that before encapsulation by protein, some RNA chains are degraded at a regular rate at or to some definite point dictated by the secondary structure of the RNA. Base ratios of M and B-RNA of BPMV although similar, are not identical. This difference implies that M component is missing a piece of RNA in contrast to the possibility where M particles could be of two types, each containing one-half fragments of B-RNA. It is worth noting that there are no particles containing one-third pieces of RNA which might be expected to occur from the indirect size estimates of squash mosaic virus RNA (Mazzone, et al. 1962)—although this could simply mean that the one-third fractions are digested before being coated.

The third suggestion is that the synthesis of M-RNA is somehow initiated by B-RNA, but that, once started, M-RNA can be made independently of B-RNA. Paul (1963) observed that the nucleoprotein component ratios of true broad bean mosaic virus, while constant in any one host, differed between hosts, indicating that the multiplication of two sizes of RNA may depend on the host. Schneider & Diener (1966) point out that different rates of increase for M and B components of tobacco ringspot virus occur early after infection. Implicit in the interpretation of any difference in component ratios is the possibility that one nucleoprotein can be preferentially disrupted *in vivo* to give an appearance of a difference in the rate of formation which, in fact, does not occur. It is doubtful that there is a turn-over, at normal temperatures, of completed BPMV nucleoprotein particles. M and B particles are produced in equal amounts at the same rate, but this does not mean that all strains must behave similarly, since mutants are known in which M nucleoprotein predominates.

ALFALFA (LUCERNE) MOSAIC VIRUS

Multicomponent systems are not confined to spherical viruses and alfalfa mosaic virus (AMV) is an example of an anisometric virus which is accompanied by defective nucleoproteins which can interact biologically with the virus. Purified preparations of AMV contain at least four serologically and electrophoretically similar nucleoproteins, called bottom, middle, top_b and top_a, which sediment at 99, 89, 73 and 68S, respectively (Bancroft & Kaesberg, 1960; Kelley & Kaesberg, 1962). The three fastest sedimenting components are rodlets with lengths of 58, 48 and 36 mμ and all contain about 21 % RNA, but differ in absolute amounts of RNA (Gibbs, Nixon & Woods, 1963). They may conveniently be regarded as anisometric equivalents of a multi-component spherical virus. Top_a component is rather different from the other components since it is isometric, contains 13·5 % RNA which has a base ratio that is similar but not identical to that of the other components (Rauws, Jaspars & Veldstra, 1964) and can easily be separated from the others because it alone is soluble in 0·05 M $MgCl_2$ (Kelley & Kaesberg, 1962). The molecular weights of bottom and top_a, which have similar if not identical amino acid compositions (Kelley & Kaesberg, 1962; Jaspars & Moed, 1966), are $7·4 \times 10^6$ and $3·5 \times 10^6$ respectively, bottom containing $1·6 \times 10^6$ and top_a, $0·5 \times 10^6$ molecular weight units of RNA (Kelley & Kaesberg, 1962). The various components appear to contain primarily single pieces of RNA (Gillaspie & Bancroft, 1965*b*).

The above characteristics are, with the exception of the polymorphism, basically similar to those found with BPMV, as apparently is the potentiating effect which the addition of middle and the combined top components have on the infectivity of the bottom component particles (Wood & Bancroft, 1965). In contrast to this activation effect, a concentration-dependent reduction of the infectivity of pure but unfractioned AMV has been observed if the top components are added at concentrations of 17 to 70 times that of the virus. Turnip yellow mosaic virus bottom component will also inhibit AMV infection, although not as efficiently as the AMV top component, whereas artificial top component of turnip yellow mosaic virus (containing no RNA), or serum albumin, are relatively ineffective (Moed, 1966).

The stability of the extracted RNA from the various components seems to differ, bottom component RNA being least and top_a-RNA most stable (Gillaspie & Bancroft, 1965b; van Vloten-Doting & Jaspars, 1967). This stability may play a part in the observation of Bosch, van Kippenberg, Voorma & van Ravenswaay Claasen (1966) that most of the messenger activity of RNA from unfractionated AMV resided in a fraction sedimenting at about 12S—the rate of top_a-RNA. It can be calculated that this small RNA, which contains about 1400 nucleotides, has enough information for perhaps two or three polypeptides, one of which is apparently coat protein (van Ravenswaay Claasen, van Leeuwen, Duijts & Bosch, 1967). If top_a-RNA was used as a messenger in a cell-free *Escherichia coli* system and the resulting radioactive polypeptides were co-precipitated with carrier coat protein in $0.05 M$ $MgSO_4$, 25 of the 29 tryptic peptides stainable with ninhydrin were also radioactive. Detailed biological studies of the effect of top_a-RNA on the infectivity of that derived from the bottom component have been made by van Vloten-Doting & Jaspars (1967). They found that the small RNA, which was not infectious by itself, caused 7–10 times more lesions to appear when added to B-RNA, than when B-RNA was assayed by itself. Further, a degradation product of B-RNA which sedimented at the same rate as top_a-RNA also enhanced the infectivity of B-RNA. They considered that the enhancement effect either resulted from top_a-RNA supplying needed supplemental information so that B-RNA could multiply, or that it resulted from the assay plant somehow being stimulated to produce lesions not ordinarily seen. Whatever the mechanism, it is clear that a small molecule carrying only limited information can be effective in enhancing infectivity.

TOBACCO RATTLE VIRUS

General description

Tobacco rattle virus (TRV) and its strains exhibit types of defectiveness and interactions unlike those previously described. Early investigations of TRV centred on what were originally called the multiplying (M) and non-multiplying (NM) forms of this virus based on differences in transmissibility (Cadman & Harrison, 1959). Both forms multiply, but the multiplying type could be transmitted mechanically by ordinary methods, whereas the other form could be transmitted efficiently only after phenol extraction of infected leaves (Sänger & Brandenburg, 1961). Cadman (1962) concluded that the NM form probably was composed of naked nucleic acid, protected perhaps by the nuclei with which it appeared to be associated. It was defective in that it could not elicit the production of coat protein, which is apparently not made, at least in amounts detectable serologically. This type of defect in which nucleic acid is not coated has been observed, probably for a different reason, with certain mutants of tobacco mosaic virus (Siegel, 1965; Siegel, Zaitlin & Sehgal, 1962). In the latter case, defective coat proteins are made which contain amino acid substitutions that apparently affect quaternary structure so that combination with nucleic acid does not occur (Siegel, Hills & Markham, 1966).

Purified preparations of the multiplying form of TRV were usually found to contain two principal classes of rod-shaped particles. The long ones of serotype 1 have a length of 188 mμ regardless of the strain and the short ones, a length of from 43 to 114 mμ depending on the strain (Harrison & Woods, 1966). Although detailed investigations have not been published on the separated components, it has generally been assumed on the basis of electrophoretic and serological data (Harrison & Nixon, 1959) that the protein of the two components is identical for any one strain. The nucleic acid, which comprises about 5 % by weight of each of the components, has also been assumed to be about the same although again, no detailed studies have been published.

Infectivity and interactions

Investigations on the infectivity of the long and short particles were made by Harrison & Nixon (1959) who concluded that only the former were infectious on beans. This observation was supported by Lister (1966) who made the additional discovery that the long and short particles could interact in rather a remarkable way. The long particles of the multiplying or stable variant form, isolated after sucrose density-gradient centrifugation, mainly caused lesions on tobacco containing the NM or unstable variant form of the virus. However, if suitable numbers of short particles were present with the long particles of the same strain, many more lesions than expected contained the multiplying form of the virus. This observation was recently confirmed by Frost, Harrison & Woods (1967), who also showed that the RNA of the short particles is responsible for the effect. In addition, Lister (1966) found that if short particles plus phenol extracts of plants previously infected with the NM form of the same or a closely related strain were rubbed on plants, stable virus was produced. Plants inoculated with only the phenol extract produced the NM form of the virus. Specificity of the interaction was established by the observation that distantly related strains would not interact. Apparently the addition of short particles, not infective by themselves, to long ones or to their unstable progeny, resulted in the production of coat protein able to enclose particles of both lengths. The addition of short particles to long ones did not increase the infectivity of the latter.

The results with TRV imply that the nucleic acids of the short and long particles are different. A number of degradation and precursor schemes can be entertained to explain the origin of short particles, but none seems likely. For example, the possibilities that short particles are strain specific degradation products of a certain population of long particles, or their RNA before encapsulation, or that they or their nucleic acids are precursors of long particles, seem rather unlikely because the long and short RNAs appear to have different functions. The situation is apparently different from that found with the multi-component spherical viruses since infection by the long particles of TRV does not result in the production of the entire system. It follows that, taking the function of short particles at face value, tissue infected with NM should only contain one type of viral RNA. Although possibly technically difficult, this point could be established. If no short RNA were found to be present, it could mean that if long particles or NM extracts were rubbed on to tobacco in the presence of short particles, the short particle RNA would multiply concomitantly with, but independently of, long RNA after some sort of initial induction—perhaps of the same sort as mentioned for tobacco necrosis virus (see next section). Alternatively, if short particle RNA were present in the NM form, then the ideas about TRV may have to be revised.

A direct approach to test the deduction that the RNA of short particles carries the information required for coat production would be to use strains of TRV in mixture experiments. Closely related strains, at least, seem to be able to interact. If the short particles of one strain, differentiable from others by some easily monitored characteristic such as serological reaction, electrophoretic mobility, or tryptic peptide map, were added to the long particles or the NM form of another strain, then the protein of the progeny should be that of the introduced short particles. If it were not, then the possibility that long particles do carry the information for coat protein specification but are unable to use it in the absence of short particles would have to be considered. Another more involved test would be to see if the RNA of the short particles could produce coat protein in cell-free systems, in contrast to the RNA of the long particles which should not. There is certainly ample RNA in any of the short particles to code for the protein coat. Offord (1966) has calculated that the long particles contain about 7100 nucleotides and that short ones of about 75 mμ in length contain 2400. The chemical unit of the protein coat of TRV is composed of 218 amino acids (Offord & Harris, 1965) which would only require 654 nucleotides by current theory and which present evidence suggests would be equivalent to a short particle length of about 20 mμ. Thus, even the shortest naturally occurring top component particles, associated with the BEL strain, for example, contain enough information for at least two polypeptides, and some, which are a little over 100 mμ long, contain as many nucleotides as found in several small spherical plant viruses which multiply independently. Consequently, the short particles, when added to the long ones, have the potentiality to do much more than to code for coat protein, and additional effects such as clear cut symptomatological or host range differences are a definite possibility (Lister, 1967).

TOBACCO NECROSIS AND SATELLITE VIRUSES

General description

The isolates comprising the tobacco necrosis virus (TNV) group, defined by Babos & Kassanis (1963) as a collection of strains all serologically related to TNV strain A, may affect, with one exception, the behaviour in plants of a relatively small nucleoprotein called the satellite virus (SV). The relationships between SV and TNV, have been described in detail by Kassanis & Nixon (1961) and Kassanis (1962, 1964, 1966).

When suitable isolates are purified and examined in the analytical centrifuge or after sucrose density-gradient centrifugation, three populations with sedimentation coefficients of 50S, 116S and 222S are observed. The 50S or SV particles are small spheres with a diameter of about 17 mμ and contain about 20% RNA. These small spheres can aggregate in twelves to form the particles comprising the 222S fraction (Kassanis & Nixon, 1961). The 116S or TNV species contains particles with about 20% RNA and have a diameter of approximately 27 mμ. Preparations containing mixtures of the SV and TNV may also be fractionated by preparative electrophoresis, SV having an isoelectric point near neutrality, where TNV is negatively charged (Kassanis, 1962). These observations clearly show that two physically distinct types of particles are present in suitably derived preparations and serological experiments show the SV and TNV particles to be completely unrelated. Having a mixture of plant viruses is not a noteworthy occurrence, but what lends distinction to the mixture in this case, is that it has been impossible to demonstrate the multiplication of SV in the absence of TNV. In other words, the multiplication of the small virus appears to be dependent on the presence of the larger one, which can multiply by itself.

The small particles are unlike any previously described and do not represent a type of middle component. They conceivably could be derived as preparative artifacts from TNV—particularly from rather unstable isolates such as TNV strain B. However, careful examination of such particles in various stages of degradation has failed to reveal SV particles. Nor do SV particles occur by themselves in detectable amounts in leaves of healthy plants, so they are not derived from host organelles.

Biological characteristics

Preparations containing both TNV and SV produce two sizes of lesions on infected leaves. When serial single lesion subcultures from large lesions are made, they are found to contain mainly 116S particles which, by themselves, are infectious like any other virus. Bulk preparations originating from small lesions, however, contain both the SV and TNV particles as well as the SV aggregates. If the 50S particles or their aggregates are isolated and rubbed on leaves, no lesions are produced and multiplication of SV cannot be detected. The satellite situation is different from several described cases where the multiplication of one virus, which can multiply by itself, affects the multiplication of an unrelated virus, which can also multiply by itself (Kassanis, 1963). If the 50S particles are added to the 116S ones, small lesions are produced whose numbers in relation to large lesions are roughly proportional to the ratio of particle types in the inoculum (Kassanis, 1962). If about 10–30% of small particles are added to large ones, most of the lesions will be large and if the particle ratio is reversed, most will be small. SV can be applied to leaves at least 2 days prior to TNV and still be induced to multiply, or SV can be activated by TNV which has already been multiplying for 2 days. The possibility that the function of TNV is to cause SV to discard its protein coat before multiplying has been discounted, for SV–RNA does not multiply by itself when applied to leaves but will in the presence of TNV (Kassanis, 1962). What probably happens during mixed infection is that if a large and small particle enter the same or neighbouring cells, the small particle multiplies—as a consequence of the proximity of a larger particle which also multiplies, but the SV for some reason retards the multiplication of the progeny of the large particle and the size of the lesion is reduced. If enough SV is added with TNV, the amount of TNV, as measured serologically, may be less by a factor of at least 16 than when SV is absent. Strain S of TNV usually only produces local lesions when inoculated with SV, but when freed from SV, multiplies systemically.

The infection process is also affected by environment, only large lesions forming on plants kept at 26°, whereas both forms occur at lower temperatures (Kassanis, 1962). Also, the species of plant used is important, only large lesions being found in *Nicotiana glutinosa* L. with inoculum producing both types of lesion on tobacco or bean.

The question arises as to the specificity of the association which allows SV to multiply. A variety of plant viruses other than TNV did not activate SV, whereas a number of TNV strains (A, B, S in serotype A; E in serotype B) did, although to different extents (Kassanis & Nixon, 1961; Babos & Kassanis, 1963). Thus, the ability of TNV to activate SV —and we are dealing with a single strain of SV—is distinct from the coat-producing function of TNV. Indeed, unstable strains of TNV which are unable to make coat protein (Kassanis & Welkie, 1963), or at least that which will enclose TNV nucleic acid, will activate SV. Furthermore, TNV strain D, which is similar to strain E, will not activate SV, the point being that the potentiality to activate SV may be divorced from all functions normally required by TNV to multiply.

Function of satellite virus RNA

Although, as far as I am aware, TNV has not yet been characterized in any detail, SV has, and consideration of its properties helps in speculating as to what may be occurring in the plant. The molecular weight of SV, obtained from sedimentation-diffusion measurements, is 1.97×10^6. The virus contains 20 % RNA existing as a single molecule containing about 1200 nucleotides (Reichmann, 1964). If all nucleotides were required to code for a single coat protein, the chemical unit of the virus should contain about 400 amino acids—a much larger value than usually found with plant viruses. Reichmann (1964) determined that the smallest possible chemical subunit molecular weight, based on one residue of half cystine, was 13,000. Tryptic digests of the protein, however, resulted in three times more peptides than predicted on the half cystine, suggesting a larger protein containing about 372 amino acid residues with a molecular weight of 39,000. The ratio of DNP-ϵ-amino lysine:DNP-α-amino alanine and end group analyses also indicated that the larger molecular weight, which does not agree with an earlier estimate (Reichmann, Rees, Symons & Markham, 1962), was correct. Interestingly, the N-terminal group was not acetylated, although it may be in protein made *in vitro* (Reichmann, Chang, Faiman & Clark, 1966). Reichmann (1964) suggested that SV may be constructed of 42 morphological units which were equated with structure units in this case. Studies on the structure of SV and its protein subunit have yet to be completed.

Evidence that the RNA could code for its own coat was obtained using SV-RNA as messenger in an *E. coli* cell-free system. The coincidence of radioactive spots from protein made *in vitro* and ninhydrin staining spots of carrier SV protein suggested that the RNA did indeed code for coat protein and little else (Clark, Chang, Spiegelman & Reichmann, 1965). This result suggests that TNV may provide all information, other than that needed for coat production, for SV to multiply. The relationship between TNV and SV which allows the latter to multiply is not known. A possibly somewhat analogous system with the appealing feature of an enzyme (RNA replicase) which apparently only recognizes its own RNA has been described (Haruna & Spiegelman, 1965). Perhaps the simplest assumption that can be made to explain the TNV–SV interaction, since it is quite specific, is that a TNV replicase exists which not only recognizes its own RNA, but that of SV also.

ACKNOWLEDGEMENTS

I wish to thank Drs Bosch, Bruening, Fulton, Jaspars, Lister, Reichmann and Tolin, for sending me manuscripts of their papers prior to publication.

REFERENCES

AGRAWAL, H. O. (1964). Identification of cowpea mosaic virus isolates. *Meded. Landbouwhogeschool Wageningen*, **64–5**, 1.

ARONSON, A. I. & BANCROFT, J. B. (1962). Density heterogeneity in purified preparations of broad bean mottle virus. *Virology*, **18**, 570.

BABOS, P. & KASSANIS, B. (1963). Serological relationships and some properties of tobacco necrosis virus strains. *J. gen. Microbiol.* **32**, 135.

BANCROFT, J. B. (1962). Purification and properties of bean pod mottle virus and associated centrifugal and electrophoretic components. *Virology*, **16**, 419.

BANCROFT, J. B. & KAESBERG, P. (1960). Macromolecular particles associated with alfalfa mosaic virus. *Biochim. biophys. Acta*, **39**, 519.

BOCKSTAHLER, L. E. & KAESBERG, P. (1965). Infectivity studies of bromegrass mosaic virus RNA. *Virology*, **27**, 418.

BOSCH, L., VAN KIPPENBERG, P. H., VOORMA, H. O. & VAN RAVENSWAAY CLAASEN, J. C. (1966). Protein synthesis by cell-free extracts of *E. coli* programmed with plant viral RNA, p. 275. In *Viruses of Plants*. Ed. A. B. R. Beemster and J. Dijkstra. Amsterdam: North Holland Publishing Co.

BRAKKE, M. K. (1960). Density gradient centrifugation and its application to plant viruses. *Adv. Virus Res.* **7**, 193.

BRUENING, G. & AGRAWAL, H. O. (1967). Infectivity of a mixture of cowpea mosaic virus nucleoprotein components. *Virology*, **32**, 306.

CADMAN, C. H. (1962). Evidence for association of tobacco rattle virus nucleic acid with a cell component. *Nature, Lond.* **193**, 49.

CADMAN, C. H. & HARRISON, B. D. (1959). Studies on the properties of soil-borne viruses of the tobacco-rattle type occurring in Scotland. *Ann. appl. Biol.* **47**, 542.

CLARK, J. M., JUN., CHANG, A. Y., SPIEGELMAN, S. & REICHMANN, M. E. (1965). The *in vitro* translation of a monocistronic message. *Proc. natn. Acad. Sci. U.S.A.* **54**, 1193.

DIENER, T. O. & SCHNEIDER, I. R. (1966). The two components of tobacco ringspot virus nucleic acid: origin and properties. *Virology*, **29**, 100.

FROST, R. R., HARRISON, B. D. & WOODS, R. D. (1967). Apparent symbiotic relationship between particles of tobacco rattle virus. *J. gen. Virol.* **1**, 777.

FULTON, R. W. (1959). Purification of sour cherry necrotic ringspot and prune dwarf viruses. *Virology*, **9**, 522.

FULTON, R. W. (1962). The effect of dilution on necrotic ringspot virus infectivity and the enhancement of infectivity by non-infective virus. *Virology*, **18**, 477.

FULTON, R. W. (1967). Purification and some properties of tobacco streak and Tulare apple mosaic viruses. *Virology*, **32**, 153.

GIBBS, A. J., NIXON, H. L. & WOODS, R. D. (1963). Properties of purified preparations of lucerne mosaic virus. *Virology*, **19**, 441.

GILLASPIE, A. G. & BANCROFT, J. B. (1965*a*). The rate of accumulation, specific infectivity and electrophoretic characteristics of bean pod mottle virus in bean and soybean. *Phytopathology*, **55**, 906.

GILLASPIE, A. G. & BANCROFT, J. B. (1965*b*). Properties of ribonucleic acid from alfalfa mosaic virus and related components. *Virology*, **27**, 391.

HARRISON, B. D. & NIXON, H. L. (1959). Separation and properties of tobacco rattle virus with different lengths. *J. gen. Microbiol.* **21**, 569.

HARRISON, B. D. & WOODS, R. D. (1966). Serotypes and particle dimensions of tobacco rattle viruses from Europe and America. *Virology*, **28**, 610.

HARUNA, I. & SPIEGELMAN, S. (1965). Specific template requirements of RNA replicases. *Proc. natn. Acad. Sci. U.S.A.* **54**, 579.

HASELKORN, R. (1962). Studies on infectious RNA from turnip yellow mosaic virus. *J. molec. Biol.* **4**, 357.

HASELKORN, R. (1966). Physical and chemical properties of plant viruses. *Ann. Rev. Pl. Physiol.* **17**, 137.

HELMS, K. & MCINTYRE, G. A. (1962). Studies on size of lesions of tobacco mosaic virus on Pinto bean. *Virology*, **18**, 535.

HULETT, H. R. & LORING, H. S. (1965). Effect of particle length distribution on infectivity of tobacco mosaic virus. *Virology*, **25**, 418.

JASPARS, E. M. J. & MOED, J. R. (1966). The complexity of alfalfa mosaic, p. 188. In *Viruses of Plants*. Ed. A. B. R. Beemster and J. Dijkstra. Amsterdam: North Holland Publishing Co.

VAN KAMMEN, A. (1967). Purification and properties of the components of cowpea mosaic virus. *Virology*, **31**, 633.

KAPER, J. M. (1960). Preparation and characterization of artificial top component from turnip yellow mosaic virus. *J. molec. Biol.* **2**, 425.

KASSANIS, B. (1962). Properties and behaviour of a virus depending on its multiplication for another. *J. gen. Microbiol.* **27**, 477.

KASSANIS, B. (1963). Interactions of viruses in plants. *Adv. Virus Res.* **10**, 219.

KASSANIS, B. (1964). Properties of tobacco necrosis virus and its association with satellite virus. *Annls Inst. Phytopath. Benaki, N.S.*, **6**, 7.

KASSANIS, B. (1966). Properties and behaviour of satellite virus. In *Viruses of Plants*, p. 177. Ed. A. B. R. Beemster and J. Dijkstra. Amsterdam: North Holland Publishing Co.

KASSANIS, B. & NIXON, H. L. (1961). Activation of one tobacco necrosis virus by another. *J. gen. Microbiol.* **25**, 459.

KASSANIS, B. & WELKIE, G. W. (1963). The nature and behaviour of unstable variants of tobacco necrosis virus. *Virology*, **21**, 540.

KELLEY, J. J. & KAESBERG, P. (1962). Biophysical and biochemical properties of top component *a* and bottom component of alfalfa mosaic virus. *Biochim. biophys. Acta*, **61**, 865.

KLECZKOWSKI, A. (1950). Interpreting relationships between the concentrations of plant viruses and numbers of local lesions. *J. gen. Microbiol.* **4**, 53.

KODAMA, T. & BANCROFT, J. B. (1964). Some properties of infectious ribonucleic acid from broad bean mottle virus. *Virology*, **22**, 23.

LISTER, R. M. (1966). Possible relationship of virus-specific products of tobacco rattle virus infections. *Virology*, **28**, 350.

LISTER, R. M. (1967). A symptomatological difference between some unstable and stable variants of pea early browning virus. *Virology*, **31**, 739.

MARKHAM, R. & SMITH, K. M. (1949). Studies on the virus of turnip yellow mosaic. *Parasitology*, **39**, 330.

MATTHEWS, R. E. F. (1960). Properties of nucleoprotein fractions isolated from turnip yellow mosaic virus preparations. *Virology*, **12**, 521.

MATTHEWS, R. E. F., BOLTON, E. T. & THOMPSON, H. R. (1963). Kinetics of labelling of turnip yellow mosaic virus with P^{32} and S^{35}. *Virology*, **19**, 179.

MAZZONE, H. M., INCARDONA, N. L. & KAESBERG, P. (1962). Biochemical and biophysical properties of squash mosaic virus and related macromolecules. *Biochim. biophys. Acta*, **55**, 164.

MOED, J. R. (1966). Onderzoekingen over Alfalfa Mosaic Virus. *Ontmantelings-vraagstuk*. Department of Biochemistry, University of Leiden, The Netherlands.

OFFORD, R. E. (1966). Electron microscopic observations on the substructure of tobacco rattle virus. *J. molec. Biol.* **17**, 370.

OFFORD, R. E. & HARRIS, J. I. (1965). The protein subunit of tobacco rattle virus. *Fed. European Biochem. Soc. 2nd Meeting*, abstracts, p. 216. Vienna: Verlag der Weiner Medizinischen Akademic.

PAUL, H. L. (1963). Untersuchungen über das Echte Ackerbohnenmosaik-Virus. *Phytopathol. Z.* **49**, 161.

RAUWS, A. G., JASPARS, E. M. J. & VELDSTRA, H. (1964). The base compositions of ribonucleic acids from alfalfa mosaic virus components. *Virology*, **23**, 283.

VAN RAVENSWAAY CLAASEN, J. C., VAN LEEUWEN, A. B. J., DUIJTS, G. A. H. & BOSCH, L. (1967). *In vitro* translation of alfalfa mosaic virus RNA. *J. molec. Biol.* **23**, 535.

REICHMANN, M. E. (1964). The satellite tobacco necrosis virus: A single protein and its genetic code. *Proc. natn. Acad. Sci. U.S.A.* **52**, 1009.

REICHMANN, M. E., CHANG, A. Y., FAIMAN, L. & CLARK, J. M., JUN. (1966). The satellite tobacco necrosis virus in studies of genetic coding. *Cold Spring Harb. Symp. quant. Biol.* **31**, 139.

REICHMANN, M. E., REES, M. W., SYMONS, R. H. & MARKHAM, R. (1962). Experimental evidence for the degeneracy of the nucleotide triplet code. *Nature, Lond.* **195**, 999.

RICE, R. V., LINDBERG, G. D., KAESBERG, P., WALKER, J. C. & STAHMANN, M. A. (1955). The three components of squash mosaic virus. *Phytopathology*, **45**, 145.

SÄNGER, H. L. & BRANDENBURG, E. (1961). Über die Gewinnung von infectiösem Press-saft aus 'Wintertyp'-Pflanzen des Tabak-Rattle-Virus durch Phenol-extraktion. *Naturwissenschaften*, **48**, 391.

SCHNEIDER, I. R. & DIENER, T. O. (1966). The correlation between the properties of virus-related products and the infectious component during the synthesis of tobacco ringspot virus. *Virology*, **29**, 92.

SEMANCIK, J. S. & BANCROFT, J. B. (1964). Further characterization of the nucleo-protein components of bean pod mottle virus. *Virology*, **22**, 33.

SIEGEL, A. (1965). Defective plant viruses, p. 113. In *Perspectives in Virology*, IV. Ed. M. Pollard. New York: Harper and Row, Inc.

SIEGEL, A., HILLS, G. J. & MARKHAM, R. (1966). *In vitro* and *in vivo* aggregation of the defective PM2 tobacco mosaic virus protein. *J. molec. Biol.* **19**, 140.

SIEGEL, A., ZAITLIN, M. & SEHGAL, O. P. (1962). The isolation of defective tobacco mosaic virus strains. *Proc. natn. Acad. Sci. U.S.A.* **48**, 1845.

STACE-SMITH, R., REICHMANN, M. E. & WRIGHT, N. S. (1965). Purification and properties of tobacco ringspot virus and two RNA-deficient components. *Virology*, **25**, 487.

VAN VLOTEN-DOTING, L. & JASPARS, E. M. J. (1967). Purification and characterization of two ribonucleic acid components from alfalfa mosaic virus. Enhancement of infectivity by combination. (In manuscript.)

WOOD, H. A. & BANCROFT, J. B. (1965). Activation of a plant virus by related incomplete nucleoprotein particles. *Virology*, **27**, 94.

DEPENDENCE AMONG RNA-CONTAINING ANIMAL VIRUSES

J. SVOBODA

Institute of Experimental Biology and Genetics, Czechoslovak Academy of Sciences, Prague, Czechoslovakia

INTRODUCTION

The problem of dependence among animal viruses containing RNA has become apparent within the last few years. This is due in no small measure to progress in studies on RNA oncogenic viruses. On a model of Rous chicken sarcoma virus (RSV) especially, data have been presented showing that completion of the maturation process of the Bryan strain of this virus is dependent on other viruses of the avian leucosis complex. On the other hand, when the same model was studied in mammals, it was shown that the maturation cycle is also determined by the host cell. When at least the genome of Rous virus replicates in the mammalian cell, the viral maturation process may take place when the otherwise non-infectious genome passes into a normal chicken cell as a result of contact. The functions of the genome are then expressed in the chicken cell and formation of the virion of RSV takes place.

Dependence on both given factors, according to recent data, is also typical for Moloney's mouse sarcoma virus (MSV) and probably other similar cases will be found, especially among tumour viruses.

The principal aim of this paper is to summarize present knowledge concerning dependence among oncogenic RNA viruses, to demonstrate their relationship to other viral models and to endeavour to evaluate their significance for viral replication and for the mechanisms of the oncogenic action of viruses.

INTERFERENCE BETWEEN VIRUSES OF THE LEUCOSIS COMPLEX

A quantitative method for titration of RSV *in vitro* was developed, which was of fundamental value in promoting studies of interference between viruses of the avian leucosis complex. This method, described by Temin & Rubin (1958), is based on a linear relationship between virus concentration and its ability to form foci of transformed cells on monolayers of chicken fibroblasts. Thus it was for the first time possible to assay the transforming activity of the oncogenic virus *in vitro*.

The first significant dependence among viruses of the avian leucosis complex to be found was interference between RSV and the resistance inducing factor (RIF) (Rubin, 1960). It was shown that RIF is a lymphomatosis virus which is vertically transmitted into chicken embryos, where it multiplies. If a tissue culture from an embryo containing RIF is prepared, then the virus rapidly multiplies without cell transformation taking place. But the RIF present in the cells produces a marked interference to RSV, so that these cells are at least 40 times more resistant to RSV infection (Rubin, 1961). From the kinetics of superinfection of RIF-infected cells by RSV it was judged that a single infectious particle of RSV may superinfect these cells. However, the probability of this superinfection is greatly reduced.

In its physico-chemical properties, RIF resembles chicken tumour viruses and cross-reacts immunologically with RSV (Friesen & Rubin, 1961). This cross-reactivity is explained by the later finding that the Bryan RSV strain is defective (see p. 251). The finding that antigenically RIF corresponds to visceral lymphomatosis virus again demonstrates that RIF is a strain of lymphomatosis virus.

Vogt & Rubin (1963) found interference between avian myeloblastosis virus (AMV) and RSV and made use of this test to work out a quantitative method for titration of AMV *in vitro*.

Likewise, Rous-associated virus (RAV) interferes with RSV. But particularly for viruses of this group it has been shown that interference appears only where the RAV and the RSV are antigenically similar. Where they are dissimilar, the infecting efficiency of RSV is, on the contrary, enhanced (Hanafusa, 1965).

DEFECTIVENESS OF RSV

The discovery that Bryan's high titre strain of RSV (BH–RSV) is defective (Hanafusa, Hanafusa & Rubin, 1963) was of basic significance for comprehension of the interaction of Rous virus with other viruses of the avian leucosis complex. Rubin & Vogt (1962) first showed that in stocks of BH–RSV there is an excess of another virus, designated by the authors as RAV (Rous-associated virus). This virus corresponds morphologically and antigenically to the RSV used, but in chickens it induces erythroblastosis, and in chicken fibroblasts, though multiplying well, it does not cause formation of foci of transformed cells.

Temin (1962, 1963) was able to obtain non-virus-producing Rous chicken cells, at a low multiplicity of infection and under experimental conditions which did not permit a spread of the Bryan's standard strain of RSV (BS–RSV) used for infection. These non-producing (NP) cells exhibited the morphological characteristics of transformed cells and proved to be malignant after transfer to chickens. When they were superinfected with different viruses of the avian leucosis complex, formation of Rous virus was induced. The viral genome is regularly transferred in a population of NP cells, since it has been possible to induce RSV formation in all clonal lines. Physico-chemical treatments which serve as inducing factors in lysogenic bacteria are not effective in the case of NP cells.

Defectiveness of BH–RSV was postulated by Hanafusa *et al.* (1963) to explain the results of experiments in which cells were infected by BH–RSV alone without RAV. The authors achieved this by exposing chick fibroblasts to high dilution of virus stocks containing both BH–RSV and RAV, so that foci arose following infection with RSV alone. The foci obtained were protected against subsequent infection by RAV with antiserum and agar overlay. Cells from these foci did not produce or contain infectious RSV, even when repeatedly passaged, and were designated by the authors as non-producing (NP) cells. If NP cells were infected with RAV, however, both RAV and RSV production commenced the following day.

These results led to the conclusion that RSV alone is not able to produce infectious progeny—in other words, it is a defective virus. Only if the RSV-infected cells are, in addition, superinfected with some other non-defective virus of the avian leucosis complex, does production of the RSV virion commence.

The RSV virion formed in this manner has a viral coat antigen corresponding to the particular helper virus used (Hanafusa, Hanafusa & Rubin, 1964; Vogt, 1964), irrespective of whether a virus-neutralizing or an immunofluorescence method is used to detect the virion coat protein. RSV is therefore designated RSV (RAV$_1$), RSV (RAV$_2$), etc., according to the helper coat. The interference pattern of RSV is also determined by the helper virus, since interference takes place only between antigenically closely related viruses (see p. 250).

For RSV it has been shown in several papers that chicken cells have a genetically determined susceptibility (or resistance) to infection by this virus. The fact that BH–RSV was found to be defective made it possible to study this question experimentally, because of the differing affinity of two Rous-associated viruses (RAV$_1$ and RAV$_2$) for certain lines of chicken cells.

Thus RAV$_1$ multiplies both in cells of genotype K and K/2, whereas RAV$_2$ effectively infects only K cells. Since RSV (RAV$_1$) and RSV (RAV$_2$) also follow this pattern, it was concluded that the helper virus determines the host range of RSV (Hanafusa, 1965; Rubin, 1965).

Genetic resistance of cells derived from different lines of chickens to infection by viruses of the avian leucosis group was employed by Vogt (1965) and Vogt & Ishisaki (1965) for classification of these viruses. At the same time it was found that viruses which fall into the same group according to these criteria are also antigenically related and have the same interference pattern.

The elegant experiments of Hanafusa & Hanafusa (1966a) showed that, in the case of mammalian cells, oncogenic activity of defective RSV is also determined by the viral coat provided by the helper virus. If Bryan's RSV received an antigenic coat from RAV$_{50}$ (RAV$_{50}$ being the helper virus isolated from a Schmidt-Ruppin strain of RSV which is highly oncogenic for mammals) then it acquired the same degree of oncogenic potency for hamsters as Schmidt-Ruppin RSV.

The helper virus probably controls some early steps of viral infection such as adsorption and penetration of the virus (Hanafusa & Hanafusa, 1966b). For instance, RSV (RAV$_2$) which does not multiply in chicken cells of genotype K/2, likewise does not multiply when these cells are simultaneously infected with RAV$_1$, which does multiply in K/2 cells. On the assumption that the RSV genome (RAV$_2$) penetrated the cell, it might be expected that RAV$_1$ would permit formation of RSV (RAV$_1$) by providing an antigenic coat. However, no RSV activity was found in mixed infected cells.

Such experiments demonstrating that the helper virus determines the host range of the defective RSV are also in accordance with the interpretation that the helper virus controls some early steps of infection.

Production of infectious BH–RSV, in view of its defectiveness, depends on whether the cell is infected simultaneously with RSV and the helper virus. It was shown that infection with a single particle of each of these viruses is sufficient to induce virus production. Kinetics of RSV production correspond to the growth kinetics of the particular helper virus which is being used. The rate of RSV production in NP cells superinfected with helper virus is the same as in chicken cells infected simultaneously with RSV and helper virus. This suggests that synthesis of the viral coat by helper virus determines the rate of maturation of RSV in the cell and, furthermore, that the mere presence of the RSV genome in an NP cell does not increase this rate.

When chicken cells were first infected with one helper virus (e.g. RAV_1) and later reinfected with RSV (RAV_2), then the initial growth rate of RSV (RAV_1) exceeded that of the respective helper virus. This type of experiment may be carried out because the two helpers do not mutually interfere. Hanafusa & Hanafusa (1966b) explain this finding as follows: the increased rate of RSV production may be due to an excess of helper coats for RSV genomes in the helper-infected cell. The authors also admit the possibility that the helper may have initiated some process stimulating RSV production in the cell.

The dependence of defective RSV on supply of the viral antigen by helper virus is connected with the phenomenon of phenotypic mixing. In this case, however, both the antigenically different viruses infecting the same cell produce coat protein, which is used at random for formation of the virion, so that a certain proportion of viral particles contains antigens of both parent viruses.

According to Hanafusa et al. (1964) 'random withdrawal of antigenic components by both viral genomes may be the same' in the case of phenotypic mixing and helper dependence of defective RSV, even though there is a marked difference between both systems in that the antigenic coat for the defective RSV is supplied by the helper virus only.

Cases of phenotypic mixing when the genome of one virus is coated only by the antigens of the other virus are interesting in their relationship to RSV. This extreme case of phenotypic mixing found in phages has been designated as genomic masking (Yamamoto & Anderson, 1961). A high proportion of genomic masking was observed in guanidine-requiring polio virus (Wecker & Lederhilger, 1964). From this aspect, RSV appears as an extreme case of genomic masking, where the virus is coated by antigens of another virus only.

The stability of BH–RSV defectiveness seemed to suggest that part of the viral genome is either missing or is inactivated. However, it is possible that it is only repressed and, under certain conditions, there may be derepression. For example, there have been recent claims that infectious particles can be detected from NP cells under special and perhaps highly sensitive assay conditions (Weiss, 1967). Therefore, different strains of cells should be tested in order to find out whether some synthesis of defective BH–RSV takes place in the absence of the helper virus. The part of the genetic information that controls the antigenic specificity of the outer viral coat must be involved. The question remains, whether this defect allows formation of virion structures detectable by electron microscopy. According to Dougherty & Di Stefano (1965) and Di Stefano & Dougherty (1965), morphologically normal particles corresponding to RSV particles are also produced in NP cells. Particles of such morphology were not found in non-infected control cells. Assuming that these particles are defective, as suggested by the above experiments, then the defect need not necessarily prevent production of morphologically normal particles. It will, however, be necessary to clarify, by other than morphological methods, the relationship of such particles to defective RSV.

It should be mentioned that certain functions of the defective RSV genome are independent of the helper virus. Temin (1960) showed that a BS–RSV population contains several kinds of mutants (morphs), each of which determines the morphological character of the cells it transforms. NP cells, obtained through infection with a certain RSV morph and superinfected with a helper virus, produced RSV which controlled the morphology of the transformed cells in similar manner to that of the original virus (Temin, 1962). The properties of RSV are thus responsible for determining the morphology of the transformed cell independently of the helper virus.

Likewise, production of soluble group-specific complement-fixing (CF) antigen (Armstrong, Okuyan & Huebner, 1964; Huebner *et al.* 1964) is independent of the helper virus, because it also takes place in NP cells (Vogt, Sarma & Huebner, 1965; Dougherty & Di Stefano, 1965). This antigen is especially interesting because it has been shown to be part of the inner component of the AMV virion (Bauer & Schäfer, 1965, 1966; Payne, Solomon & Purchase, 1966).

The nature of the CF antigen is, as yet, unknown. Dougherty & Di Stefano (1966) found it in chicken embryonic tissue free of known avian leucosis viruses. According to their suggestions leucosis viruses merely derepress synthesis of normal chicken antigens which are then revealed in the complement-fixation reaction.

Since in experiments so far, inbred chickens fully compatible for transplantation antigens were not used, it was not possible to study in detail the tumour specific transplantation antigen of NP cells.

So far, characteristic defectiveness has with certainty been described only for Bryan's strain of RSV. Other strains, such as Schmidt-Ruppin (SR–RSV) and Prague (PR–RSV) (Goldé & Vigier, 1966; Dougherty & Rasmussen, 1964; Hanafusa, 1964) were not qualitatively defective, and production of NP cells could not be found when these strains were used to infect chicken cells. However, superinfection with RAV of chicken cells earlier infected with SR–RSV leads to 50–100-fold increase of virus production (Hanafusa & Hanafusa, 1966*b*). This increase is the result of the formation of SR–RSV (RAV). Thus, this is a case of phenotypic mixing. These results, on the other hand, indicate that the genome of the helper virus is capable of initiating the synthesis of antigenic determinants of the virion more effectively than the SR–RSV genome, so that these determinants are utilized in the mixed infected cell for SR–RSV virion formation.

According to Rubin (1964) this implies that SR–RSV is defective in a quantitative sense, and its production in the infected cell, as compared with that of the helper virus, is limited by the supply of coat protein. If such a cell is superinfected with helper virus, a sufficient amount of coat protein is synthesized. As for transforming activity, according to Rubin, SR–RSV causes a less marked cell transformation of chicken cells, but on the other hand this RSV strain is highly oncogenic for mammals, for whose embryo cells it causes morphologically distinct alterations *in vitro*, as does another non-defective strain: PR–RSV.

Possibly the dependence of SR–RSV on other viruses of the leucosis complex will be detectable when heterologous avian hosts and cells are used. Thus Shipman & Levine (1965) found that tumours induced with SR–RSV in Japanese quail do not contain virus infectious for this avian species. But if the Japanese quail are preinfected with other viruses of the leucosis complex, then cell-free extracts of these tumours are tumorigenic for them. One possible explanation of these experiments would be that for SR–RSV infection, quail cells must be simultaneously infected by another leucosis virus acting as helper. This would be analogous to the defectivity of Moloney sarcoma virus in mice (see further). The question of the interaction of defective RSV with helper virus in the heterologous avian cell deserves more attention. The role of helper virus in the phenomenon described by Rauscher, Reyniers & Sacksteder (1964) still remains to be clarified. These authors found that BH–RSV produces infectious progeny in tumours induced in adult Japanese quail, but not in tumours produced on the CAM.

It is not improbable that quite new types of dependence and helper viruses will be found if investigations are carried out on different kinds of heterologous avian cells.

Not only defectiveness which may be compensated by helper virus is involved. Experiments of Goldé & Latarjet (1966) showed that after irradiation of SR–RSV with high doses of X-rays, some of the virus particles gave rise to transformed NP cells, which do not, however, become virus producers after infection with helper virus. Probably this treatment caused a more profound alteration of those functions of the viral genome responsible for synthesis of the virion that are not compensated by the action of the helper virus.

It will unquestionably be very interesting to find out whether in such NP cells further characteristic markers connected with cell transformation are retained, such as specific tumour transplantation antigen and CF antigen, or whether any one of these becomes lost.

DEFECTIVENESS IN THE GROUP OF
MOUSE LEUKAEMIC VIRUSES

Using Moloney's mouse leukaemia virus, Harvey (1964) was successful in obtaining a new virus causing sarcomas in mice, rats and hamsters. This virus will be referred to as mouse sarcoma virus (MSV). Hartley & Rowe (1966) showed that MSV is capable of transforming mouse embryo cells *in vitro*. With a low dilution of MSV the number of foci obtained in embryonic fibroblasts falls in proportion to the dilution factor. However, with higher dilutions, the transforming activity of the virus falls in proportion to the square of the dilution factor. This led them to the conclusion that two viruses were necessary for the formation of foci. Experimental confirmation was forthcoming for this assumption when an excess of Moloney's mouse leukaemia virus was added to MSV. In that case the focus-forming activity of the virus preparation was always proportional to its concentration. Of all the other mouse leukaemia viruses tested, only Rauscher's virus had a similar effect.

These results show that MSV (like BH–RSV) is defective. However, there is a significant difference between the defectiveness of these two viruses. MSV is dependent on a helper virus for cell transformation and presumably also for virus production. In contrast, defective BH–RSV transforms the cell without the aid of a helper virus, which is, however, necessary for infectious RSV to form in the transformed cells.

The basis of these differences has not yet been studied, nor has the role of the helper virus in maintaining the character of MSV-transformed cells.

So far, defectivity has been found in the group of viruses of the avian and murine leucosis complex. In both cases, viruses which caused sarcoma formation were always involved. These virus groups have been most widely studied and are in fact representative of the avian and mammalian tumour viruses containing RNA. It must be remembered that defectiveness may occur in other RNA oncogenic virus groups and may be one of the reasons why these viruses elude detection. This naturally is also true of human malignancies, and especially of sarcomas.

Experience so far has shown that even in experimental systems defectiveness escaped attention for a long time. This was because the defective virus was mixed with helper virus, so that without a suitable sensitive test to dissociate the two, it was impossible to detect defectiveness.

So far data are missing that would show whether and how often defective viruses may be the cause of spontaneous tumours. For example, defectiveness might be an essential state of some oncogenic viruses or it may arise as a laboratory artifact, only after repeated passaging. These questions are not only theoretically significant but are also related to the etiology of human tumours.

THE DEFECTIVE MATURATION CYCLE OF
RSV IN MAMMALIAN CELLS

When it was found that RSV attacks mammals and causes lethal haemorrhagic cystic disease (Zilber & Kryukova, 1957; Svet-Moldavsky, 1957) the possibility was open to study the reproductive cycle of this virus in the mammalian cell. From results obtained to date, it may be concluded that RSV in the majority of haemorrhagic cysts is not detectable, though it was found in some cases after relatively long latency when tissues from cysts had been transferred to chickens (Svet-Moldavsky & Skorikova, 1960; Svoboda & Grozdanovic, 1960; Munroe & Southam, 1964). By means of fluorescein-labelled antiviral antibodies, Kryukova & Obukh (1964a) and Kryukova (1966) also occasionally found viral antigen in the cells of cyst walls and in organs of animals with cysts.

The study of the behaviour of the RSV genome in the mammalian cell and of conditions under which this virus may mature was made possible when RSV-induced tumours of proved RSV origin were obtained in mammals.

The first tumour studied in which RSV aetiology was definitely confirmed was the XC sarcoma (Svoboda, 1960, 1961). It was demonstrated that after transfer to chickens in the form of intact tissue suspensions this sarcoma gives rise to Rous sarcomas containing a normal amount of RSV, which is neutralized with anti-RSV antisera. The only effective components of XC sarcoma are intact cells, named virogenic cells, whereas extracellular materials, as well as destroyed cells, were inactive after transfer to chickens (Svoboda, 1962). In a parallel study, Simkovic, Valentova & Thurzo (1962) and Simkovic, Svoboda & Valentova (1963, 1965) found production of infectious RSV only when XC cells were co-cultivated with chicken fibroblasts.

The necessity for close contact of virogenic cells with chicken cells was shown by further experiments in which both types of cells, whether during cultivation *in vivo* or *in vitro*, were divided by a semi-permeable membrane (Chyle, 1964). Under these conditions there was no virus formation, although the membranes used were permeable for RSV and macromolecular components but not for cells. On the basis of these findings, XC cells, according to the character of virus-cell interaction, have been classified as belonging to the cell-productive type of inter-action (Svoboda, 1964).

In other papers (Chyle, Klement & Svoboda, 1963; Svoboda, Chyle, Simkovic & Hilgert, 1963) it was shown that infectious RSV is not detectable by preparation methods from gramme quantities of XC tumour or in high-speed sediments from culture fluid. Similarly, virus-neutralizing antibodies do not form in rats bearing or having rejected this tumour. Likewise, formation of RSV is not induced by physico-chemical treatment, inductive on lysogenic phages, nor by various physiological growth conditions of cells *in vitro*. Not only the original XC cell population but also all derived clones evince virogenic activity which suggests that the greater part, if not the whole, of the cell population is virogenic. This, coupled with the finding that the RSV genome is retained in a population of XC cells as long as 7 years in passage, shows that the genome is regularly transmitted in the cellular population.

Similar findings were obtained in rat cells transformed *in vitro* with PR–RSV (Svoboda & Chyle, 1963). In this case, too, all tested clones and subclones were virogenic (Svoboda, Vesely, Vrba & Jiranek, 1965).

Further papers dealt with tests for determining whether there is a relationship between virogeny of XC cells and RSV defectiveness. For this purpose Svoboda (1964), Simkovic (1964), Vigier & Svoboda (1965) and Simkovic *et al.* (1965) tested the inductive action of avian myelo-blastosis virus, RAV_1, infectious PR–RSV and Gross mouse leukaemia virus, as possible helper virus of mammalian origin. All these types of treatment proved ineffective. Mixed cultivation of XC cells with RIF-free chicken cells also leads to RSV formation. This, together with the earlier-mentioned finding of non-defectiveness of PR–RSV, shows that RSV formation resulting from contact and co-cultivation of PR–RSV-transformed mammalian cells with chicken cells does not occur as a result of chicken cells being a source of known helper virus. Thus virogeny inducible by helper virus and virogeny inducible by chicken cells are apparently different phenomena.

The effect of contact and co-cultivation of virogenic mammalian cells with chicken cells was interpreted as follows (Svoboda *et al.* 1963; Svoboda, 1964; Simkovic, 1964):

The RSV genome present in virogenic cells is labile and a complete degradation takes place when the cell is destroyed. If contact is made with the chicken cell, the viral genome (provirus) may pass into the chicken cell, where the whole vegetative reproductive cycle may take place. A mutual exchange of cellular material between two types of cells, or fusion of cells, may lead to the same effect.

At the same time, transfer of the genome of non-defective RSV from the mammalian to the chicken cell need not always lead to virus production, as may be inferred from experiments by Simkovic (1964) who co-cultivated chicken cells with X-irradiated XC cells. After mixed cultivation during which the amount of XC cells fell below the threshold quantity necessary for tumour formation in chickens, Simkovic obtained tumours in some chickens only when he injected them with a suspension of live co-cultivated cells, and not with cell-free material. A possible explanation of these results is that the RSV genome transferred to chicken cells was capable of transformation but could not evoke a complete vegetative cycle of the virus.

A situation very similar to that of PR–RSV was found in the case of another non-defective Rous virus: SR–RSV. Thus in rat tumours induced by SR–RSV, Rous sarcomas were obtained only if the chickens were inoculated with a suspension of tumour tissue (Ahlström & Jonsson, 1962) or with a suspension of live tumour cells (Klement, Chyle & Svoboda, 1963). This finding was also proved valid for tumours induced with the same virus in hamsters (Ahlström & Forsby, 1962; Huebner *et al.* 1964) or cells transformed *in vitro* (Vesely & Svoboda, 1965) for guinea-pig tumours (Ahlström, Bergman & Ehrenberg, 1963) for hamster BHK cells transformed with SR–RSV (Macpherson, 1965) and for human and bovine cells transformed with Engelbreth–Holm strain of RSV and SR–RSV (Stenkvist, 1966). In SR–RSV-transformed monkey cells *in vitro*, the virus was demonstrated only after mixed cultivation with chicken cells (Jensen, Girardi, Gilden & Koprowski, 1964).

Rat cells transformed with BS–RSV *in vitro* also produce RSV only after transfer to chickens (Febvre, Rothschild, Arnoult & Haguenau, 1964).

The study of mammalian cells transformed with defective RSV gave interesting results. Such studies were carried out by Hanafusa & Hanafusa (1966 *a*) on hamster cells transformed with BH–RSV (RAV$_{50}$), by Sarma, Vass & Huebner (1966) on cells explanted from hamster tumours induced with BH–RSV (Rabotti, Raine & Sellers, 1965) and by Vigier (1966) on a line of hamster cells BHK 21/13 transformed with BH–RSV.

All these authors obtained the following identical results:

Transformed hamster cells, if superinfected with helper viruses, or if co-cultivated with chicken cells without helper viruses, do not produce infectious RSV. But if helper virus is added to mixed cultures then infectious virus is produced.

The most probable explanation for these findings is the following. In the first stage, the viral genome is transferred to the chicken cell by mechanisms which have been discussed earlier. The chicken cell thus becomes an NP tumour cell containing defective RSV genome. Production of RSV virion begins immediately this cell is superinfected with helper virus, compensating for the defectiveness of the RSV.

This pattern assumes that NP cells are produced as a result of co-cultivation of virogenic mammalian cells with chicken cells. So far, however, they have been demonstrated only by Sarma et al. (1966) probably thanks to the fact that these authors transferred mixed cultures to chickens, when NP cells gave rise to tumours. Thanks to such selection even a small number of NP cells may be detected which otherwise would be hard to identify microscopically in mixed cultures.

Because of the possible transfer of the viral genome to the chicken cell, there is an interesting finding made by Teyssie, Rothschild & Febvre (1965), that production of specific antigen detectable by the fluorescence method (probably corresponding to CF antigen, see below) passes from rat cells transformed by BS–RSV to chicken cells as a result of mixed cultivation.

Recently more direct evidence concerning the importance of at least cytoplasmic contact between the virogenic mammalian cell and the chicken cell for the initiation of virus production has accumulated. Okada (1962) and Schneeberger & Harris (1966) have shown that u.v.-inactivated Sendai virus produces the formation of cytoplasmic bridges and cellular fusion between cells. When the mixture of virogenic mammalian and normal chicken cells is treated first with Sendai virus and afterwards plated, the frequency of the transfer of non-infectious RSV genome from mammalian to chicken cells is clearly increased (Svoboda, Machala & Hlozanek, 1967; Vigier, 1967). This was demonstrated both by shortening of latency of RSV appearance and increased RSV titres in Sendai-treated mixed cultures. Similarly, the number of chicken cells in the Sendai virus-treated mixed cultures which become transformed is increased 40 times (Svoboda et al. 1967).

An analogous situation to that of mammalian cells transformed with defective RSV was observed by Huebner *et al.* (1966) for hamster tumour cells obtained using defective MSV (Moloney). In this case, too, the virus was detected only after mixed cultivation of these cells with mouse fibroblasts and only if mouse leukaemia virus acting as helper virus was added to the mixed cultures. Moloney's, Rauscher's and Friend's viruses were all effective. Serological specificity of the MSV was determined by virus-neutralization and corresponds to the helper virus used. Similarly, it was possible to achieve virus rescue *in vivo*—by application of hamster tumour cells and helper virus to mice. The MSV released induced sarcomas in such treated mice.

Virogenic mammalian tumours were also studied from the aspect of their antigenicity. In these tumours a specific tumour transplantation antigen was found in several laboratories (Sjögren & Jonsson, 1963; Harris & Chesterman, 1964; Koldovsky & Bubenik, 1964; Radzichovskaya, 1964).

Different tumours induced by the same strain of RSV, and also tumours induced by different strains of RSV all cross-react antigenically (Bubenik *et al.* 1967; Svoboda, 1967). This antigen differs from coat antigen of the virion (Jonsson & Sjögren, 1965; Svoboda, 1967). It is questionable whether its production is coded by the viral genome. In any case, it represents a specific fingerprint of the transforming action of RSV on the mammalian cell. Both in chicken NP cells and in mammalian virogenic cells obtained after application of non-defective RSV (Huebner *et al.* 1964) or defective RSV (Casey *et al.* 1966), CF antigen was found, representing—as already mentioned— the inner antigenic component of the virion.

It remains to be seen whether further virus-specific antigens also appear in virogenic cells, and what is the relationship of the antigen detected by immunofluorescence (Teyssie *et al.* 1965; Klement, Svoboda, Sovova & Zemek, 1966) to both these antigens, and especially to the CF antigen.

In addition to tumour cells in which the presence of the RSV genome can be determined by means of contact of virogenic cells with sensitive chicken cells, 'non-productive' tumours that arise after application of RSV are also known. In these tumours it has not been possible to induce virus production. This is true of tumours obtained in rats after application of Carr–Zilber strain of RSV (CZ–RSV) (Svet-Moldavsky, 1958; Kryukova, 1960), of fibromatous nodes obtained by Zilber & Kryukova (1958) in rabbits, of mouse tumours induced by CZ–RSV (Morgunova & Kryukova, 1962) and of some tumours induced in rats by PR–RSV (Svoboda, 1962). Since PR–RSV at least, is non-defective, the negative result cannot be ascribed to the absence of helper virus in the chicken cells used.

It is also interesting that tumours induced with PR–RSV which formed after long latency in rats (Svoboda, 1962), and long-growing tumours induced with SR–RSV in guinea-pigs (Ahlström *et al.* 1963), do not give rise to RSV after transfer to chickens. Similarly, viral activity was not found in passages of a guinea-pig tumour (Ahlström *et al.* 1963) and in passages of a rat tumour induced with SR–RSV (Harris & Chesterman, 1964), although in primary tumours both virogenic and non-virogenic tumour cells were present. Non-virogenic cells may acquire a selective advantage during growth *in vivo*, either because they do not contain specific tumour antigen or for some other reason.

Experiments by Macpherson (1965), who found formation of non-virogenic revertants from virogenic hamster BHK cells transformed by SR–RSV *in vitro*, indicate that in some cases a similar phenomenon might arise in tumours induced *in vivo*.

The reason for negative findings of RSV genome in some tumours may only be found on the level of the methods used. It is not improbable that certain virogenic mammalian cells react very weakly with chicken cells and the frequency of transfer of the non-infectious genome is very small. For this reason it will be necessary to try out long-term co-cultivation and also some types of treatment provoking cell-fusion as, for instance, treatment with the Sendai virus.

The possibility of utilizing different physical and chemical methods of induction has not yet been explored extensively. In experiments with XC cells mentioned earlier, as well as in those carried out by Shevlyagin (1964), such methods did not lead to production of infectious RSV. However, Kryukova (1961), after X- and u.v.-irradiation of a homo-genized tissue suspension of rabbit fibromatous nodes, obtained Rous sarcoma production in some cases after transfer to chickens, whereas non-irradiated material was inactive. The mechanism of the action of such treatment, ascribed to inactivation of an unknown viral inhibitor or perhaps to induction, still remains to be clarified.

In addition to the possibility that some non-productive tumour cells do not contain the viral genome, it is necessary to take into account a further possibility: that only the part of the viral genome responsible for transforming activity is present in these cells. If such a genome lacks the ability to control synthesis of parts of the virion other than the antigenic determinants, then even superinfection of the chicken cells, into which this genome is transferred by the helper virus, will not lead to virion production.

Theoretically, the transfer of this genome to the chicken cell would result in chicken cells being transformed but lacking capacity to form virions, even if superinfected with helper virus.

That such cases may exist is shown by Bubenik et al. (1967) and Bubenik & Bauer (1967) who found, after application of PR–RSV to mice, tumours in which specific tumour transplantation antigen, but not CF antigen, was detectable. Thus it is possible that in this case the RSV genome is not capable of directing synthesis of the virion's internal antigens.

It remains to be clarified to what extent these cases are the result of the influence of a heterologous cellular environment on the RSV genome or the result of selection of viral mutants which, in spite of an incomplete genetic make-up, display an ability to transform mammalian cells.

In some cases it has been possible to obtain convincing evidence that mammalian tumours induced with RSV may produce infectious virus. This virus-productive interaction has been described in connection with several hamster tumours induced with PR–RSV in hamsters (Svoboda & Klement, 1963; Klement & Vesely, 1965). Infectious virus was detected in some passages of these tumours. The virus production found was low: approximately five orders lower than in chicken sarcomas.

Infectious virus production was observed in passages of rat tumour induced with chicken sarcoma virus B 77 (Altaner & Svec, 1966). Virus recovered from rat tumours cross-reacts with virus-neutralizing antisera against original chicken sarcoma virus, but cross-reactivity is decreased. Virus particles having the morphology of avian tumour viruses were found by Rabotti, Bucciarelli & Dalton (1966) in dog brain tumours induced with two strains of RSV. Infectious RSV was, however, not recovered from tumour tissue. The relationship of observed particles to RSV remains to be clarified by additional methods.

Infectious virus was also found in one rabbit sarcoma induced with CZ–RSV (Kryukova, 1966). In other tumours of this type it was not possible to confirm virus production even if tumour tissue was transferred to chickens. In spite of this, virus-neutralizing antibodies were detected in carriers and viral antigen was detected by antiviral fluorescein-labelled antisera (Kryukova & Obukh, 1964b; Kryukova, 1966). This demonstrates that at least some of the antigenic determinants of the viral outer coat are synthesized in the carriers without a complete virion being formed, or—and this is in agreement with the other findings—that mature virus does form but is inhibited by an unknown mechanism sensitive to irradiation.

Formation of small amounts of RSV was found by Shevlyagin & Martirosyan (1966) but only around the 20th day after infection of mouse fibroblasts with CZ–RSV and SR–RSV. In some of the cells during this period, antigen synthesis, localized in the cytoplasm and detected by fluorescein-labelled chicken antiviral antibodies, was found. The same experiments, carried out with hamster cells, did not show virus production even though a similar antigen was found (Martirosyan & Shevlyagin, 1966). Since in neither mouse nor hamster cells was a characteristic transformation of infected cultures observed by the authors, these results imply a possibility that a complete, or incomplete, maturation cycle of RSV may take place in certain mammalian cells without transformation of the culture.

Similarly, synthesis of small amounts of virus was observed shortly after the infection of bovine lung fibroblasts with Engelbreth–Holm strain of RSV (Stenkvist, 1966).

It is not yet known which are the factors that determine whether virus will be produced in the mammalian cell. Results to date show that one of the important factors could be the type of heterologous cell used, which is best shown in the case of PR–RSV. Of course other factors have to be taken into account, such as the state of the infected cell, unknown helper viruses, multiplicity of infection, etc.

In some cases a situation may arise where virus is produced for a limited period after transformation, but later is detectable only if contact is established between the virogenic cell and the chicken cell. Such a situation was observed in rat cells transformed by PR–RSV *in vitro* (Svoboda *et al.* 1965) and shows that various types of interaction may pass into one another.

The mechanism whereby the maturation process of RSV reaches different degrees in different mammalian cell systems is not known. A possible explanation is that certain enzymes and metabolic pathways necessary for virion synthesis are lacking in the mammalian cells. This certainly occurs in bacteria infected with viruses. For instance, T-even phages grown on mutants of *Escherichia coli*, defective in glucosylating enzymes, create a progeny whose DNA contains little or no glucose (Hattman & Fukasawa, 1963; Shedlovsky & Brenner, 1963). Such phages later grow normally only in some strains of bacteria. After passage on such hosts they acquire normal DNA.

On the other hand, it is not excluded that in mammalian cells there are repressors that inhibit the function of certain parts of the viral genome, or that genes necessary to ensure correct functioning of the viral genome are missing. From this aspect, findings with the RNA-containing host-dependent bacteriophage f2 are interesting. These mutants grow only on 'permissive' bacteria, which contain certain suppressor genes (Zinder & Cooper, 1964). The genes renew the function of mutated phage cistrons and thus enable the complete phage replication cycle to be realized. If a 'non-permissive' bacterial host is infected with mutant Su-1, normal-size phage particles are formed that are not viable, while in the permissive bacteria viable phages are produced. Further analysis carried out by Lodish, Horiuchi & Zinder (1965) showed that non-viability of Su-1 mutants, because late-functioning phage product is not formed, is necessary perhaps for phage absorption, or assembly of RNA and protein into intact virions.

In contrast to this, the Su-3 mutant does not form non-infectious particles after infection of non-permissive hosts. Its presence, however, is detectable because it rescues Su-1 mutants (Valentine, Engelhardt & Zinder, 1964). Thus, if non-permissive bacteria are infected by both mutants, then, thanks to complementation, both give rise to infectious progeny.

Experimental material obtained so far on virogenic mammalian cells does not give any data in confirmation of this supposition. Although tests for the presence of virus-inactivating substances in virogenic cells were negative (Chyle et al. 1963), it does not follow that intracellular repressors functioning at a certain stage of viral synthesis are necessarily absent. The study of the suppressive action of cellular genes is, for the present, outside the range of experimental possibilities afforded by the cytogenetics of the animal cell.

The problem of an incomplete reproductive cycle of RSV in mammalian cells should be seen also in connection with data on the abortive multiplication of myxoviruses.

A whole range of data exists to show that infection of various types of cells by myxovirus has, as a consequence, the synthesis of antigenic components of these viruses, but not virion formation. In some cases, only certain antigenic components of the virus were synthesized. Thus, synthesis of particular antigenic components of fowl-plague virus was found to be dependent, respectively, on the time of application of *para*-fluorophenylalanine (Zimmermann & Schäfer, 1960) and on the length of treatment with ethyleneiminoquinone (Scholtissek & Rott, 1964).

To what extent the activity of these substances is related to repression of RSV virion formation in the mammalian cell remains to be clarified.

CONCLUSION

Virion formation of some RNA-containing oncogenic viruses in the virogenic cell is dependent on two factors. The first is the 'helper' virus, which in the case of infection of the homologous cell by a defective virus provides an outer coat for this virus, thus permitting it to mature. The second factor is contact with the sensitive ('helper') cell, permitting transfer of the viral genome from the cell, where certain functions are not expressed, to a cellular environment in which all its functions may be realized. Some cases are known in which a combination of both factors is necessary before virion formation can take place.

A further knowledge of the mechanisms which play a role in both 'helper' effects will open further ways for the investigation of animal viruses.

These mechanisms are also important in viral oncogenesis. They show that simple attempts to isolate oncogenic agents from cell-free tumour preparations may fail because dependence of oncogenic virus formation on the helper virus, the cell, or both, has not been taken into account.

ACKNOWLEDGEMENT

The author expresses his gratitude to Dr R. J. C. Harris for the revision of the manuscript.

REFERENCES

AHLSTRÖM, C. G., BERGMAN, S. & EHRENBERG, B. (1963). Neoplasms in guinea pigs induced by an agent in Rous chicken sarcoma. *Acta path. microbiol. scand.* **58**, 177.

AHLSTRÖM, C. G. & FORSBY, N. (1962). Sarcomas in hamsters after injection with Rous chicken tumor material. *J. exp. Med.* **115**, 839.

AHLSTRÖM, C. G. & JOHNSSON, N. (1962). Induction of sarcoma in rats by a variant of Rous virus. *Acta path. microbiol. scand.* **54**, 145.

ALTANER, C. & SVEC, F. (1966). Virus production in rat tumors induced by chicken sarcoma virus. *J. natn. Cancer Inst.* **37**, 745.

ARMSTRONG, D., OKUYAN, M. & HUEBNER, R. J. (1964). Complement-fixing antigens in tissue cultures of avian leucosis viruses. *Science*, **144**, 1584.

BAUER, H. & SCHÄFER, W. (1965). Isolierung eines gruppenspezifischen Antigens aus dem Hühner-Myeloblastose-Virus (BAI-Stamm A). *Z. Naturf.* **20***b*, 815.

BAUER, H. & SCHÄFER, W. (1966). Origin of group-specific antigen of chicken leukosis viruses. *Virology*, **29**, 494.

BUBENIK, J. & BAUER, H. (1967). Antigenic characteristics of the interaction between Rous sarcoma virus and mammalian cells. Complement-fixing and transplantation antigens. *Virology*, **31**, 489.

BUBENIK, J., KOLDOVSKY, P., SVOBODA, J., KLEMENT, V. & DVORAK, R. (1967). Induction of tumours in mice with three variants of Rous sarcoma virus and studies on the immunobiology of these tumours. *Folia biol., Praha*, **13**, 29.

CASEY, M. J., RABOTTI, G. F., SARMA, P. S., LANE, W. T., TURNER, H. C. & HUEBNER, R. J. (1966). Complement-fixing antigens in hamster tumors induced by the Bryan strain of Rous sarcoma virus. *Science*, **151**, 1086.

CHYLE, P. (1964). Analysis of the virus-producing interaction of rat Rous sarcoma (XC and MR₅) with the fowl cell. *Folia biol., Praha*, **10**, 359.

CHYLE, P., KLEMENT, V. & SVOBODA, J. (1963). Attempts to induce formation of Rous sarcoma virus in cells of tumour XC. *Folia biol., Praha*, **9**, 92.

DI STEFANO, H. & DOUGHERTY, R. (1965). Cytological observations of 'nonproducer' Rous sarcoma cells. *Virology*, **27**, 360.

DOUGHERTY, R. & DI STEFANO, H. (1965). Virus particles associated with 'non-producer' Rous sarcoma cells. *Virology*, **27**, 351.

DOUGHERTY, R. & DI STEFANO, H. (1966). Lack of relationship between infection with avian leukosis virus and the presence of COFAL antigen in chick embryos. *Virology*, **29**, 586.

DOUGHERTY, R. & RASMUSSEN, R. (1964). Properties of a strain of Rous sarcoma virus that infects mammals. *Natn. Cancer Inst. Monogr.* **17**, 337.

FEBVRE, H., ROTHSCHILD, L., ARNOULT, J. & HAGUENAU, F. (1964). *In vitro* malignant conversion of rat embryonic cell lines with the Bryan strain of Rous sarcoma virus. *Natn. Cancer Inst. Monogr.* **17**, 459.

FRIESEN, B. & RUBIN, H. (1961). Some physicochemical and immunological properties of an avian leucosis virus (RIF). *Virology*, **15**, 387.

GOLDÉ, A. & LATARJET, R. (1966). Dissociation, par irradiation, des fonctions oncogène et infectieuse du virus de Rous, souche de Schmidt-Ruppin. *C. r. hebd. Séanc. Acad. Sci., Paris*, **262**, 420.

GOLDÉ, A. & VIGIER, P. (1966). Non-défectivité du virus de Rous de la souche de Prague. *C. r. Acad. Sci., Paris*, **262**, 2793.

HANAFUSA, H. (1964). Nature of the defectiveness of Rous sarcoma virus. *Natn. Cancer Inst. Monogr.* **17**, 543.

HANAFUSA, H. (1965). Analysis of the defectiveness of Rous sarcoma virus. III. Determining of a new helper virus on the host range and susceptibility to interference of RSV. *Virology*, **25**, 248.

HANAFUSA, H. & HANAFUSA, T. (1966a). Determining factor in the capacity of Rous sarcoma virus to induce tumours in mammals. *Proc. natn. Acad. Sci. U.S.A.* **55**, 532.

HANAFUSA, H. & HANAFUSA, T. (1966b). Analysis of defectiveness of Rous sarcoma virus. IV. Kinetics of RSV production. *Virology*, **28**, 369.

HANAFUSA, H., HANAFUSA, T. & RUBIN, H. (1963). The defectiveness of Rous sarcoma virus. *Proc. natn. Acad. Sci. U.S.A.* **49**, 572.

HANAFUSA, H., HANAFUSA, T. & RUBIN, H. (1964). Analysis of the defectiveness of Rous sarcoma virus. II. Specification of RSV antigenicity by helper virus. *Proc. natn. Acad. Sci. U.S.A.* **51**, 41.

HARRIS, R. J. C. & CHESTERMAN, F. C. (1964). Growth of Rous sarcoma in rats, ferrets and hamsters. *Natn. Cancer Inst. Monogr.* **17**, 321.

HARTLEY, J. W. & ROWE, W. P. (1966). Production of altered cell foci in tissue culture by defective Moloney sarcoma virus particles. *Proc. natn. Acad. Sci. U.S.A.* **55**, 780.

HARVEY, J. J. (1964). An unidentified virus which causes the rapid production of tumours in mice. *Nature, Lond.* **204**, 1104.

HATTMAN, S. & FUKASAWA, T. (1963). Host-induced modification of T-even phages due to defective glucosylation of their DNA. *Proc. natn. Acad. Sci. U.S.A.* **50**, 297.

HUEBNER, R. J., ARMSTRONG, D., OKUYAN, M., SARMA, P. S. & TURNER, H. C. (1964). Specific complement-fixing viral antigens in hamster and guinea-pig tumors induced by the Schmidt-Ruppin strain of avian sarcoma. *Proc. natn. Acad. Sci. U.S.A.* **51**, 742.

HUEBNER, R. J., HARTLEY, J. W., ROWE, W. P., LANE, W. T. & CAPPS, W. I. (1966). Rescue of the defective genome of Moloney sarcoma virus from a non-infectious hamster tumor and the production of pseudotype sarcoma viruses with various murine leukemia viruses. *Proc. natn. Acad. Sci. U.S.A.* 56, 1164.

JENSEN, F. C., GIRARDI, A. J., GILDEN, R. V. & KOPROWSKI, H. (1964). Infection of human and simian tissue cultures with Rous sarcoma virus. *Proc. natn. Acad. Sci. U.S.A.* 52, 53.

JONSSON, N. & SJÖGREN, H. O. (1965). Further studies on specific transplantation antigens in Rous sarcoma of mice. *J. exp. Med.* 118, 403.

KLEMENT, V., CHYLE, P. & SVOBODA, J. (1963). Comparison of the biological properties of tumours formed in rats after the administration of two variants of Rous sarcoma. *Folia biol., Praha,* 9, 412.

KLEMENT, V., SVOBODA, J., SOVOVA, V. & ZEMEK, J. (1966). Antigenicity of mammalian tumour cells transformed by Rous sarcoma virus. *Symposium on the mutational process. Genetic Variation in somatic cells. Academia, Prague,* p. 325.

KLEMENT, V. & VESELY, P. (1965). Tumour induction with the Rous sarcoma virus in hamsters and production of infectious Rous sarcoma virus in an heterologous host. *Neoplasma,* 12, 147.

KOLDOVSKY, P. & BUBENIK, J. (1964). Occurrence of tumours in mice after inoculation of Rous sarcoma and antigenic changes in these tumours. *Folia biol., Praha,* 10, 81.

KRYUKOVA, I. N. (1960). Findings on haemorrhagic disease induced with Rous virus in rats. (In Russian.) *Vop. Virus.* 5, 602.

KRYUKOVA, I. N. (1961). An attempt to demask Rous virus in mammalian tissues. *Vop. Virus.* 6, 313.

KRYUKOVA, I. N. (1966). On two types of interaction of Rous sarcoma virus (Carr strain) with mammalian cells *in vivo. Acta virol.* 10, 440.

KRYUKOVA, I. N. & OBUKH, I. B. (1964a). Distribution of infectious Rous virus and viral antigen in infected rat and mouse organisms. *Vop. Onkol.* 10, 3.

KRYUKOVA, I. N. & OBUKH, I. B. (1964b). Distribution of viral antigens in the internal organs and fibromatose nodes of rabbits injected with Rous virus. (In Russian.) *Vop. Onkol.* 10, 60.

LODISH, H. F., HORIUCHI, K. & ZINDER, N. D. (1965). Mutants of the bacteriophage f2. V. On the production of non-infectious phage particles. *Virology,* 27, 139.

MACPHERSON, I. (1965). Reversion in hamster cells transformed by Rous sarcoma virus. *Science,* 148, 1731.

MARTIROSYAN, D. M. & SHEVLYAGIN, V. Y. (1966). Relationship of Rous virus to guinea-pig and hamster cells *in vitro. Folia biol., Praha,* 12, 248.

MORGUNOVA, T. D. & KRYUKOVA, I. N. (1962). Mouse sarcomas induced by Rous virus. (In Russian.) *Vop. Virus.* 7, 367.

MUNROE, J. S. & SOUTHAM, C. M. (1964). Oncogenicity of two strains of chicken sarcoma virus for rats. *J. natn. Cancer Inst.* 32, 591.

OKADA, Y. (1962). Analysis of giant polynuclear cell formation caused by HVJ from Ehrlich's ascites tumor cells. *Expl. Cell Res.* 26, 98.

PAYNE, F. E., SOLOMON, J. J. & PURCHASE, H. G. (1966). Immunofluorescent studies on group-specific antigen of the avian sarcoma-leukosis viruses. *Proc. natn. Acad. Sci. U.S.A.* 55, 341.

RABOTTI, G. F., BUCCIARELLI, E. & DALTON, A. J. (1966). Presence of particles with the morphology of viruses of the avian leukosis complex in meningeal tumors induced in dogs by Rous sarcoma virus. *Virology,* 29, 684.

RABOTTI, G. F., RAINE, W. A. & SELLERS, R. L. (1965). Brain tumors (gliomas) induced in hamsters by Bryan's strain of Rous sarcoma virus. *Science,* 147, 504.

RADZICHOVSKAYA, R. (1964). Observation on induction of resistance to Rous sarcoma cell antigens in hamsters. *Nature, Lond.* **204**, 393.

RAUSCHER, F. J., REYNIERS, J. A. & SACKSTEDER, M. R. (1964). Response or lack of response of apparently leucosis-free Japanese Quail to avian tumor viruses. *Natn. Cancer Inst. Monogr.* **17**, 211.

RUBIN, H. (1960). A virus in chick embryos which induces resistance *in vitro* to infection with Rous sarcoma virus. *Proc. natn. Acad. Sci. U.S.A.* **46**, 1105.

RUBIN, H. (1961). The nature of a virus-induced cellular resistance to Rous sarcoma virus. *Virology*, **13**, 200.

RUBIN, H. (1964). Virus defectiveness and cell transformation in the Rous sarcoma. *J. cell. comp. Physiol.* **64** (suppl. **1**), 173.

RUBIN, H. (1965). Genetic control of cellular susceptibility to pseudotypes of Rous sarcoma virus. *Virology*, **26**, 270.

RUBIN, H. & VOGT, P. K. (1962). An avian leucosis virus associated with stocks of Rous sarcoma virus. *Virology*, **17**, 184.

SARMA, P. S., VASS, W. & HUEBNER, R. J. (1966). Evidence for the *in vitro* transfer of defective Rous sarcoma virus genome from hamster tumor cells to chick cells. *Proc. natn. Acad. Sci. U.S.A.* **55**, 1435.

SCHNEEBERGER, E. E. & HARRIS, H. (1966). An ultrastructural study of interspecific cell fusion induced by inactivated Sendai virus. *J. Cell Science*, **1**, 401.

SCHOLTISSEK, C. & ROTT, R. (1964). Behaviour of virus-specific activities in tissue cultures infected with myxoviruses after chemical changes of the viral ribonucleic acid. *Virology*, **22**, 169.

SHEDLOVSKY, A. & BRENNER, S. (1963). A chemical basis for the host-induced modification of T-even bacteriophages. *Proc. natn. Acad. Sci. U.S.A.* **50**, 300.

SHEVLYAGIN, V. J. (1964). Hamster tumours induced by Rous sarcoma virus (Carr strain). (In Russian.) *Vop. Virus.* **5**, 533.

SHEVLYAGIN, V. J. & MARTIROSYAN, D. M. (1966). Reproduction of Rous virus in mouse embryonic tissue *in vitro. Folia biol., Praha*, **12**, 184.

SHIPMAN, C. & LEVINE, A. S. (1965). Tumor production in the Japanese Quail by the Schmidt-Ruppin strain of Rous sarcoma virus. *Virology*, **27**, 637.

SIMKOVIC, D. (1964). Interaction between mammalian tumor cells induced by Rous virus and chicken cell. *Natn. Cancer Inst. Monogr.* **17**, 351.

SIMKOVIC, D., SVOBODA, J. & VALENTOVA, N. (1963). Clonal analysis of line XC⁻ᵗᶜ rat tumour cells (derived from tumour XC) grown *in vitro. Folia biol., Praha*, **9**, 82.

SIMKOVIC, D., SVOBODA, J. & VALENTOVA, N. (1965). Induction of formation and release of infectious Rous virus by cells of rat tumour XC *in vitro. Folia biol., Praha*, **11**, 350.

SIMKOVIC, D., VALENTOVA, N. & THURZO, V. (1962). An *in vitro* system for the detection of the Rous sarcoma virus in the cells of the rat tumour XC. *Neoplasma*, **9**, 104.

SJÖGREN, H. O. & JONSSON, N. (1963). Resistance against isotransplantation of mouse tumors induced by Rous sarcoma virus. *Expl. Cell Res.* **32**, 618.

STENKVIST, B. (1966). Evidence for early synthesis of Rous sarcoma virus in bovine fibroblasts. *Acta Universitatis Upsaliensis. Abstract of Uppsala Dissertations in Medicine* **30**.

SVET-MOLDAVSKY, G. J. (1957). Development of multiple cysts and of haemorrhagic affections of internal organs in albino rats treated during the embryonic or new-born period with Rous sarcoma virus. *Nature, Lond.* **180**, 1299.

SVET-MOLDAVSKY, G. J. (1958). Sarcoma in albino rats treated during the embryonic stage with Rous virus. *Nature, Lond.* **182**, 1452.

270 J. SVOBODA

SVET-MOLDAVSKY, G. J. & SKORIKOVA, A. S. (1960). The pathogenicity of Rous sarcoma virus for mammals. Detection of virus and of antigenic substances of Rous sarcoma in the cyst-haemorrhagic disease of albino rats. *Acta Virol.* **4**, 47.

SVOBODA, J. (1960). Presence of chicken tumour virus in sarcoma of the adult rat inoculated after birth with Rous sarcoma virus. *Nature, Lond.* **186**, 980.

SVOBODA, J. (1961). The tumorigenic action of Rous sarcoma in rats and the permanent production of Rous virus by the induced rat sarcoma XC. *Folia biol., Praha*, **7**, 46.

SVOBODA, J. (1962). Further findings on the induction of tumours by Rous sarcoma in rats and on the Rous virus-producing capacity of one of the induced tumours (XC) in chicks. *Folia biol., Praha*, **8**, 215.

SVOBODA, J. (1964). Malignant interaction of Rous virus with mammalian cells *in vivo* and *in vitro*. *Natn. Cancer Inst. Monogr.* **17**, 277.

SVOBODA, J. (1967). Antitumour immunity against RSV-induced tumours in mammals. *Ser. Cancerologica*, **2**, 12.

SVOBODA, J. & CHYLE, P. (1963). Malignization of rat embryonic cells by Rous sarcoma virus *in vitro*. *Folia biol., Praha*, **9**, 329.

SVOBODA, J., CHYLE, P., SIMKOVIC, D. & HILGER, T. I. (1963). Demonstration of the absence of infectious Rous virus in rat tumour XC, whose structurally intact cells produce Rous sarcoma when transferred to chicks. *Folia biol., Praha* **9**, 77.

SVOBODA, J. & GROZDANOVIC, J. (1960). Notes on the role of immunological tolerance in the induction of haemorrhagic disease in young rats. *Folia biol., Praha*, **6**, 27.

SVOBODA, J. & KLEMENT, V. (1963). Formation of delayed tumours in hamsters inoculated with Rous virus after birth and finding of infectious Rous virus in induced tumour P_1. *Folia biol., Praha*, **9**, 403.

SVOBODA, J., MACHALA, O. & HLOZANEK, I. (1967). Influence of Sendai virus on RSV formation in mixed culture of virogenic mammalian cells and chicken fibroblasts. *Folia biol., Praha*, **13**, 155.

SVOBODA, J., VESELY, P., VRBA, M. & JIRANEK, M. (1965). Biological properties of rat cells transformed *in vitro* by Rous sarcoma virus. *Folia biol., Praha*, **11**, 251.

TEMIN, H. M. (1960). Control of cellular morphology in embryonic cells infected with Rous sarcoma virus *in vitro*. *Virology*, **10**, 182.

TEMIN, H. M. (1962). Separation of morphological conversion and virus production in Rous sarcoma virus infection. *Cold Spring Harb. Symp. quant. Biol.* **27**, 407.

TEMIN, H. M. (1963). Further evidence for a converted non-virus producing state of Rous sarcoma virus-infected cells. *Virology*, **20**, 235.

TEMIN, H. M. & RUBIN, H. (1958). Characteristics of an assay for Rous sarcoma virus and Rous sarcoma cells in tissue culture. *Virology*, **6**, 669.

TEYSSIE, A., ROTHSCHILD, L. & FEBVRE, H. (1965). Transmission in vitro d'un antigène spécifique par association de cellules tumorales de rat induites par le virus du sarcome de Rous à des cellules normales de poulet. *C. r. hebd. Séanc. Acad. Sci., Paris*, **261**, 4925.

VALENTINE, R. C., ENGELHARDT, D. L. & ZINDER, N. D. (1964). Host-dependent mutants of the bacteriophage f2. II. Rescue and complementation of mutants. *Virology*, **23**, 195.

VESELY, P. & SVOBODA, J. (1965). Malignant transformation of Syrian hamster embryonic cells with Rous virus of the Schmidt-Ruppin strain *in vitro*. *Folia biol., Praha*, **11**, 78.

VIGIER, P. (1966). Persistance du génome du virus de Rous dans les cellules de hamster converties in vitro par un virus non défectif et un virus défectif. *C. r. hebd. Séanc. Acad. Sci., Paris*, **262**, 2554.

VIGIER, P. (1967). Persistance du génome du virus Rous dans des cellules du hamster converties in vitro, et action du virus Sendai inactive sur sa transmission aux cellules de poule. *C. r. hebd. Séanc. Acad. Sci., Paris*, **264**, 422.

VIGIER, P. & SVOBODA, J. (1965). Étude en culture de la production du virus de Rous par contact entre les cellules du sarcome XC du rat et les cellules d'embryon de poule. *C. r. hebd. Séanc. Acad. Sci., Paris*, **261**, 4281.

VOGT, P. K. (1964). Fluorescence microscopic observations on the defectiveness of Rous sarcoma virus. *Natn. Cancer Inst. Monogr.* **17**, 523.

VOGT, P. K. (1965). A heterogeneity of Rous sarcoma virus revealed by selectively resistant chick embryo cells. *Virology*, **25**, 237.

VOGT, P. K. & ISHISAKI, R. (1965). Reciprocal patterns of genetic resistance to avian tumor viruses in two lines of chickens. *Virology*, **26**, 664.

VOGT, P. K. & RUBIN, H. (1963). Studies in the assay and multiplication of avian myeloblastosis virus. *Virology*, **19**, 92.

VOGT, P. K., SARMA, P. S. & HUEBNER, R. J. (1965). Presence of avian tumor virus group-specific antigen in non-producing Rous sarcoma cells of the chicken. *Virology*, **27**, 233.

WECKER, E. & LEDERHILGER, G. (1964). Genomic masking produced by double-infection of HeLa cells with heterotypic poliovirus. *Proc. natn. Acad. Sci. U.S.A.* **52**, 705.

WEISS, R. (1967). Spontaneous virus production from 'non-virus producing' Rous sarcoma cells. *Virology*, **32**, 719.

YAMAMOTO, N. & ANDERSON, T. F. (1961). Genomic masking and recombination between serologically unrelated phages P22 and P221. *Virology*, **14**, 430.

ZILBER, L. A. & KRYUKOVA, I. N. (1957). Haemorrhagic disease in rats caused by Rous sarcoma virus. (In Russian.) *Vop. Virus.* **4**, 239.

ZILBER, L. A. & KRYUKOVA, I. N. (1958). Fibromatosis in rabbits caused by Rous sarcoma virus. (In Russian.) *Vop. Virus.* **3**, 166.

ZIMMERMANN, T. & SCHÄFER, W. (1960). Effect of *p*-fluorophenylalanine on fowl plague virus multiplication. *Virology*, **11**, 676.

ZINDER, N. D., & COOPER, S. (1964). Host-dependent mutants of the bacteriophage f2. I. Isolation and preliminary classification. *Virology*, **23**, 152.

DEPENDENCE AND COMPLEMENTATION AMONG ANIMAL VIRUSES CONTAINING DEOXYRIBONUCLEIC ACID

F. RAPP*

Department of Virology and Epidemiology, Baylor University College of Medicine, Houston, Texas 77025

The recognition that some deoxyribonucleic acid (DNA)-containing viruses replicate in certain cells only in the presence of an unrelated helper virus is a recent advance in animal virology. This has served to focus renewed attention on the molecular events underlying virus replication and has emphasized that application of the techniques used in genetics, biology, biochemistry, biophysics and immunology will be necessary to resolve the many questions raised. In addition, the studies already conducted have important implications to viral oncogenesis and, thus, to the whole problem of the regulation of cell growth and division.

This chapter will review the more extensively studied examples of dependence and complementation between DNA-containing animal viruses. The first section includes systems in which one virus can multiply well in the host cell but the second virus can multiply only in the presence of the first, a situation termed non-reciprocal complementation. The second section will cover reciprocal complementation, a term used to denote examples in which neither virus can multiply in the host cell in the absence of the other. No further subdivisions can be made at this time because the exact mechanism of complementation in most of the systems has not been elucidated.

Complementation has been defined (Butel & Rapp, 1966a) as 'the functional interaction between two viruses, such that replication occurs under otherwise inhibitory conditions'. This definition will be taken to exclude examples of both recombination (in which some of the progeny produced are genetically different from the parental viruses) and phenotypic mixing (which is a temporary situation because the foreign capsids encasing the progeny genome are lost in the next replicative cycle).

In contrast to most examples of complementation which have been well-defined with bacterial viruses, the examples included in this chapter occur between unrelated viruses, rather than between viruses

* American Cancer Society Professor of Virology.

which are genetically similar. However, it should be pointed out that some examples of complementation between closely related DNA-containing animal viruses are known for members of the poxvirus group (see review by Fenner & Sambrook, 1964) and for herpes simplex virus (Roizman & Aurelian, 1965).

NON-RECIPROCAL COMPLEMENTATION

Human adenoviruses and simian virus 40

It had been generally believed that human adenoviruses could adapt to growth in monkey kidney cells following serial passage in these cells. This interpretation was called into question when O'Conor, Rabson, Berezesky & Paul (1963) observed that simultaneous infection of simian cells with simian papovavirus SV40 and adenovirus type 12 resulted in an increase in the number of cells in the culture in which adenovirus particles could be readily visualized by electron microscopy. A subsequent survey of various 'adapted' strains of human adenoviruses revealed that the majority were contaminated with SV40 information, based either on the detection of SV40 tumour (T) antigen following inoculation of simian cell cultures or by the isolation of complete, infectious SV40. It appeared in some instances that phenotypic mixing had occurred with complete SV40 genomes being encased in adenovirus capsids at each passage level (see review by Rapp & Melnick, 1966). One of the adenovirus type 7 strains, however, was found to be carrying defective SV40 genetic material (Huebner, Chanock, Rubin & Casey, 1964; Rapp, Melnick, Butel & Kitahara, 1964; Rowe & Baum, 1964) and this virus will be described in detail in the section on reciprocal complementation. Thus far, there have been no reports conclusively showing that a human adenovirus has been adapted to simian cells without the aid and continuous carriage of some foreign genetic information—which does not necessarily have to be SV40 (Butel, Rapp, Melnick & Rubin, 1966).

Subsequent studies demonstrated that human adenoviruses undergo only an abortive cycle in the simian cells. During single infection, the adenovirus tumour (or T) antigen (an early protein whose biological role remains obscure) is synthesized (Feldman, Butel & Rapp, 1966; Malmgren, Rabson, Carney & Paul, 1966), but only in the presence of SV40 is adenovirus capsid protein detected. The induction of the adenovirus T antigen clearly showed that the abortive cycle is not due to faulty adsorption, penetration or uncoating of the adenovirion.

Analysis by analytical ultracentrifugation of DNA extracted from infected green monkey kidney cells revealed that adenovirus DNA was synthesized during the abortive infection (Rapp, Feldman & Mandel, 1966). In fact, there appears to be as much adenovirus DNA synthesized during the abortive cycle as during the productive cycle in the presence of SV40. In addition, high levels of thymidine kinase are induced in the simian cells during the abortive cycle of the human adenoviruses (E. Bresnick & F. Rapp, unpublished observations). These observations indicate that the helper effect of SV40 concerns a molecular event following the biosynthesis of adenovirus nucleic acid.

The exact nature of the SV40 helper effect has not yet been determined. One possibility is that SV40 represses the formation of an inhibitor normally synthesized by monkey kidney cells which prevents the multiplication of human adenoviruses. Attempts in our laboratory to abort synthesis of such a hypothetical inhibitor with ultraviolet light have failed. Another possibility is that SV40 inhibits the formation of monkey kidney cell interferon, much as parainfluenza type 3 prevents the synthesis of interferon in calf kidney cells so that Newcastle disease virus can replicate (Hermodsson, 1963). However, attempts by Dr Butel in our laboratory to demonstrate an interferon-like inhibitor present during abortive infection of simian cells by human adenoviruses have thus far also failed.

Therefore, at this time, it seems more reasonable to postulate that SV40 contributes something that is normally lacking in simian cells but which is required by the human adenoviruses to complete their replicative cycle. This factor must either be present in human cells or the adenoviruses must be capable of inducing the synthesis of the factor in human, but not in monkey cells. It is possible that the factor is involved in the translation of adenovirus messenger RNA, such as a transfer RNA.

Differences appear to exist among the abortive cycles of various human adenovirus serotypes in monkey cells. Growth curves constructed from sequential harvests of simian cells infected with adenovirus types 2, 5, 6, 14, 16, or 21 show an initial drop in titre which remains low over a 72-hr. period (Butel & Rapp, 1967; Rapp & Jerkofsky, 1967). In contrast, cultures infected with adenovirus types 1 or 7 show the same initial drop in titre which, in turn, is followed by a subsequent increase in infectious virus, usually within 48 hr. after infection (Rapp & Jerkofsky, 1967; Jerkofsky & Rapp, 1967). The final yield in infectious virus closely parallels the amount of input virus. This relationship was found to hold over a 10,000-fold difference in level of input virus. This phenomenon is under investigation, and preliminary results suggest that simian cells are 'leaky' for some strains of human adenoviruses, a finding analogous to those previously made with conditional lethal mutants in certain bacterial virus systems. This suggestion implies that there are at least minor differences in the biosynthetic phases of the replicative cycles of various adenovirus serotypes.

Human adenoviruses and simian adenoviruses

Initial experiments with other viruses seemed to indicate that the helper effect on human adenoviruses in monkey cells was specific for SV40. Papovaviruses of other species (human wart and rabbit papilloma) and herpes simplex and measles viruses all failed to substitute for SV40 (Feldman *et al.* 1966). However, it has recently been found that co-infection of the monkey cells with simian adenoviruses (SV15, SA7) will enable the human adenoviruses to complete their replicative cycle (Naegele & Rapp, 1967). Both SV15 and SA7 are effective helper viruses and it is reasonable to assume that other simian adenoviruses can probably be substituted as helpers.

The increase in the yield of human adenoviruses from simian cells is dependent upon the time of addition of helper SV15. The maturation of the human adenoviruses is delayed until the end of the replicative cycle of SV15. Therefore, the block encountered by the human adenoviruses appears to be related to a late step in the growth cycle of the simian virus. Pre-infection of the monkey cells with a human adenovirus has no effect on the replication of SV15. These results conclusively demonstrate that the enhancement of replication of human adenoviruses in simian cells is not an exclusive property of SV40 virus.

The steps in the adenovirus replicative cycle in human cells are presented graphically in Figure 1. In the abortive infection of simian cells, synthesis of the intranuclear adenovirus T antigen indicates that steps 1–5 are successfully completed. Since adenovirus-specific DNA can be detected, step 6 must also occur in the abortive cycle. The block might be at either step 7 or 8. However, virus-specific messenger RNA has been isolated from adenovirus-transformed hamster cells (Fujinaga & Green, 1966). Since the transformed cells also appear to be abortively infected, late transcription may also be occurring in simian cells. This has not been proved, however, by actual extraction of virus messenger RNA from abortively infected monkey cells.

Step 8 does not occur because virus capsid antigen cannot be detected. Although the possibility exists that defective capsid protein is made that has not been detected by the immunological reagents currently available, late translation in the abortive cycle is either faulty or does not occur at all. Helper viruses SV40, SV15 and SA7 permit the human adenoviruses to effectively complete step 8 and to carry the cycle to completion (step 10).

As a result of this broadened viewpoint it seems probable that many other examples of virus-helper systems will be discovered. Chany & Brailovsky (1967) have already reported that adenovirus 12 enhances

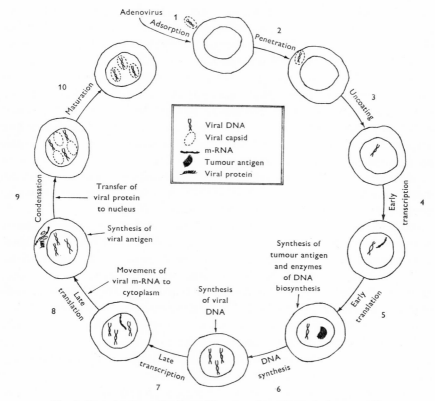

Fig. 1. Steps in the replication of human adenoviruses in human cells.

the replication of rat virus in rat embryo fibroblasts, although adenovirus 12 itself undergoes an abortive replicative cycle in these cells. The mechanism appears to be a reversal of the inhibitory effect of interferon. The authors postulate that the adenovirus induces the synthesis of an active substance they named 'stimulon'. The adenovirus had no such stimulating effect on the replication of vaccinia, herpes or vesicular stomatitis viruses.

No helper system has yet been reported which involves the interaction of a DNA- with an RNA-containing virus.

Adenoviruses and adeno-associated satellite viruses

Increased emphasis on research with human adenoviruses and the more careful inspection of adenovirus stocks for possible contamination with SV40 led to the observation with the electron microscope of small, 20 mμ particles (Pl. 1) closely associated with many adenoviruses (Melnick, Mayor, Smith & Rapp, 1965; Hoggan, Blacklow & Rowe, 1966; Archetti, Bereczky & Bocciarelli, 1966; Atchison, Casto & Hammon, 1965, 1966). These particles have been called adeno-associated or adeno-associated satellite viruses.

The small satellite particles have been found in preparations of various human adenoviruses, including types 3, 4, and 7 and in stocks of simian adenoviruses SV11 and SV15. A number of investigators have noted that the satellite viruses do not replicate in the absence of the helper adenovirus and are, therefore, defective. Thus far, other viruses have failed to substitute for the adenovirus as helpers for the satellite viruses. However, human, simian, and canine adenoviruses can complement the satellite particles, as long as the helper adenovirus can replicate in the host cell (Blacklow, Hoggan & Rowe, 1967).

Atchison *et al.* (1965) first showed that the small particles were serologically distinct from the adenovirus and this has been confirmed in a number of laboratories. At the present time, there appear to be three or four serotypes of the satellite virus (Hoggan *et al.* 1966; Melnick & Parks, 1966). Studies with type 1 satellite virus by Rose, Hoggan & Shatkin (1966) yielded strong evidence for a double-stranded molecule for satellite DNA with a molecular weight of 3.6×10^6 daltons and a guanine plus cytosine content of 52 moles %.

Biophysical studies with various satellite serotypes indicate that these viruses have a buoyant density ranging from 1·38 to 1·45 g./ml. in caesium chloride. The virus particles have been reported to possess cubic symmetry of the icosahedral type apparently with 12 morphological subunits (Mayor, Jamison, Jordan & Melnick, 1965). Smith, Gehle & Thiel (1966), however, concluded that the symmetry of satellite virus might be represented by some form of lattice-like capsomere arrangement similar to that proposed for reovirus (Vasquez & Tournier, 1964) rather than a knob-like arrangement of capsomeres.

A recent study of DNA from type 4 satellite virus (Parks, Green, Piña & Melnick, 1967) revealed a guanine plus cytosine content of 54–61 moles %, based on the kinetics of thermal dissociation. These experiments also yielded a profile of thermal dissociation characteristic of a double-stranded molecule. Examination of isolated DNA in the electron microscope showed an absence of circular forms. The larger strands were approximately $1\cdot4$–$1\cdot6$ μ in length, indicative of a molecular weight of about $2\cdot9 \times 10^6$ daltons.

The buoyant density of nucleic acid extracted from type 4 satellite virus was $1\cdot721$ g./ml. in caesium chloride, which also corresponds to a guanine plus cytosine content of $62\cdot2$ moles % (Parks, Green, Piña & Melnick, 1967). Since Rose et al. (1966) obtained a molecular weight of $3\cdot6 \times 10^6$ daltons for DNA from the type 1 satellite, there appear to be significant differences between type 1 and type 4 satellite viruses. Examination of both serotypes in the same laboratory would greatly aid in resolution of this problem.

Ability to separate the satellite virus from its helper adenovirus by density gradient centrifugation allowed experiments to determine the role of the helper virus in the replication of the satellite particles. Generally, detection and assay of these particles have been either by electron microscopy (where individual satellite particles can be counted) or by serologic tests such as complement fixation to detect satellite antigens. Recently, a test based on the detection of satellite antigens by the immunofluorescence technique has been used to follow replication of the defective virus (Ito, Melnick & Mayor, 1967). These techniques have revealed, for satellite serotype 4 in the presence of its helper simian adenovirus SV15, that the kinetics of satellite production closely follows the kinetics of replication of its helper adenovirus (Parks, Melnick, Rongey & Mayor, 1967). Both undergo an eclipse period of 12–16 hr. which is followed by the replication of both viruses. The eclipse period for satellite particles was shortened to 4–6 hr. by pre-infection of the cultures with the adenovirus about 12 hr. before addition of the satellite virus. Thus, satellite virus replication seems dependent upon a relatively late event in the adenovirus replication cycle.

It has also been noted by a number of investigators (Hoggan et al. 1966; Smith et al. 1966; Parks, Melnick, Rongey & Mayor, 1967; Casto, Atchison & Hammon, 1967) that in addition to the enhancement of the satellite by the adenovirus, reduced yields of adenovirus are often obtained in cultures co-infected with both viruses. This coincides with a decreased synthesis of adenovirus complement-fixing antigen. With satellite type 4 and SV15, the eclipse period of the adenovirus was not affected and attempts to demonstrate an interferon-like substance also failed. The interference effect seems to operate in all cells in the culture, since infective centre assays revealed that most of the cells will produce the adenovirus in the presence of satellite virus but that the number of particles produced per cell is reduced. The interference noted may be cell dependent since interference of human adenoviruses in human kidney cells is not significant although the human KB cells and monkey kidney cells yield less human or simian adenovirus respectively, if co-infected with satellite virus. The mechanism of the enhancement of satellite replication by adenovirus and the interference of satellite particles with the replication of its helper adenovirus awaits clarification. This will undoubtedly require information on the molecular events underlying the role of the adenovirus in aiding the replication of its defective satellite virus.

RECIPROCAL COMPLEMENTATION

Defective simian virus 40 and human adenoviruses

A strain of adenovirus type 7, adapted to growth in monkey kidney cells for vaccine purposes, was observed in 1964 to induce tumours in hamsters which were subsequently found to contain a specific SV40-induced antigen (Huebner et al. 1964). This same virus was found to induce the synthesis of SV40 T antigen in a variety of cells in tissue culture (Rapp et al. 1964; Rowe & Baum, 1964). No SV40 capsid antigen could be detected and all biological activity of the preparation was abolished by treatment with adenovirus 7 antiserum. It was suggested that a 'hybrid' of the unrelated viruses, SV40 and adenovirus, had evolved. The term 'hybrid' was used with reservation, however, because the actual physical state of the two nucleic acids was not known.

The growth characteristics of this unique virus were analysed in detail. Plaque formation by the 'hybrid' virus in human embryonic kidney cells followed one hit kinetics. Progeny derived from the plaques in human cells were found to have lost the determinant for SV40 T antigen and could no longer grow in simian cells (Boeyé, Melnick & Rapp, 1965). Plaque formation in green monkey kidney cells, however, was found to follow two hit kinetics, revealing that the interaction of two virus particles was required in order to initiate a plaque (Rowe & Baum, 1965; Boeyé, Melnick & Rapp, 1966). The simultaneous addition of large amounts of non-adapted, non-cytopathic human adenovirus to the monkey kidney cells used in the assay converted plaque formation by the 'hybrid' virus to single hit kinetics. This implied that one of the two necessary particles must be a human adenovirion. The progeny from the simian cell plaques invariably retained the ability to induce the synthesis of SV40 T antigen (Boeyé et al. 1965). Therefore, it was felt that the particle containing the SV40 genetic material was the second essential element required for plaque formation.

The phenomenon of altered kinetics of plaque formation by this virus in the presence of helper adenovirus is presented graphically in Figure 2. In the absence of additional helper virus, the number of observed plaques in simian cells falls as the square of the dilution factor, resulting in a continual decrease in the calculated titre of the 'adapted' virus stock. In the presence of helper virus, however, the calculated titre of the 'adapted' virus is higher and remains constant over a series of dilutions.

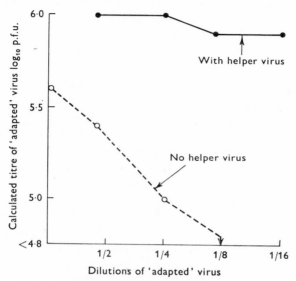

Fig. 2. Kinetics of plaque formation in monkey cells for human adenoviruses carrying unrelated defective genomes. Following an initial dilution of 1:100 (to reduce expected plaques to countable numbers in 60 mm. Petri dish cultures), twofold dilutions of the virus were seeded on to monolayers of green monkey kidney cells in the absence and presence of helper adenovirus. Enough helper virus was added to make certain that every cell in the culture was infected with at least one infectious unit of the helper virus. Following adsorption of the viruses at 37° C for one hour, nutrient agar was added to the cell cultures. The cultures were then incubated at 37° C in an atmosphere of 5 % CO_2. A second agar overlay was added seven days after inoculation of the virus. Plaques were read between the 10th and 14th day following inoculation of the viruses.

Under the conditions employed, only cells co-infected with an adenovirus and with a particle carrying defective determinants will give rise to plaques. In the absence of added helper adenovirus, the titre of the parent virus drops as the square of the dilution. When all the cells are infected with a helper adenovirion, one-hit kinetics are obtained and the titre calculated from the number of plaques obtained at various dilutions remains constant.

The particle carrying the defective SV40 genome was named PARA, the letters abbreviated from 'particle aiding (and aided by) the replication of adenovirus' (Rapp, Butel & Melnick, 1965). An assay system for this particle was devised in which the green monkey kidney cells were co-infected with non-adapted human adenovirus. This assay system was based on the observation previously described that plaque formation by the 'hybrid' virus was enhanced and followed single-hit kinetics in the presence of helper adenovirus.

An interesting observation based on this assay system was that maximum titres of PARA were obtained when the dilution of helper virus allowed a multiplicity of only about five physical particles of adenovirus per monkey cell (Butel, Melnick & Rapp, 1966). This input corresponded to 0·02–0·1 plaque-forming units of helper virus per cell. Obviously, then, there were many adenovirus particles in the stock which had failed to form a plaque during titration of the helper virus stock in human cells, but were, nevertheless, biologically able to interact with PARA in simian cells.

Any human adenovirus serotype could function as helper virus in the plaque assay for PARA. When the progeny from terminal plaques were analysed, it was found that they were all antigenically similar to the serotype of the helper virus (Rapp et al. 1965; Rowe, 1965; Butel & Rapp, 1966a). The SV40 determinants for T antigen and ability to enhance adenovirus growth in monkey cells had been transferred from the parental adenovirus type 7 to the helper virus, e.g. adenovirus type 2. This was not a case of phenotypic mixing because adenovirus 7 did not appear upon subsequent passage; the new PARA-adenovirus populations bred true in monkey cells. The term 'transcapsidation' was used (Rapp et al. 1965) to describe the conversion of the foreign determinants in PARA in a stable fashion from one adenovirus population to another. The phenomenon of transcapsidation is shown diagrammatically in Figure 3.

One additional test was necessary to show, unequivocally, that the capsid of PARA had been changed. It was possible that the neutralization of growth capacity had been due merely to neutralization of the helper virus. Therefore, a PARA-adenovirus population was neutralized with homologous antiserum and then assayed in monkey kidney cells in the presence of heterologous helper adenovirus. If PARA were not neutralized with the homologous adenovirus antiserum, plaques would form in the presence of the provided helper virus. Since plaques failed to form, the results indicated that PARA was actually encased in an adenovirus capsid and that this capsid was changed during transcapsidation (Rapp et al. 1965; Butel & Rapp, 1966a).

The addition of PARA to populations of human adenoviruses enabled those viruses to replicate in simian cells (Rapp et al. 1965; Butel & Rapp, 1966a; Rapp & Jerkofsky, 1967). Close analysis of the replication of PARA-adenoviruses in monkey cells revealed that both components replicated in unison after a latent period of 16–24 hr. with maximum titres reached by 48 hr. post-infection (Butel & Rapp, 1966a; Rapp & Jerkofsky, 1967). This growth of the PARA-adenoviruses in simian cells is an example of reciprocal complementation because PARA is defective for coat protein and is dependent upon the helper adenovirus for provision of a capsid. In turn, PARA serves the same function as complete SV40 in enabling human adenoviruses to replicate in monkey cells.

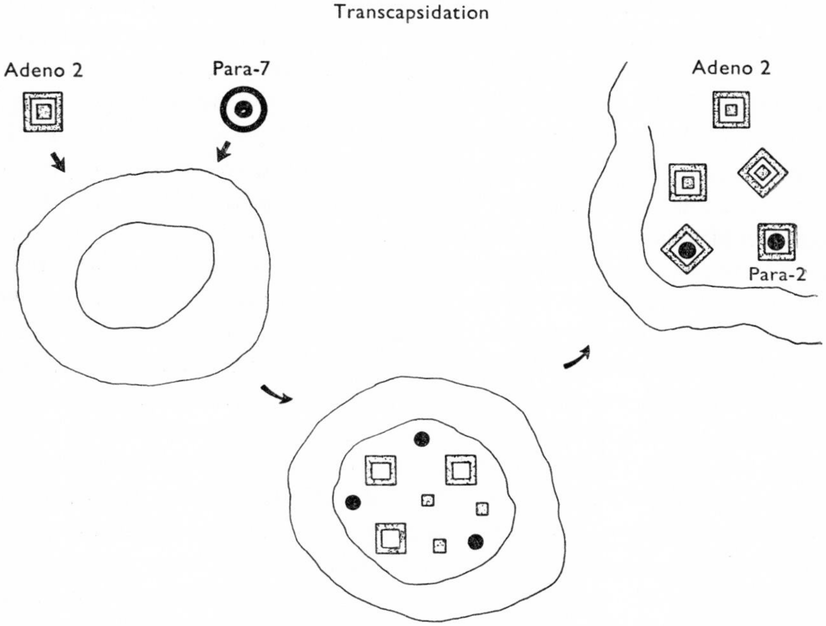

Fig. 3. The phenomenon of transcapsidation. Simian cells infected with adenovirus type 2 and co-infected with a defective SV 40 genome in an adenovirus type 7 capsid (PARA-7) replicate the type 2 adenovirus. The progeny defective SV 40 genome, however, become encased in an adenovirus type 2 capsid (PARA-2). Such a transcapsidant population breeds true.

Experiments were devised to determine what other SV40 genetic markers were carried by PARA, in addition to the information for T antigen synthesis. Transplantation rejection tests using weanling hamsters revealed that PARA could immunize the animals as effectively as complete SV40 so that they would reject transplants of SV40-transformed cells (Rapp, Tevethia & Melnick, 1966). In addition, transcapsidant populations (PARA-adenovirus 2 and PARA-adenovirus 12) also effectively immunized hamsters against challenge with SV40-transformed cells (Rapp, Butel, Tevethia & Melnick, 1967). Since these populations were derived from single plaques and possess the ability to induce SV 40 T antigen, it appears that the genetic information for the SV 40 tumour and transplantation antigens are linked in the PARA genome. Further supporting evidence for the concept of genetic linkage of the two markers is that the PARA-adenovirus 7 progeny obtained from plaques in human cells no longer carry the information for T antigen and are not able to induce SV 40 transplantation immunity.

PARA also carries the necessary information needed to transform either hamster or human cells *in vitro* (Black & Todaro, 1965). A study of the *in vivo* oncogenic potential of PARA revealed that when PARA was added to non-tumourigenic adenovirus type 2, the virus became oncogenic (Rapp, Butel, Tevethia, Katz & Melnick, 1966; Rapp, Melnick & Levy, 1967; Black & White, 1967; Igel & Black, 1967). This observation has been extended to other non-oncogenic serotypes of adenovirus, as well (F. Rapp & M. A. Jerkofsky, unpublished observations). Animals bearing tumours induced by the PARA-adenoviruses usually respond with antibodies against SV 40 and adenovirus T antigens. The tumours induced by the mixed populations vary histopathologically (Rapp, Melnick & Levy, 1967). Some resemble SV 40 giant cell-containing fibrosarcomas, some appear to be adenovirus-like undifferentiated epithelioid tumours, and still others contain elements of both types of tumours.

The preceding observations have important implications in the field of viral oncology. They show that viruses can acquire tumourigenic properties by the acquisition of defective portions of genomes of unrelated oncogenic viruses. Such defective genomes cannot be detected unless specific markers are available by which to identify the parent tumour virus. Only a knowledge of the SV 40 T antigen system, plus the availability of the necessary immunological reagents, enabled the identification of PARA and PARA-induced tumours.

All attempts to physically separate PARA from the helper adenovirions present in the same stock have failed. Both types of particles band at the same buoyant density after centrifugation to equilibrium (Butel & Rapp, 1966b; Rowe, Baum, Pugh & Hoggan, 1965). Both PARA and adenovirus are inactivated at elevated temperature at similar rates (Butel & Rapp, 1966b; Rowe *et al.* 1965; Boeyé *et al.* 1966). However, PARA was found to be more sensitive to inactivation by ultraviolet light than was the helper adenovirus or complete SV 40 (Butel & Rapp, 1966b), suggesting that PARA may contain more nucleic acid than either SV 40 or adenovirus.

The studies of the properties of defective PARA raised the question of whether any adenovirus DNA is contained in the particle. The presence of such additional DNA could explain both the density and ultraviolet light inactivation results. Rowe & Pugh (1966) reported, based on immunofluorescence staining patterns with specific hamster antibody, that heterologous transcapsidant populations of PARA are able to induce the synthesis of adenovirus 7 T antigen. Recently, it has been reported that the SV 40 and adenovirus DNA of the 'hybrid' virus are non-dissociable (Baum *et al.* 1966) because density gradient centrifugation failed to separate the nucleic acids as measured by nucleic acid hybridization techniques. These findings indicate that the two types of DNA are linked by an alkali-stable bond and that the particles may correctly be called hybrids by biophysical criteria. However, attempts to separate adenovirus DNA from PARA DNA by density gradient centrifugation in caesium chloride following extraction of DNA from either the virus or infected simian cells have failed (Rapp & Khare, 1967).

It can readily be seen from the foregoing data that the PARA-adenovirus interaction is very complex. It may be unique or it may be the prototype of other examples of defective viruses. The techniques devised in the course of the study of this particular population have already resulted in the discovery of a different type of defective particle which will be described in the following section.

Monkey cell-adapting component and human adenoviruses

The elimination of SV 40 from vaccine strains of adenoviruses caused resumption of attempts to 'adapt' the human viruses to simian cells. Various strains of adenoviruses 'adapted' to simian cells were therefore characterized. An adenovirus type 7 strain, capable of replication in simian cells and free from any SV 40 determinants, was then carefully analysed.

Using the assay systems devised for PARA-adenoviruses, it was readily determined that plaque formation by the 'adapted' adenovirus 7 followed single hit kinetics in human embryonic kidney cells, but demonstrated a 2-particle requirement in simian cells (Butel, Rapp, Melnick & Rubin, 1966). The addition of non-adapted human adenovirus as helper virus in the monkey kidney cell assay system converted the kinetics of plaque formation by the 'adapted' virus to a one-particle requirement and also increased the titre of the virus (refer to Fig. 2).

Heterologous serotypes of human adenoviruses were used as helper viruses and the terminal plaques were found to consist antigenically of only the helper virus (Butel, Rapp, Melnick & Rubin, 1966; Butel &

Rapp, 1967). The progeny from these plaques were able to replicate in simian cells as well as if SV40 were present. The particle with which the helper adenovirus had interacted to initiate plaque formation was named the monkey cell-adapting component, or MAC (Butel, Rapp, Melnick & Rubin, 1966). MAC and the adenovirus replicated in unison during joint infection after a latent period of 16 hr. and attained maximum titres 40 hr. after infection (Butel & Rapp, 1967). MAC could be removed by successive plaque purifications of the carrier adenovirus populations in human cells. The derived progeny were then unable to replicate in monkey cells, or replicated poorly, unless SV40 was added to the cultures (Butel & Rapp, 1967).

Neutralization tests showed that MAC was encased in adenovirus capsids and could be transcapsidated from one adenovirus serotype to another (Butel, Rapp, Melnick & Rubin, 1966; Butel & Rapp, 1967). It could not replicate in the absence of helper adenovirus because it was dependent on the adenovirus for provision of capsid protein. In return, MAC complemented adenovirus by substituting for SV40 as helper virus for enhancement of adenovirus growth in simian cells.

Both MAC and adenovirus have a buoyant density of 1·34 g./ml. at equilibrium in caesium chloride (J. S. Butel & F. Rapp, unpublished observations). Therefore, this defective particle could not be physically separated from the helper adenovirions. MAC, like PARA, was found to be more sensitive to ultraviolet light than either adenovirus or SV40 (J. S. Butel & F. Rapp, unpublished observations).

Neither MAC-adenovirus 2 nor MAC-adenovirus 7 could confer transplantation immunity against SV40-transformed cells in weanling hamsters (Rapp, Butel, Tevethia & Melnick, 1967). These results eliminated the possibility that this SV40 marker is incorporated in the MAC genome. Studies in collaboration with Dr Sherman Weissman revealed that DNA extracted from MAC-infected GMK cells could not hybridize with SV40 complementary RNA. Therefore, there is no evidence that MAC is defective SV40.

Some properties of the two defective viruses known to be associated with human adenoviruses (PARA and MAC) are compared in Table 1. Even though they have many properties in common, as described above and in the preceding section, there is no direct evidence that MAC is derived from SV40. In view of the recent finding that simian adenoviruses can enhance the replication of human adenoviruses in monkey cells, it is possible that MAC is actually a defective simian adenovirus. There are no reagents available at this time to identify it as such.

To date, MAC has not been found to be oncogenic. It is important, however, as an example of a foreign defective genome carried by human adenoviruses. It demonstrates that such an association of a defective agent with adenoviruses is not specific for SV 40 determinants. It was detected only by its effect on the growth properties of the carrier adenovirus. At this time, no specific antigens have been detected to identify the origin of the monkey cell-adapting component.

Table 1. *Comparison of properties of defective viruses*
PARA (SV 40) and MAC

Property	PARA	MAC
Enhances human adenoviruses in GMK cells	+	+
Dependent on helper adenovirus for replication	+	+
Encased in adenovirus capsids	+	+
Can be transcapsidated to other adenoviruses	+	+
Removed by plaque purification in HEK cells	+	+
Buoyant density = 1·34 g./ml. in CsCl	+	+
More sensitive to u.v. than adenovirus or SV 40	+	+
Induces synthesis of SV 40 V antigen	0	0
Induces synthesis of SV 40 T antigen	+	0
Confers SV 40 transplantation immunity	+	0
DNA hybridizes *in vitro* with SV 40-specific RNA	+	0

GMK = green monkey kidney cells; HEK = human embryonic kidney cells; V antigen = virus (capsid) antigen; T antigen = tumour antigen; u.v. = ultraviolet light; + = possesses property; 0 = does not possess property.

SPECULATIONS

The examples of complementation between unrelated defective and complete genomes described in this chapter are perhaps the first of many such systems to be uncovered amongst DNA-containing animal viruses. Their importance at the present time lies in the fact that such interactions allow viruses to 'adapt' to cells in which they are otherwise unable to replicate. They also permit the propagation of defective viruses under conditions in which they would otherwise be lost. Though the examples presented here may represent laboratory artifacts, it is possible that similar phenomena occur under natural conditions.

The properties of SV 40 virus, human adenoviruses, and adenovirus populations carrying the defective genomes described in this chapter are summarized in Table 2. It is obvious that there are many differences in the systems under investigation. Thus, PARA carries the SV 40 determinants for the tumour and transplantation antigens and renders adenoviruses carrying those determinants oncogenic, while MAC does not carry markers readily recognizable as belonging to SV 40. Recognition of PARA or MAC defective viruses is based on the presence of SV 40 markers or a two-particle requirement for plaque formation in

Table 2. *Properties of simian virus 40, human adenovirus type 7, and adenovirus populations plus defective genomes (PARA, MAC, satellite)*

Property	Simian virus 40		Human adenovirus 7		PARA-adenovirus 7		MAC-adenovirus 7		Satellite + human adenovirus	
	GMK	HEK	GMK	HEK	GMK	HEK	GMK	HEK	GMK	HEK
Biological										
Synthesis of Ad T antigen	0	0	+	+	+	+	+	+	+	+
Synthesis of Ad DNA	0	0	+	+	+	+	+	+	+	+
Synthesis of Ad V antigen	0	0	0	+	+	+	+	+	0	+
Synthesis of Ad virions	0	0	0	+	+	+	+	+	0	+
Synthesis of SV 40 T antigen	+	Rare	0	0	+	0	0	0	0	0
Synthesis of SV 40 V antigen	+	Rare	0	0	0	0	0	0	0	0
SV 40 transplantation antigens	Positive		Negative		Positive		Negative		Negative	
Synthesis of SV 40 virions	+	Rare	0	0	0	0	0	0	0	0
SV 40 helper effect	0	0	+	0	0	0	0	0	+	0
No. of particles for plaque formation	1	—	—	1	2	1	2	1	—	1
Oncogenic	Strongly		Weakly		Strongly*		Weakly		Satellite is negative	
Biophysical										
Morphology	SV 40		Adeno		Adeno		Adeno		Satellite and adeno	
Coat protein	SV 40		Adeno		Adeno		Adeno		Satellite and adeno	
Buoyant density (CsCl) Particle	1·320 g./ml.		1·340 g./ml.		1·340 g./ml.		1·340 g./ml.		1·38–1·43 g./ml.†	
DNA	1·700 g./ml.		1·713 g./ml.		1·713 g./ml.		1·713 g./ml.		Varies	

* Some non-oncogenic adenoviruses are oncogenic when carrying PARA.   † Varies with satellite serotype. Adenovirus is 1·34 g./ml.

GMK = green monkey kidney cells; HEK = human embryonic kidney cells; T antigen = tumour antigen; V antigen = virus (capsid) antigen; Ad = adenovirus; DNA = deoxyribonucleic acid; + = positive; 0 = negative.

simian cells. However, the defective adeno-associated satellite viruses were first detected by electron microscopy and now the known antigenic types can be demonstrated by a variety of serological techniques. Of the defective determinants described, only the adeno-associated satellite virus can induce synthesis of its own coat protein; PARA and MAC are dependent upon the adenovirus for the provision of a capsid.

It is therefore obvious that different molecular events are involved in the enhancement and complementation of the viruses described. Unfortunately, at the present time, none of the systems has been completely characterized. The enhancement of human adenoviruses in simian cells by SV15 appears to involve a late step in the replication of SV15; the human adenovirus can apparently carry out the early steps (including synthesis of virus-specific DNA) without a helper virus. However, both PARA and MAC, which appear to be unable to cause synthesis of capsid protein (and therefore late steps in the replicative cycle) are effective helpers for the human adenovirus in the simian cells. It is therefore possible that the assistance provided by these various genomes is either different or involves a step in virus replication unknown at the present time. This may, for example, involve a transfer mechanism from the nucleus to the cytoplasm or the reverse. However, it is certain that it finally involves a step in translation of messenger RNA specifying for late proteins. The effect of these determinants may also be directly on the host cell, such as the derepression of a gene that allows the adenovirus to complete its cycle of replication. Presumably, such a gene is either not repressed in human cells or can be derepressed by the adenovirus in those cells.

The reason why the adeno-associated satellite viruses are defective is completely unknown at the present time. It is not even known whether they are able to induce synthesis of virus-specific DNA in the absence of the helper virus; the effect of the helper adenovirus for its defective satellite may, therefore, involve a step in DNA biosynthesis. This would be different from the other systems described here, since the human adenoviruses are capable of causing the synthesis of large amounts of their own DNA in simian cells.

Other lines of evidence strongly support the assumption that DNA-containing viruses that are oncogenic can transform cells only in systems in which the viruses undergo an abortive replicative cycle. This has been demonstrated for the papovaviruses as well as for the adenoviruses. Support for this is also obtained by the fact that PARA (which contains only defective SV40) can effectively transform cells *in vitro* and cause the induction of tumours when inoculated into newborn animals.

The replication of such defective determinants and their continued propagation in the presence of viruses that can complement their replication adds an important parameter to the study of viral oncogenesis. It is tempting to speculate that many of the oncogenic viruses are carrying such defective determinants and that it is these genomes that ultimately transform the cells. With most virus systems, large input multiplicities must be used to obtain a low frequency of transformation. It is quite possible that under these circumstances, only defective viruses are able to convert a normal cell into a cancer cell. An understanding of the molecular events underlying replication of these defective genomes, as well as the help they give in allowing other viruses to replicate, will ultimately be very useful in delineating their role in cell transformation. For example, PARA, which cannot induce synthesis of SV40 capsid proteins, can transform cells and this alone demonstrates conclusively that at least one late event in the cycle of replication of SV40 is not required for transformation. It would be of importance, for example, if a defective SV40 unable to induce the synthesis of the SV40 tumour antigen could be developed and still found to be oncogenic. In this way, the SV40 determinants necessary to transform cells could be delineated.

Although the studies cited in this chapter have made only moderate progress in characterizing the molecular events involved in complementation between unrelated DNA-containing viruses, the techniques available to the molecular biologist should resolve many of the questions posed by the discovery of the various systems described.

ACKNOWLEDGEMENTS

The work reported from the author's laboratory involves contributions made by many colleagues. Dr Joseph L. Melnick deserves credit for supplying many useful suggestions and for continued support of these studies. Dr Janet S. Butel has not only contributed many critical ideas and experiments, but has also helped in the preparation of this article. For this, I am especially grateful.

The original work cited from the author's laboratory was supported in part by Public Health Service research grants CA-04600 and CA-10036 from the National Cancer Institute, research grant AI-05382, and training grant 5T1AI74 from the National Institute of Allergy and Infectious Diseases, National Institutes of Health.

REFERENCES

ARCHETTI, I., BERECZKY, E. & BOCCIARELLI, D. S. (1966). A small virus associated with the simian adenovirus SV 11. *Virology*, **29**, 671.

ATCHISON, R. W., CASTO, B. C. & HAMMON, W. McD. (1965). Adenovirus-associated defective virus particles. *Science, N.Y.* **149**, 754.

ATCHISON, R. W., CASTO, B. C. & HAMMON, W. McD. (1966). Electron microscopy of adenovirus-associated virus (AAV) in cell cultures. *Virology*, **29**, 353.

BAUM, S. G., REICH, P. R., HYBNER, C. J., ROWE, W. P. & WEISSMAN, S. M. (1966). Biophysical evidence for linkage of adenovirus and SV40 DNA's in adenovirus 7-SV40 hybrid particles. *Proc. natn. Acad. Sci. U.S.A.* **56**, 1509.

BLACK, P. H. & TODARO, G. J. (1965). *In vitro* transformation of hamster and human cells with the adeno 7-SV40 hybrid virus. *Proc. natn. Acad. Sci. U.S.A.* **54**, 374.

BLACK, P. H. & WHITE, B. J. (1967). *In vitro* transformation by the adenovirus-SV40 hybrid viruses. II. Characteristics of the transformation of hamster cells by the adeno 2-, adeno 3-, and adeno 12-SV40 viruses. *J. exp. Med.* **125**, 629.

BLACKLOW, N. R., HOGGAN, M. D. & ROWE, W. P. (1967). Immunofluorescent studies of the potentiation of an adenovirus-associated virus by adenovirus 7. *J. exp. Med.* **125**, 755.

BOEYÉ, A., MELNICK, J. L. & RAPP, F. (1965). Adenovirus-SV40 'hybrids': Plaque purification into lines in which the determinant for the SV40 tumor antigen is lost or retained. *Virology*, **26**, 511.

BOEYÉ, A., MELNICK, J. L. & RAPP, F. (1966). SV40-adenovirus 'hybrids': presence of two genotypes and the requirement of their complementation for viral replication. *Virology*, **28**, 56.

BUTEL, J. S., MELNICK, J. L. & RAPP, F. (1966). Detection of biologically active adenovirions unable to plaque in human cells. *J. Bact.* **92**, 433.

BUTEL, J. S. & RAPP, F. (1966a). Replication in simian cells of defective viruses in an SV40-adenovirus 'hybrid' population. *J. Bact.* **91**, 278.

BUTEL, J. S. & RAPP, F. (1966b). Inactivation and density studies with PARA-adenovirus 7 (SV40-adenovirus 'hybrid' population). *J. Immun.* **97**, 546.

BUTEL, J. S. & RAPP, F. (1967). Complementation between a defective monkey cell-adapting component and human adenovirus in simian cells. *Virology*, **31**, 573.

BUTEL, J. S., RAPP, F., MELNICK, J. L. & RUBIN, B. A. (1966). Replication of adenovirus type 7 in monkey cells: a new determinant and its transfer to adenovirus type 2. *Science, N.Y.* **154**, 671.

CASTO, B. C., ATCHISON, R. W. & HAMMON, W. McD. (1967). Studies on the relationship between adeno-associated virus type 1 (AAV-1) and adenoviruses. I. Replication of AAV-1 in certain cell cultures and its effect on helper adenovirus. *Virology*, **32**, 52.

CHANY, C. & BRAILOVSKY, C. (1967). Stimulating interaction between viruses (stimulons). *Proc. natn. Acad. Sci. U.S.A.* **57**, 87.

FELDMAN, L. A., BUTEL, J. S. & RAPP, F. (1966). Interaction of a simian papovavirus and adenoviruses. I. Induction of adenovirus tumor antigen during abortive infection of simian cells. *J. Bact.* **91**, 813.

FENNER, F. & SAMBROOK, J. F. (1964). The genetics of animal viruses. *Ann. Rev. Microbiol.* **18**, 47.

FUJINAGA, K. & GREEN, M. (1966). The mechanism of viral carcinogenesis by DNA mammalian viruses: Viral-specific RNA in polyribosomes of adenovirus tumor and transformed cells. *Proc. natn. Acad. Sci. U.S.A.* **55**, 1567.

HERMODSSON, S. (1963). Inhibition of interferon by an infection with parainfluenza virus type 3 (PIV-3). Virology, **20**, 333.

HOGGAN, M. D., BLACKLOW, N. R. & ROWE, W. P. (1966). Studies of small DNA viruses found in various adenovirus preparations: Physical, biological, and immunological characteristics. *Proc. natn. Acad. Sci. U.S.A.* **55**, 1467.

HUEBNER, R. J., CHANOCK, R. M., RUBIN, B. A. & CASEY, M. J. (1964). Induction by adenovirus type 7 of tumors in hamsters having the antigenic characteristics of SV40 virus. *Proc. natn. Acad. Sci. U.S.A.* **52**, 1333.

IGEL, H. J. & BLACK, P. H. (1967). *In vitro* transformation by the adenovirus-SV40 hybrid viruses. III. Morphology of tumors induced with transformed cells. *J. exp. Med.* **125**, 647.

292   F. RAPP

ITO, M., MELNICK, J. L. & MAYOR, H. D. (1967). An immunofluorescence assay for studying replication of adeno-satellite virus. *J. gen. Virol.* **1**, 199.

JERKOFSKY, M. A. & RAPP, F. (1967). Differences in the abortive cycle of various human adenoviruses in simian cells. *Bact. Proc.* p. 137.

MALMGREN, R. A., RABSON, A. S., CARNEY, P. G. & PAUL, F. J. (1966). Immunofluorescence of green monkey kidney cells infected with adenovirus 12 and with adenovirus 12 plus simian virus 40. *J. Bact.* **91**, 262.

MAYOR, H. D., JAMISON, R. M., JORDAN, L. E. & MELNICK, J. L. (1965). Structure and composition of a small particle prepared from a simian adenovirus. *J. Bact.* **90**, 235.

MELNICK, J. L., MAYOR, H. D., SMITH, K. O. & RAPP, F. (1965). Association of 20-millimicron particles with adenoviruses. *J. Bact.* **90**, 271.

MELNICK, J. L. & PARKS, W. P. (1966). Identification of multiple defective and noncytopathic viruses in tissue culture. In *Relazione al VI Cong. Intern'l. di Patol. Clinica*, Rome, October, 1966.

NAEGELE, R. F. & RAPP, F. (1967). Enhancement of the replication of human adenoviruses in simian cells by simian adenovirus SV15. *J. Virol.* **1**, 838.

O'CONOR, G. T., RABSON, A. S., BEREZESKY, I. K. & PAUL, F. J. (1963). Mixed infection with simian virus 40 and adenovirus 12. *J. natn. Cancer Inst.* **31**, 903.

PARKS, W. P., GREEN, M., PIÑA, M. & MELNICK, J. L. (1967). Physicochemical characterization of adeno-associated satellite virus type 4 and its nucleic acid. *J. Virol.* **1**, 980.

PARKS, W. P., MELNICK, J. L., RONGEY, R. & MAYOR, H. D. (1967). Physical assay and growth cycle studies of a defective adeno-satellite virus. *J. Virol.* **1**, 171.

RAPP, F., BUTEL, J. S. & MELNICK, J. L. (1965). SV40-adenovirus 'hybrid' populations: transfer of SV40 determinants from one type of adenovirus to another. *Proc. natn. Acad. Sci. U.S.A.* **54**, 717.

RAPP, F., BUTEL, J. S., TEVETHIA, S. S., KATZ, M. & MELNICK, J. L. (1966). Antigenic analysis of tumors and sera from animals inoculated with PARA-adenovirus populations. *J. Immun.* **97**, 833.

RAPP, F., BUTEL, J. S., TEVETHIA, S. S. & MELNICK, J. L. (1967). Comparison of ability of defective foreign genomes (PARA and MAC) carried by human adenoviruses to induce SV40 transplantation immunity. *J. Immun.* **99**, 386.

RAPP, F., FELDMAN, L. A. & MANDEL, M. (1966). Synthesis of virus deoxyribonucleic acid during abortive infection of simian cells by human adenoviruses. *J. Bact.* **92**, 931.

RAPP, F. & JERKOFSKY, M. (1967). Replication of PARA (defective SV40)-adenoviruses in simian cells. *J. gen. Virol.* **1**, 311.

RAPP, F. & KHARE, G. P. (1967). DNA biosynthesis in monkey kidney cells infected with PARA (SV40)-adenoviruses. *Proc. Soc. exp. Biol. Med.* **126** (in the Press).

RAPP, F. & MELNICK, J. L. (1966). Papovavirus SV40, adenovirus and their hybrids: transformation, complementation, and transcapsidation. *Prog. med. Virol.* **8**, 349.

RAPP, F., MELNICK, J. L., BUTEL, J. S. & KITAHARA, T. (1964). The incorporation of SV40 genetic material into adenovirus 7 as measured by intranuclear synthesis of SV40 tumor antigen. *Proc. natn. Acad. Sci. U.S.A.* **52**, 1348.

RAPP, F., MELNICK, J. L. & LEVY, B. (1967). Correlation of immunology and histopathology of tumors induced by defective SV40-adenovirus hybrids. *Am. J. Path.* **50**, 849.

RAPP, F., TEVETHIA, S. S. & MELNICK, J. L. (1966). Papovavirus SV40 transplantation immunity conferred by an adenovirus-SV40 hybrid. *J. natn. Cancer Inst.* **36**, 703.

PLATE I

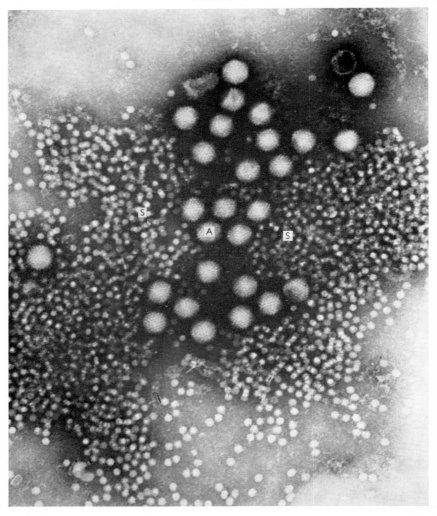

ROIZMAN, B. & AURELIAN, L. (1965). Abortive infection of canine cells by herpes simplex virus. I. Characterization of viral progeny from co-operative infection with mutants differing in capacity to multiply in canine cells. *J. molec. Biol.* **11**, 528.

ROSE, J. A., HOGGAN, M. D. & SHATKIN, A. J. (1966). Nucleic acid from an adeno-associated virus: Chemical and physical studies. *Proc. natn. Acad. Sci. U.S.A.* **56**, 86.

ROWE, W. P. (1965). Studies of adenovirus-SV40 hybrid viruses. III. Transfer of SV40 gene between adenovirus types. *Proc. natn. Acad. Sci. U.S.A.* **54**, 711.

ROWE, W. P. & BAUM, S. G. (1964). Evidence for a possible genetic hybrid between adenovirus type 7 and SV40 viruses. *Proc. natn. Acad. Sci. U.S.A.* **52**, 1340.

ROWE, W. P. & BAUM, S. G. (1965). Studies of adenovirus SV40 hybrid viruses. II. Defectiveness of the hybrid particles. *J. exp. Med.* **122**, 955.

ROWE, W. P., BAUM, S. G., PUGH, W. E. & HOGGAN, M. D. (1965). Studies of adenovirus SV40 hybrid viruses. I. Assay system and further evidence for hybridization. *J. exp. Med.* **122**, 943.

ROWE, W. P. & PUGH, W. E. (1966). Studies of adenovirus-SV40 hybrid viruses. V. Evidence for linkage between adenovirus and SV40 genetic materials. *Proc. natn. Acad. Sci. U.S.A.* **55**, 1126.

SMITH, K. O., GEHLE, W. D. & THIEL, J. F. (1966). Properties of a small virus associated with adenovirus type 4. *J. Immun.* **97**, 754.

VASQUEZ, C. & TOURNIER, P. (1964). New interpretation of the reovirus structure. *Virology*, **24**, 128.

EXPLANATION OF PLATE 1

Electron photomicrograph of SV15 simian adenovirus (A) and adeno-associated satellite virus type 4 (S). Similar to photographs in Parks, Melnick, Rongey & Mayor (1967). Magnification, × 150,000.

HOST-CONTROLLED RESTRICTION AND MODIFICATION OF BACTERIOPHAGE

W. ARBER

Laboratoire de Biophysique, Université de Genève, Switzerland

INTRODUCTION

Viral properties have frequently been reported to undergo non-heritable changes upon passage through certain host strains. But very little was known about the molecular mechanisms forming the basis of such host-controlled modifications, when we started in 1960 to investigate one particular virus–host system. This work, carried out with bacteriophage λ and a few host strains all derived from *Escherichia coli*, soon revealed the existence of a highly specific recognition mechanism able to screen infecting and intracellular DNA molecules for absence or presence on these molecules of a host-specific stamp. This has more recently been identified as nucleotide methylation. It is the scope of my contribution to this symposium to lay out the crucial experiments and arguments that lead to the present understanding of this system, which may be interpreted as serving the cell as a defence mechanism against infection with foreign genetic material.

Definitions and notations

Since it will not be possible to present here all the various aspects of host-controlled modification of bacteriophage, I shall confine myself to discussing only some experiments carried out with the well-known bacteriophage λ and with the class of small, filamentous, male-specific phages that include phages fd (Hoffmann-Berling, Marvin & Dürwald, 1963), f1 (Zinder, Valentine, Roger & Stoeckenius, 1963), M13 (Hofschneider, 1963) and F12 (see Arber, 1966).

The bacterial host strains will be classified here according to the type of specific modification that they provide to the DNA: strains giving the modification typical for *E. coli* K12 will be denoted K, strains giving the same modification as *E. coli* B will be denoted B, strains giving no detectable modification (e.g. *E. coli* C) will be denoted O. The genetic determinants of K and B host specificity are located on the bacterial chromosome linked to the threonine markers (Boyer, 1964; Colson, Glover, Symonds & Stacey, 1965; Wood, 1966). Two independent episomic elements are known to carry their own system of host specificity determinants: phage P1 (Lederberg, 1957; Arber & Dussoix, 1962) and a resistance transfer factor, called RTF-2 (Arber & Morse, 1965). Combinations of one or both of these episomes with the strains O, K or B yield strains with one, two or three different host specificity systems. A P1-lysogenic strain K(P1), for example, provides K- as well as P1-specific modification. Phage λ grown in such a strain is called λ.K(P1). The notation of the last host, together with the symbol of the phage and separated from it by a dot, usually denotes the type of host-specific modification carried by a phage stock. A direct check of host specificity is provided by the determination of the efficiency of plating (e.o.p.) of a given phage stock on a series of different host strains.

Table 1. *Efficiency of plating (e.o.p.) of λ variants on different host strains of* E. coli

Phage	\multicolumn{6}{c}{Host}
	O	K	B	O (RTF-2)	O (P1)	K (P1)
λ.O	1	4×10^{-4}	10^{-4}	10^{-2}	2×10^{-5}	10^{-7}
λ.K	1	1	10^{-4}	10^{-2}	2×10^{-5}	2×10^{-5}
λ.B	1	4×10^{-4}	1	10^{-3}	2×10^{-5}	10^{-7}
λ.O (RTF-2)	1	4×10^{-4}	10^{-4}	1	2×10^{-5}	10^{-7}
λ.O (P1)	1	4×10^{-4}	10^{-4}	10^{-2}	1	4×10^{-4}
λ.K (P1)	1	1	10^{-4}	10^{-2}	1	1

Experimental e.o.p. values of phage λ obtained under standardized conditions (Arber & Dussoix, 1962) are given in Table 1. One notes that in general the e.o.p. is one, if a phage variant is assayed on bacteria that had already served for its growth before. Under such conditions practically every phage particle gives rise to the formation of a plaque. Plating of phage on most other bacterial strains gives low but still quite characteristic e.o.p. values. In these latter cases, the phage is said to be restricted. A somewhat more subtle definition will become obvious in the presentation of the experimental results. Host strains requiring more than one type of modification on the infecting phage display an increased factor of restriction, as is illustrated also in Table 1 with host strain K(P1): it plates λ.O with a probability of 10^{-7} only as compared to the e.o.p. of 2×10^{-5} for λ.K and of 4×10^{-4} for λ.O(P1). Strain O is characterized as restricting none of the phage variants involved, while it does not provide the phage with any of the host specificities tested, i.e. phage λ.O is restricted in all strains except O.

DNA AS CARRIER OF HOST SPECIFICITY

Breakdown of restricted DNA

Whenever an infecting phage λ is restricted a majority of the infected cells does not succeed in producing phage progeny. Dussoix & Arber (1962) showed that this failure is explained by an efficient control mechanism that enables the host to degrade most of the restricted DNA molecules upon or quite soon after their injection. We used ^{32}P-labelled λ phage for infection of either non-restricting or restricting host bacteria. Phage adsorbed efficiently on both types of hosts. But from the moment when DNA penetration occurred the following striking difference was observed: the complexes formed by the restricting host cells started to yield a rapidly increasing amount of acid soluble ^{32}P label, while no such breakdown products appeared with the non-restricting host-virus complexes. Despite this abortive infection the restricting host cells were not killed and did not even show any delay in their growth.

Restriction is never complete as shown indeed by Table 1: a low and for each system quite characteristic proportion of infected cells seems to be leaky in the restriction and allows the phage to reproduce. If part or all of progeny phages are also modified, the formation of a phage plaque on the restricting indicator is ensured. The probability of such leakiness is somewhat dependent on physiological conditions (Bertani & Weigle, 1953) as well as on the multiplicity of infection (Paigen & Weinfeld, 1963).

It seems thus that restricting cells have a remarkably efficient mechanism to check the host specificity of infecting phage and if the right modification is absent the phage DNA is degraded. We still do not know the sequence of processes leading to this breakdown. One of the most attractive ideas is that a very specific restriction enzyme transforms the DNA in such a way as to make it accessible to exonucleases that are present in *E. coli* cells. The enzymic apparatus involved in restriction is not induced by the phage infection: conditions that inhibit protein synthesis do not inhibit breakdown of restricted phage DNA (Dussoix, 1964). Restriction is not dependent on initiation of replication of the infecting DNA, as shown for example by the breakdown of restricted λ DNA upon infection of immune, λ-lysogenic strains (Lederberg & Meselson, 1964). Various lines of evidence indicate that the decisive act in restriction occurs rather early in the infection process: most of the DNA breakdown products appear within a few minutes after DNA injection; rescue of markers from restricted phage genomes by superinfecting, non-restricted phage rapidly loses its efficiency (Dussoix & Arber, 1962); and finally, restriction is overcome in co-operative infection within the first 3 min. following phage penetration (Weinfeld & Paigen, 1964). It has not yet been possible to get solid information as to the location in the cell where restriction is exerted. Does it act on the cell surface like a border patrol while DNA penetrates into the cell or does it randomly check on DNA molecules in the cytoplasm? The first of these mechanisms seems to better explain certain observations, but none of these could really prove it.

Non-hereditary nature of the modification

Selection of modified phage growing on a restricting host has operationally much in common with selection of host range mutants from a phage strain; that is phages which do not grow on a bacterial plating strain. Host-modified phage, however, is easily distinguished from a host range mutant by its behaviour upon replating on the original, non-modifying host. The host range mutant maintains its capacity to grow on the second host strain, but the host-modified phage does not. Hence modification is not hereditary. Still, the following series of experiments prove that it is the DNA molecule that undergoes the host-controlled modification and stably carries this characteristic.

Transfer of parental host specificity jointly with the
phage DNA molecule in one cycle growth

Arber & Dussoix (1962) grew heavy, density-labelled phage λ.K(P1) for one single cycle on non-lysogenic K cells in non-labelled medium. We then centrifuged the phage progeny to equilibrium in a CsCl density gradient, collected density fractions and assayed each fraction for phage plaques on K and K(P1) indicator bacteria (Fig. 1). Most phages of course were non-labelled and they only grew on K but not on K(P1): the K host had not provided them with the P1-specific modification nor did they inherit it from the infecting λ.K(P1) parents. A very definite and sharp peak of phage growing on K(P1) appeared, however, in a region corresponding to the buoyant density of phage built with new, unlabelled protein coats and semi-conserved DNA (one parental, density-labelled strand associated with a newly synthesized, unlabelled strand). This finding showed that the P1-specific modification of the parental λ.K(P1) phage is intimately associated with the DNA molecule. The host specificity characteristic obviously does not interfere with the replication of DNA, and upon intracellular growth of the phage, only phages that inherit parental DNA will also inherit parental host

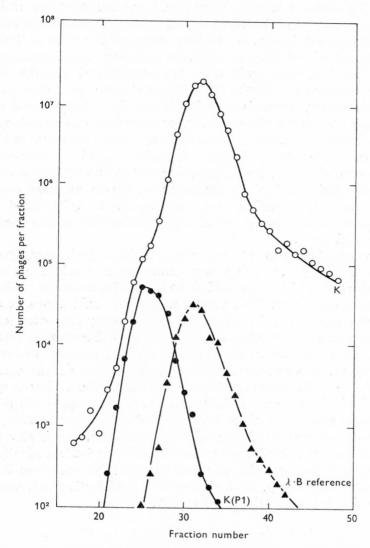

Fig. 1. CsCl density gradient distribution of one cycle lysate obtained by growth of heavy (deuterated) λ.K (P 1) on strain K after infection at a multiplicity of 0·019 phages per cell. The tube was spun for 16 hr. at 22,000 rev./min. in swinging bucket SW 39 of Spinco model L ultracentrifuge. A total number of 82 fractions was collected, which were then assayed on K (P 1) (●) and K (○). Phage λ.B had been added to the tube as density reference (▲). From Arber & Dussoix (1962), by permission of Academic Press.

specificity. Another labelling technique, incorporation of ^{32}P at high specific activity into the parental DNA, causing phage inactivation by the radioactive disintegration, entirely confirmed the result of the first experiment (Arber & Dussoix, 1962). We then grew λ. K on strain O and found that K-specific modification too is associated with the DNA (Arber, Hattman & Dussoix, 1963). This and more recent experiments by Kellenberger, Symonds & Arber (1966) show that parental host specificity can be transfered serially for more than one growth cycle, always jointly with parental DNA. But it appeared that only about half of the one cycle progeny phages with semi-conserved DNA molecules still grow on the first, parental host, the other half do not overcome the restriction and their DNA is broken down. We do not know how to explain this situation and can only speculate that it may be related with the mechanism of DNA penetration into the restricting cells (Kellenberger *et al.* 1966).

A somewhat more complex experiment of this series is reproduced in Figure 2. Heavy λ.K phage was grown on a restriction-deficient derivative of B which still provided the DNA with B-specific modification. This experiment illustrates what has been said above. It also illustrates that after multiple infection some progeny phages are found to carry a fully conserved, heavy DNA molecule (Meselson & Weigle, 1961). Such phages are accepted with full efficiency upon re-infection of the first host. Another interesting observation from the data in Figure 2 is related to the imprinting of host specific modification. The one cycle progeny of λ. K on $Br^-m_B^+$ is fully modified: not only phages with newly synthesized and with semi-conserved DNA display B-modification, but also those with conserved parental DNA molecules. This implies that the modifying reaction is independent of DNA replication. Finally, in undergoing B-specific modification, the parental DNA molecules do not lose their K-specificity. K- and B-specific modifications are exerted independently of each other on the DNA.

The distribution of host specificity sites on the DNA

Since a single DNA molecule can carry more than one type of host specificity, it is likely that specific sites exist on the DNA molecules and that modification is exerted by the host on these sites. On the other hand, it is most likely that such sites also form the recognition loci in restriction reactions. Only non-modified DNA is submitted to restriction, and again, two independent restrictions seem to be exerted independently of each other on the DNA molecules, since their effects are additive as shown in Table 1.

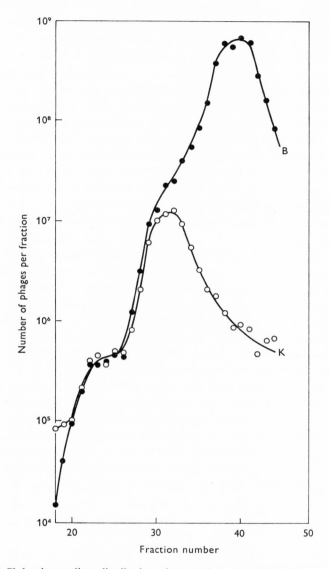

Fig. 2. CsCl density gradient distribution of one cycle lysate obtained by growth of heavy (deuterium and ^{15}N) λ.K on a restriction deficient strain Br$^-m_B^+$ after infection at a multiplicity of 4 phages per cell. Same conditions of centrifugation as in Fig. 1. Eighty-eight fractions were collected and assayed on K (○) and B (●) indicator bacteria. Fractions 22 to 25 = phage with conserved DNA; fractions 30 to 33 = phage with semiconserved DNA; fractions 38 to 41 = phage with newly synthesized DNA. From Kellenberger, Symonds & Arber (1966), by permission of Springer-Verlag.

Evidence for the existence of specificity sites was looked for in the one cycle growth experiments discussed above. Upon intracellular growth, phage λ undergoes some genetic recombination. If such an exchange takes place between a DNA molecule carrying one of the parental, density labelled strands and a light progeny molecule, the resulting recombinant will contain parental nucleotides only on a fraction of the DNA filament. Phage particles with such DNA molecules have a buoyant density intermediate between that of phages with semi-conserved DNA and that of phages with entirely light DNA. Density gradient analysis gave, however, no evidence that such recombinants contained also parental host specificity. It is then possible that there is more than one specificity site per DNA molecule and that these sites are located at different places on the phage genome.

A somewhat more sensitive method to answer this question is to use genetic markers. The experiment is then designed according to the following scheme: strain O is infected simultaneously with λ.O and with λ.K that carries a genetic marker a. The one cycle progeny is then tested for phages with parental K-specificity having acquired the marker a^+ from the λ.O parental genotype. If such recombinants are found, one likes to conclude that the site of marker a^+ is not also a site of K-specific modification and restriction. Experiments were carried out involving various genetic markers and various host specificities (Arber & Dussoix, 1962; Ihler & Meselson, 1963; M. Fluck & W. Arber, unpublished results). They indicated that certain regions of the genome could be outcrossed from the modified parent without loss of the parental modification, while others could not. It also became obvious that the extent of regions non-essential for maintenance of parental host specificity is rather limited. This permits the prediction that the number of specificity sites carried by the DNA molecule must be higher than one for all of the host specificities analysed. A precise estimation of this number, however, was not possible. We will take up this point again later in the discussion of experiments with phage fd.

Host specificity of purified DNA

The high degree of stability in the association of host specificity with the phage DNA molecule is best demonstrated with purified DNA. Dussoix & Arber (1965) extracted DNA with phenol from a number of λ phage stocks carrying various types of host specificity. We then mixed these DNA preparations with helper infected cells of a number of different host strains. Successful infection was defined by the formation of phage plaques upon plating of the host-DNA complexes with appropriate indicator bacteria. The results were quite clear: the patterns of restriction given in Table 1 for infection with phage particles were also found to hold true in the infection with purified DNA. The DNA extracted from $10^7\lambda$. K(P1) phages, for example, gave 12,700 successful infections on K and 11,000 on K(P1), while the same amount of λ. K DNA initiated 7400 infections on K, but only 76 on the restricting host K(P1). We also carried out infections with density labelled DNA and showed in density gradient analysis of the one cycle progeny phage particles that, indeed, the parental DNA molecules, although deproteinized and purified, still had carried P1-specific modification.

Conclusions

We have seen, so far, that host-specific modification is acquired by the phage in a process independent of DNA replication. This reaction produces a non-hereditary alteration on the phage DNA molecule, which, incidentally does not noticeably change any of a series of tested physico-chemical properties of the DNA. The association between host-specific modification and the DNA molecule is remarkably stable, surviving DNA extraction and purification as well as intracellular DNA replication, which is not affected by the presence of host specificity. Restriction is exerted on all those phages that do not carry specific modification. DNA breakdown results from such restrictions. It is then likely that modification and restriction affect both the same specific sites on the DNA molecule. By modification these sites would become inert against the restriction. A limited number of specificity sites seem to be distributed along the DNA molecule.

THE NATURE OF THE MODIFICATION AND RESTRICTION PROCESSES

In the last few years various lines of approach have been used in the attempt to get more insight into the processes underlying host-controlled modification and restriction. One approach resides in the study of the genetic determinants that control these reactions in the host cells. Another line is the search for *in vitro* modifying and restricting activity. Finally the product of the enzymic modification was looked for on the DNA by biochemical methods, and the number and position of specificity sites were determined on genomes of phages with small DNA content.

Genetic determinants for restriction and modification functions

We have mentioned in the introduction that genetic determinants for functions involved in restriction and modification by strains K and B are located on the bacterial chromosome linked to the threonine locus. This conclusion was reached by several authors, who worked with genetic hybrids between strains K and B (Boyer, 1964; Hoekstra & De Haan, 1965) or with restriction deficient mutants of these strains (Wood, 1965, 1966; Colson *et al.* 1965; Lederberg, 1966*a*).

The selection applied by Wood (1966) in the isolation of restriction deficient (r^-) mutants is based on the observation, that restriction and modification are exerted also on transducing phage λdg (Arber, 1964): K gal^- receptor bacteria were infected with restricted λdg. B. A few gal^+ transductants were obtained and many of them were indeed r^- mutants, particularly if the recipient cells had previously been mutagenized. A most interesting aspect appeared upon further characterization of these mutants.

Roughly half of them still gave a completely normal K-specific modification. A second class of mutants, occurring at about the same frequency as the first, was defective not only in restriction, but also in modification of λ. They have been called r^-m^- and are, according to our definition of strains, type O. But in view of their high rate of occurrence it is unlikely that they are double mutants. Various other explanations were proposed and we will come back to one of them later. A minority of mutants finally were of intermediate types, giving either no restriction combined with partial modification or partial restriction combined with complete modification. The phenotypic response of some of them turned out to be temperature dependent.

Some of these mutants were then used in genetic mapping experiments involving bacterial conjugation or phage mediated transduction. All mutants from both major classes as well as those giving intermediate responses map, as said before, close to the threonine locus. A fine structure map of the mutants is unfortunately not yet available. As far as we know, a systematic search for restricting revertants has not been undertaken yet, nor an attempt to isolate directly non-modifying mutants from a wild-type strain. Research in these lines might help to increase our knowledge of the genetic information that cells possess for controlling the restriction and modification processes.

Range of action of restriction and modification processes

We mentioned before that restriction is not observed only with phage λ. Indeed, it soon became clear that neither the actions of restriction nor those of modification are limited to λ DNA. Many other phages are subject to the same host specificity systems as λ. Last but not least, bacterial DNA itself is a substrate for these actions (Arber, 1962). This became obvious in interstrain crosses: in bacterial conjugation, for example, a mixture of Hfr K donor cells with $F^-K(P1)$ receptors is 100- to 1000-fold less fertile than any of the three other combinations Hfr $K \times F^-K$, Hfr $K(P1) \times F^-K$ and Hfr $K(P1) \times F^-K(P1)$ (Arber & Morse, 1965). Further conjugation studies involving strains with all of the host specificities mentioned above confirmed that high degrees of fertility are in general obtained only with recipient strains which do not restrict λ phage grown on the donor strain used in the conjugation. Secondly, transferred genetic characters appear unlinked in conjugation with restricting strains. Both these manifestations are presumably caused by a fragmentation of the infecting segment from the donor genome under the influence of the restriction system of the recipient strain (Pittard, 1964; Boyer, 1964; Wood, 1966). Restriction also reduces the rate of acceptance of bacterial characters on episome mediated transfer involving F' fertility factors, resistance transfer factors or transducing phages (Arber, 1964; Arber & Morse, 1965; Glover, Schell, Symonds & Stacey, 1963; Wood, 1966; Inselburg, 1966). The fundamental importance of restriction in all processes involving exchange of genetic material between different strains does not need to be emphasized further.

Restriction and modification of small bacteriophages

A number of different bacteriophage strains with small nucleic acid content have been tested for restriction by the K, the B, the P1 or the RTF-2 host specificity systems. None of these systems restricted any of three different RNA-phage strains (M. Fluck & W. Arber, unpublished results). Among the small phages with single stranded DNA molecules, $\phi X 174$ is not restricted by K or B (B. Schnegg & P. H. Hofschneider, personal communication), while the filamentous, male-specific phage fd was found by H. Hoffmann-Berling (personal communication) to be restricted and to undergo host-controlled modification in B strains (Table 2). The question then arose if the restriction and modification of fd in B is under the control of the same genes as B-specific restriction and modification of λ. This is indeed the case: transduction hybrids between K and B, whose chromosomes have only the small segment with threonine and host specificity markers from B but are otherwise K, also restrict fd (Table 2). Hybrids with mutant character $r^-m_B^+$, which no longer restrict λ, are also deficient in restriction of fd, while they produce B-modified progeny of both phages (Arber, 1966).

Table 2. *Efficiency of plating of fd and some related phages*
on male derivatives of different strains of E. coli

		Host		
		Strains with B-restriction and modification		
			Transduction hybrid	
Phage	K	B	$Kr_B^+m_B^+$	O (P 1)
fd.K	1	4×10^{-4}	7×10^{-4}	0·3
fd.B	1	1	1	0·3
fd.O (P 1)	1	4×10^{-4}	7×10^{-4}	1
f1.K	1	3×10^{-4}	6×10^{-4}	0·3
f1.B	1	1	1	0·3
f1.O (P 1)	1	3×10^{-4}	6×10^{-4}	1
M 13.K	1	$1·7 \times 10^{-2}$	$2·4 \times 10^{-2}$	0·3
M 13.B	1	1	1	0·3
F 12.K	1	$1·6 \times 10^{-2}$	$2·2 \times 10^{-2}$	1
F 12.B	1	1	1	1

Upon infection of a restricting host B the DNA of non-modified fd is broken down, as we have described for λ. After growth of fd.B on strain O, some progeny phage particles still display the parental B-specific modification, and again as found with λ, such phage particles contain parental DNA as shown by the density labelling techniques. B-specificity does not seem to dissociate from fd DNA upon purification, as evidenced by infectivity assays on spheroplasts. There is no doubt that modification and restriction functions are exerted on DNA molecules of the phage fd (Arber, 1966). It remains unanswered, however, if these functions are exerted on the single-stranded DNA or rather while the phage goes through its replicative states.

Correlation between enzymic methylation and host-controlled modification

In the search for the biochemical nature of host specificity carried by the DNA, special attention was soon given to one hypothesis, which seemed to meet all requirements dictated by the experimental findings exposed above. This hypothesis postulated that enzymic methylation of DNA bases is responsible for host-specific modification. DNA-methylating enzymes have been isolated and purified from extracts of

E. coli and other bacterial strains (Gold & Hurwitz, 1963, 1964*a*, *b*; Fujimoto, Srinivasan & Borek, 1965). Their study confirmed that enzymic methylation acts on the macromolecular DNA and that S-adenosyl-methionine serves as methyl donor. The enzymes display a rather strict species specificity in the sense that a purified enzyme preparation usually methylates only DNA from other sources, but is inactive on DNA extracted from the same bacterial strain. Supposedly all potential methylation sites on homospecific DNA are already occupied before the extraction. It was then not unreasonable to postulate that these or similar methylating enzymes could produce the host-specific modification.

With this hypothesis in mind, we tried to grow phage in absence of a methyl donor needed for the enzymic reaction (Arber, 1965*a*). A λ-lysogenic strain K(λ) *met⁻pro⁻*, requiring the amino acids methionine and proline, was irradiated with an optimal u.v. induction dose for initiation of vegetative phage production. After the first 20 min. of incubation in complete medium, one of the required amino acids was deleted and the cells then further incubated for 50–100 min. During this period DNA was synthesized at least to some extent, but protein synthesis of course was stopped. As to the enzymic methylation of nucleic acids, we supposed that enough enzyme was still in the cells but that it could only act if methionine was given to the medium, while this reaction should have been independent of presence or absence of proline. Finally, towards the end of the growth period we restored the complete medium by re-addition of methionine or proline, respectively. Protein synthesis was thus restored and progeny phage particles could mature. Plating of this progeny on both O and K indicator bacteria showed how many active phages were produced and how many also had obtained the K-specific modification. When methionine had been removed from the medium we found up to 400 times more plaque formers on O than on K. When proline had been removed, exactly equal numbers of phages were obtained on both O and K indicators. Hence we found our expectation confirmed that DNA grown in absence of methionine is incompletely supplied with host-specific modification, while other amino acids (a few others in addition to proline were tested also) did not produce a similar effect. If this experiment does not prove the methylation hypothesis—it only showed that methionine is involved in an essential step in the host-specific modification—it at least encouraged us to undertake more direct, biochemical approaches.

The most straightforward experiment was the determination of the extent of methylation found in the DNA of various strains, either bacterial DNA or DNA from phage grown in these strains. At first these experiments did not give the expected answer. Perhaps the most striking in the series of negative answers was obtained, when we compared the extent of methylation of DNA from K with that of DNA from non-modifying (r^-m^-) mutants of K: within the error of the analysis, which was about 10 %, both DNA preparations showed the same extent of methylation, some 0·35 mole % of 6-methyl-aminopurine and some 0·2 mole % of 5-methylcytosine (J. D. Smith, W. B. Wood & W. Arber, unpublished results). DNA from strain B has no measurable amount of 5-methylcytosine, while the level of 6-methylaminopurine corresponds to that of K. But it is quite obvious that this difference between methylation encountered in K and B is not correlated with host-specific modification, first, since r^-m^- mutants of B give the same extent of methylation as B, and secondly since genetic hybrids between B and K which contain the region with host specificity markers $r_K m_K$ from K, still do not produce any 5-methylcytosine. The extent of methylation of phage λ DNA seemed to be somewhat lower than that of bacterial DNA, but still no correlation appeared between methylation and host specificity type carried by the phage DNA. Other authors obtained qualitatively the same results independently of ours (Gough & Lederberg, 1966; Lederberg, 1966b; Klein, 1965).

A more encouraging situation was encountered upon analysis of methylation of DNA from phage fd (Arber & Smith, 1966). As in the experiments cited above, we used methionine requiring bacteria to grow the phage and supplied to the medium methionine carrying ^{14}C label in the methyl group. The first striking observation was that the extent of methylation in fd DNA was about 10- or 20-fold lower than that found with λ or bacterial DNA. Per phage DNA molecule, which is composed of about 7,000 nucleotides, some 1·5 to two 6-methylaminopurines were detected for fd grown on either strains O or K. fd.B, grown on the $Kr_B^+ m_B^+$ hybrid strain (see Table 2), however, was definitely more methylated: three to four 6-methylaminopurines were found per DNA molecule. The amount of 5-methylcytosine, which has not yet been measured accurately, is certainly lower than one base per DNA molecule of either fd.O or fd.B, and we decided to ignore it for the moment, since we know that all DNA molecules in the population of fd.B carry host-specific modification. It might then seem, that the fd.B DNA molecule carries, in addition to some background methylation encountered also in non-modified fd.O DNA, about two methylated adenines which might be responsible for B-specific modification. The experiments described in the next section will provide evidence that this interpretation is reasonable. We then can also conclude that the methyl group involved in host-specific modification is indeed donated by methionine or a derivative thereof, since methyl groups from other sources would not have been detected in our experiments.

Unrestricted mutants of phage fd

Since the measure of methylation in fd.B had suggested that this phage carried perhaps two host specificity sites on its DNA molecule, we looked for an independent method to confirm this expectation. If the B-specificity site is defined by a special base sequence in the DNA molecule, which seemed to us to be a reasonable assumption, mutations occurring within this sequence should supposedly be able to destroy the specific site. Loss of a specificity site may have two measurable consequences: reduction in the restriction of non-modified phage upon infection of B and less methylation of phage grown in B.

To isolate mutants of fd with reduced restriction in B the following enrichment procedure was successfully applied. The phage was first grown on O; infection of B then provided a first step of enrichment for the expected mutants. The progeny from growth in B was then re-adapted to O and used again for infection of B. Usually after three to four such double passages through hosts O and B most of the phages turned out to be mutants with the expected properties, even if no mutagen was applied to the original phage stock (Arber & Kühnlein, 1967). Much to our pleasure, many independently isolated mutants all showed about the same restriction in B: their efficiency of plating was 3 to 4×10^{-2} as compared to the e.o.p. of 7×10^{-4} of fd.O wild-type phage on the same B indicator strain. When we applied the same enrichment technique a second time to a first step mutant, we found with the same ease as in the first instance completely unrestricted phages, plating with an e.o.p. of one on B even if they had been grown before on O. We interpreted these findings as reflecting a stepwise loss of B-specificity sites by mutational changes occurring in the fd DNA molecule. Since restriction was completely lost by two steps of mutation it seems that wild-type fd DNA carries indeed two B-specificity sites, as suggested by the methylation measurement.

One unrestricted double mutant of fd was grown on B and its methylation determined by the same method as described before. It had only about 1·4 6-methylaminopurines per DNA molecule, hence the same low level as fd.O (J. D. Smith & W. Arber, unpublished results). This result strongly suggests the existence of a direct correlation between the sites on the DNA molecule where restriction is exerted and the sites of modification, i.e. the locations of those adenine nucleotides which undergo specific methylation in strain B.

It is interesting to compare the degrees of restriction of four independently isolated phage strains that are all closely related to each other. Table 2 shows that phages fd and f1 are restricted to the same degree in B. Completely unrestricted derivatives of both of them can be obtained only by accumulation of two independent mutations. Phages M 13 and F 12, however, are both less restricted by B, and their efficiency of plating is close to that found for the one-step mutants of fd. We were then not surprised when we found that M 13 and F 12 yielded completely unrestricted variants in only one mutational step (Arber & Kühnlein, 1967). It is thus likely that M 13 and F 12 have only one B-specific site on their DNA molecule.

Conclusions

The findings described above open the doors to genetic localization of specificity sites on phage DNA molecules, and mapping experiments are now in progress in our laboratory. We expect such sites to occur at random, but according to a specific nucleotide sequence requirement, within sections of the phage genome that carry information for synthesis of phage specific products. One might then expect that mutations occurring within a specificity site could affect the activity of the phage or at least its growth characteristics. Such changes were indeed observed with mutants of phage M 13, that were isolated as unrestricted on B: they gave somewhat better phage yields than wild type M 13. No other correlation between change in restriction and in physiological properties of phage has yet been observed, but only a few cases have been examined up to now.

A very crude estimation of the average occurrence of host specificity sites per nucleotide base of the DNA molecule is possible in considering our results obtained with phage fd and its relatives and with phage λ. Among four different host specificities (K, B, P1 and RTF-2) tested, fd was found to have two sites for B-specificity, perhaps one site for P1-specificity, since restriction of fd in P1 lysogenic strains is low (Table 2), but no site for the two remaining specificities. The DNA molecule of phage fd is composed of some 7,000 nucleotides. We can then estimate that in the average one particular host specificity site is found per roughly 10^4 nucleotides, assuming that the sites for each different specificity display similar construction, i.e. having about the same number of bases within the sequence of nucleotides that determine the specific site. One may expect a tendency for phages with small DNA content to lose host specificity sites by natural selection upon repeated change of host bacteria, such as we have demonstrated under experimental conditions. But consideration of the results from experiments with phage λ indicates that our estimation is reasonable. The genome of λ has some 10^5 nucleotides. We then expect to find on the DNA molecule in the order of ten sites for each of the four host specificities. Since λ is restricted in strains with either of the four specificities considered, it certainly has at least one site for each of them. The recombination experiments discussed above indicated that there is more than one site but that, on the other hand, there exist sections on the genome that are devoid of sites. Hence their number cannot be very large. The biochemical analysis finally would have revealed a number of specifically methylated bases exceeding about 20–50 per phage DNA molecule. Hence we think that the estimation of one specific site per 10^4 nucleotides is not too far from the correct value.

We can now calculate how many nucleotides form a specificity site. The four common DNA bases occur with roughly equal frequencies in both λ and *E. coli* DNA. DNA of phage fd has some 24 % adenine, 20 % guanine, 22 % cytosine and 34 % thymine (Hoffmann-Berling *et al.* 1963). But for an estimation, we can ignore this deviation from completely random occurrence of each of the nucleotides. A specific sequence of three nucleotides occurs then on the average once in a section of 64 nucleotides, while a particular sequence formed by six nucleotides appears only once every 4,000 odd nucleotides. More than 16,000 nucleotides are needed for an average occurrence of one particular sequence of seven nucleotides. It seems then that some six or seven nucleotides form our hypothetical specificity sites. Whether these bases are in a closed sequence or in some other arrangement remains to be seen.

A number of different experimental approaches might give information on the composition of these sites. First, one may try to determine biochemically the bases found next to the specifically methylated adenine. Secondly, if by chance a specificity site would be found within a gene for some major protein such as phage coat subunits, the analysis of proteins from wild-type phage and from unrestricted variants thereof could give valuable information as to the precise location as well as the composition of this specificity site. A third approach will be feasible when it becomes possible to isolate the hypothetical primary products of the restriction reaction, that is supposed to cut the restricted DNA molecules at or near the specificity sites. It might then be possible to determine the terminal bases of the scission products.

This brings us to consider two important aspects, that of the search for *in vitro* enzymic activity and that of the possible mechanisms by which the enzymes might act on the DNA. These two problems are of course interconnected. The isolation of modification and restriction enzymes would be greatly aided by knowledge of how they act on the DNA, while the availability of purified enzymes could provide solid evidence for certain specific reaction mechanisms. A broad variety of approaches have been attempted in the last few years in various laboratories in the search for the enzymes in question, but the yield of these experiments remained rather poor until now. The first positive indication of restricting activity on DNA molecules *in vitro* may have been found by Takano, Watanabe & Fukasawa (1966) in sonicated extracts from cells carrying a resistance transfer factor similar to the one described in Table 1. It may be expected that these lines of approach will soon help to increase our knowledge of the restriction as well as the modification mechanisms. Among the various hypotheses that have been proposed in view of these mechanisms, I would like to take up briefly a model (Arber, 1965b) which is based on the knowledge of the specificity site on the DNA substrate as defined above. Such sites are recognized in the modification reaction which consists of the methylation of a specifically located adenine nucleotide. On the other hand, exactly the same sites are presumably recognized in the restriction reaction, which leads to the breakdown of the non-modified DNA and which perhaps also consists of base methylation (the methyl donor methionine seems indeed to play a role in restriction, see Arber, 1965a), but leaving the DNA molecule injured and thus susceptible to further breakdown. Hence for modification as well as for restriction the specificity site around the non-methylated adenine is detected. It is then probable that the enzymes of modification and of restriction have at least this site recognition part in common. Genetic mutants affecting the activity of this recognition mechanism would then be equally defective in modification and in restriction. Such mutants could be identical with what we called above $r^- m^-$ and which might then better be called s^- (for site recognition). The enzymes in question could be composed of subunits, carrying out the site recognition, the modification and the restriction, or they could perhaps rather consist of one single gene product. In this latter case the reactions of modification or restriction could be conditioned by the conformation or localization of the enzyme in question. One might, for example, expect that modification is particularly active in the cytoplasm, while restriction could preferentially be exerted by enzymes connected in some way to structures of the cell surface. It will be exciting to follow the experiments that are in course and promise to clarify our ideas on these reactions.

REFERENCES

ARBER, W. (1962). Spécificités biologiques de l'acide désoxyribonucléique. *Path. Microbiol.* **25**, 668.

ARBER, W. (1964). Host specificity of DNA produced by *Escherichia coli*. III. Effects on transduction mediated by λ *dg*. *Virology*, **23**, 173.

ARBER, W. (1965*a*). Host specificity of DNA produced by *Escherichia coli*. V. The role of methionine in the production of host specificity. *J. molec. Biol.* **11**, 247.

ARBER, W. (1965*b*). Host-controlled modification of bacteriophage. *A. Rev. Microbiol.* **19**, 365.

ARBER, W. (1966). Host specificity of DNA produced by *Escherichia coli*. 9. Host-controlled modification of bacteriophage fd. *J. molec. Biol.* **20**, 483.

ARBER, W. & DUSSOIX, D. (1962). Host specificity of DNA produced by *Escherichia coli*. I. Host controlled modification of bacteriophage λ. *J. molec. Biol.* **5**, 18.

ARBER, W., HATTMAN, S. & DUSSOIX, D. (1963). On the host-controlled modification of bacteriophage λ. *Virology*, **21**, 30.

ARBER, W. & KÜHNLEIN, U. (1967). Mutationeller Verlust B-spezifischer Restriktion des Bacteriophagen fd. *Path. Microbiol.* (in the Press).

ARBER, W. & MORSE, M. L. (1965). Host specificity of DNA produced by *Escherichia coli*. VI. Effects on bacterial conjugation. *Genetics*, **51**, 137.

ARBER, W. & SMITH, J. D. (1966). Host-controlled modification of phage fd and its correlation with specific methylation of desoxyribonucleotides. Abstracts of papers, *IX. International Congress for Microbiology, Moscow*, p. 5.

BERTANI, G. & WEIGLE, J. J. (1953). Host controlled variation in bacterial viruses. *J. Bact.* **65**, 113.

BOYER, H. (1964). Genetic control of restriction and modification in *Escherichia coli*. *J. Bact.* **88**, 1652.

COLSON, C., GLOVER, S. W., SYMONDS, N. & STACEY, K. A. (1965). The location of the genes for host-controlled modification and restriction in *Escherichia coli* K-12. *Genetics*, **52**, 1043.

DUSSOIX, D. (1964). Contrôle par la bactérie de l'acceptation d'acide désoxyribonucléique phagique. Thèse de doctorat no. 1376, Faculté des Sciences, Université de Genève.

DUSSOIX, D. & ARBER, W. (1962). Host specificity of DNA produced by *Escherichia coli*. II. Control over acceptance of DNA from infecting phage λ. *J. molec. Biol.* **5**, 37.

DUSSOIX, D. & ARBER, W. (1965). Host specificity of DNA produced by *Escherichia coli*. IV. Host specificity of infectious DNA from bacteriophage lambda. *J. molec. Biol.* **11**, 238.

FUJIMOTO, D., SRINIVASAN, P. R. & BOREK, E. (1965). On the nature of the desoxyribonucleic acid methylases. Biological evidence for the multiple nature of the enzymes. *Biochemistry*, **4**, 2849.

GLOVER, S. W., SCHELL, J., SYMONDS, N. & STACEY, K. A. (1963). The control of host-induced modification by phage P1. *Genet. Res., Camb.* **4**, 480.

GOLD, M. & HURWITZ, J. (1963). The enzymatic methylation of the nucleic acids. *Cold Spring Harb. Symp. quant. Biol.* **28**, 149.

GOLD, M. & HURWITZ, J. (1964*a*). The enzymatic methylation of ribonucleic acid and deoxyribonucleic acid. V. Purification and properties of the deoxyribonucleic acid-methylating activity of *Escherichia coli*. *J. biol. Chem.* **239**, 3858.

GOLD, M. & HURWITZ, J. (1964*b*). The enzymatic methylation of ribonucleic acid and deoxyribonucleic acid. VI. Further studies on the properties of the deoxyribonucleic acid methylation reaction. *J. biol. Chem.* **239**, 3866.

314 W. ARBER

GOUGH, M. & LEDERBERG, S. (1966). Methylated bases in the host-modified deoxyribonucleic acid of *Escherichia coli* and bacteriophage λ. *J. Bact.* **91**, 1460.

HOEKSTRA, W. P. M. & DE HAAN, P. G. (1965). The location of the restriction locus for λ.K in *Escherichia coli* B. *Mutation Res.* **2**, 204.

HOFFMANN-BERLING, H., MARVIN, D. A. & DÜRWALD, H. (1963). Ein fädiger DNS-Phage (fd) und ein sphärischer RNS-Phage (fr), wirtsspezifisch für männliche Stämme von *E. coli*. 1. Präparation und chemische Eigenschaften von fd und fr. *Z. Naturf.* **18b**, 876.

HOFSCHNEIDER, P. H. (1963). Untersuchungen über 'kleine' *E. coli* K12 Bakteriophagen. *Z. Naturf.* **18b**, 203.

IHLER, G. & MESELSON, M. (1963). Genetic recombination in bacteriophage λ by breakage and joining of DNA molecules. *Virology*, **21**, 7.

INSELBURG, J. (1966). Phage P1 modification of bacterial DNA studied by generalized transduction. *Virology*, **30**, 257.

KELLENBERGER, G., SYMONDS, N. & ARBER, W. (1966). Host specificity of DNA produced by *Escherichia coli*. 8. Its acquisition by phage λ and its persistence through consecutive growth cycles. *Z. VererbLehre*. **98**, 247.

KLEIN, A. (1965). Mechanismen der wirtskontrollierten Modifikation des Phagen T1. *Z. VererbLehre*. **96**, 346.

LEDERBERG, S. (1957). Suppression of the multiplication of heterologous bacteriophages in lysogenic bacteria. *Virology*, **3**, 496.

LEDERBERG, S. (1966a). Genetics of host-controlled restriction and modification of deoxyribonucleic acid in *Escherichia coli*. *J. Bact.* **91**, 1029.

LEDERBERG, S. (1966b). 5-Methylcytosine in the host-modified DNA of *Escherichia coli* and phage λ. *J. molec. Biol.* **17**, 293.

LEDERBERG, S. & MESELSON, M. (1964). Degradation of non-replicating bacteriophage DNA in non-accepting cells. *J. molec. Biol.* **8**, 623.

MESELSON, M. & WEIGLE, J. J. (1961). Chromosome breakage accompanying genetic recombination in bacteriophage. *Proc. natn. Acad. Sci. U.S.A.* **47**, 857.

PAIGEN, K. & WEINFELD, H. (1963). Cooperative infection by host-modified lambda phage. *Virology*, **19**, 565.

PITTARD, J. (1964). Effect of phage-controlled restriction on genetic linkage in bacterial crosses. *J. Bact.* **87**, 1256.

TAKANO, T., WATANABE, T. & FUKASAWA, T. (1966). Specific inactivation of infectious λ DNA by sonicates of restrictive bacteria with R factors. *Biochem. biophys. Res. Commun.* **25**, 192.

WEINFELD, H. & PAIGEN, K. (1964). Evidence for a new intermediate state of the viral chromosome during cooperative infection by host-modified lambda phage. *Virology*, **24**, 71.

WOOD, W. B. (1965). Mutations in *E. coli* affecting the host-controlled modification of bacteriophage λ. *Path. Microbiol.* **28**, 73.

WOOD, W. B. (1966). Host specificity of DNA produced by *Escherichia coli*. 7. Bacterial mutations affecting the restriction and modification of DNA. *J. molec. Biol.* **16**, 118.

ZINDER, N. D., VALENTINE, R. C., ROGER, M. & STOECKENIUS, W. (1963). f1, a rod-shaped male-specific bacteriophage that contains DNA. *Virology*, **20**, 638.

LYSOGENY

R. THOMAS

*Laboratoire de Génétique, Faculté des Sciences, Université
libre de Bruxelles*

1. INTRODUCTION

Temperate bacteriophages are able to establish a permanent association
with their host cell. In this situation the viral chromosome (then called
prophage) is inserted in the continuity of the bacterial chromosome. The
resulting organism (*lysogenic* bacterium) carries hereditarily the
potentiality of producing the phage in question. This lethal character is
expressed spontaneously in a very small fraction of the population only,
but in many lines its expression can be induced in the whole population
by various agents. On the other hand, the lysogenic cell possesses a
remarkable *immunity* towards infection with the corresponding phage.
Immunity prevents the superinfecting phage from replicating and from
expressing most of its functions. The same mechanism is responsible for
preventing the prophage itself from replicating autonomously and
from expressing its functions. This immunity is quite specific. Whereas a
lysogenic cell is immune towards infection with the phage it carries as a
prophage, and most of its mutants ('homoimmune' phages), it is usually
not immune towards independently isolated temperate phages (then
called 'heteroimmune' phages). Heteroimmune superinfection usually
results in the normal development of the superinfecting phage, but
not of the prophage, which remains under the control of immunity.

Major features of lysogenic systems have been discovered in particular
by the Pasteur group (reviews: Lwoff, 1953; Jacob, 1954; Jacob &
Wollman, 1957). There have since been several outstanding reviews on
the subject (Bertani, 1958; Campbell, 1962).

This paper deals with current advances in the field, especially those
concerning the regulatory mechanisms involved in the establishment
and loss of lysogeny, and in the development of temperate phages—with
special emphasis on the coliphage λ.

2. TEMPERATE PHAGES

A number of different temperate phages have been isolated. They are
usually called after the specificity of their immunity, i.e. if one isolates

from native, or from a cross between heteroimmune phages, a strain with the immunity of, say, λ, it will be called λ.

The most extensively studied temperate phages are the coliphage λ and its relatives the so-called 'lambdoid' phages (434, 82, 21, φ80, . . .). They are inducible, yield recombinants with each other and map in the S–E part of the *Escherichia coli* chromosome as it is conventionally drawn. Other well-studied temperate phages are the coliphage P2 and the *Salmonella* phage P22.

The genetic map of bacteriophage λ

The first genetic markers to be used were morphologic markers, such as, for instance, markers affecting plaque size. They are less and less used now, with the notable exception of the so-called 'clear mutants' (see below). The most widely used mutants are those which can be selected or counter-selected: (1) Host range mutants (Appleyard, McGregor & Baird, 1956). (2) Virulent mutants, which grow even on lysogenic strains and are thus able to overcome immunity (Jacob, Wollman & Siminovitch, 1953). (3) Defective mutants (Appleyard, 1954b; Jacob, Fuerst & Wollman, 1957; Arber & Kellenberger, 1958). In spite of the fact that their genetic defect usually prevents them from producing progeny they can be perpetuated as a prophage in defective lysogenic bacteria. These bacteria are usually immune but produce very little or no phage. In addition to the '*absolute*' defectives which are unable to develop by themselves under any conditions, there are '*conditional*' defective mutants (Campbell, 1959, 1961) which are able to grow under certain conditions: temperature-sensitive (*ts*) mutants grow if the temperature is not too high and suppressor-sensitive (*sus*) mutants grow in bacterial strains with a suitable suppressor.

Defective mutations allow detection of those genes whose operation is necessary for the viral development. The isolation of over 100 *sus* mutants of λ, and complementation tests, led Campbell to identify 18 genes (labelled A to R). These genes could be mapped, either by deletion analysis with *dg* (defective, galactose-transducing) prophages (see below, §4), or by three factor crosses (Campbell, 1961). Some additional genes were identified through absolute defective (Eisen *et al.* 1966; L. Siminovitch & C. R. Fuerst, personal communication), temperature-sensitive (Brown & Arber, 1964; Harris, Mount, Fuerst & Siminovitch, 1967), or *sus* (Thomas *et al.* 1967) mutations.

Dispensable genes are in general more difficult to identify. Nevertheless it has been possible recently to discover and locate several genes involved in the integration of the prophage or in functions which are not understood as yet (Zissler, 1967; Gottesman & Yarmolinsky, 1967). Figure 1 is a map of the λ chromosome.

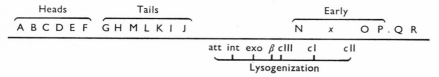

Fig. 1. Genetic map of bacteriophage λ. A–R are the genes identified by Campbell. *att*, attachment region; *int*, 'integrase'; *exo*, exonuclease; *β*, protein *β*.

The functions of the λ genes

(a) Genes involved in the lytic development

Genes A through F are involved in building the phage heads (Weigle, 1966; L. Siminovitch, personal communication). In addition, the mutants affected in these genes do not produce infective DNA (Dove, 1966) as if the 'maturation' of DNA took place during the formation of the phage heads.

Genes G to J are involved in the synthesis of the tails (Weigle, 1966; L. Siminovitch, personal communication).

N, O, P and x mutants are unable to replicate their DNA by themselves (Brooks, 1965; Eisen *et al.* 1966; Joyner, Isaacs, Echols & Sly, 1966). Since replication normally begins some minutes after infection these genes are early ones. Gene Q seems to be involved in the promotion of late functions (Dove, 1966; Dambly, Couturier & Thomas, 1968). The two terminal genes of the map are directly involved in cell lysis: their mutants produce a normal phage complement, which, however, remains in the cells, unless they are liberated by artificial lysis. The exact function of the penultimate gene (Harris *et al.* 1967) is not known. The last gene (R) is the structural gene of lysozyme (Del Campillo-Campbell & Campbell, 1965). Mutants affected in genes N, x, O, P and Q do not lyse the cells either, but this is due to secondary effects (see below, §6).

(b) Functions involved in the establishment, maintenance or loss of lysogeny

Temperate bacteriophages usually produce turbid plaques because a fraction of the cells in the lysis area have become immune. A number of mutants of P2 (see Bertani, 1954), P22 (see Levine, 1957), λ (see Kaiser, 1957) and other temperate phages produce clear plaques because they are unable to lysogenize. In the case of λ, there are at least three cistrons involved. They all map in the same region ('c' region) of the chromosome. Mutants cII and cIII lysogenize at a low rate (10^{-5}, 10^{-2}, respectively), but clones lysogenic for these phages have a normal stability. The operation of genes cII and cIII is thus required for the establishment but not for the maintenance of the lysogenic condition. In contrast, gene cI, which has been identified since as the structural gene for the immunity repressor (Jacob & Campbell, 1959; Ptashne, 1967a,b) is necessary for both the establishment and maintenance of lysogeny. There are other mutants which produce turbid plaques of a normal aspect, but which are defective in the ability of integrating and/or excising the prophage (see below, §4).

λ-specific proteins

Only a few λ-specific proteins have been identified as yet: the lysozyme (specified by gene R), whose primary structure is currently being determined (D. S. Hogness, personal communication); the immunity repressor (specified by gene cI) isolated by Ptashne (1967a,b); the λ-specific exonuclease (Korn & Weissbach, 1963, 1964; Radding, 1964; Little, 1967) whose function is not yet understood, in spite of a clear identification of its enzymatic specificity; the β protein (Radding & Shreffler, 1966), known only by its antigenic specificity.

The structural genes for the last two proteins have been located by indirect methods, although no mutation affecting these loci is known as yet (Radding, Szpirer & Thomas, 1967).

The structure of the chromosome in λ and related phages
(review: Hogness, 1966)

The chromosome of bacteriophage λ is a double-stranded DNA molecule (molecular weight: 3×10^7; about 5×10^4 nucleotide pairs; length: 16μ). λ DNA can be used to effect transformation (Kaiser & Hogness, 1960). The technique consists of infecting bacterial cells with a defective bacteriophage in the presence of DNA extracted from a non-defective strain.

When the λ chromosome is broken into pieces by shearing, it is found that different regions have a different density indicative of a different base composition (Hershey, Burgi, Frankel, Goldberg, Ingraham & Mosig, 1963). The two strands also have distinctly different base composition. Their density difference can be magnified by suitable techniques, and it is then easier to separate the two strands completely in a CsCl gradient.

Marker rescue experiments using broken molecules have established the colinearity of the molecule with the genetic map (Kaiser & Inman,

1965). In addition, the ends of the molecule must have some crucial role, since only those fragments which include either of the ends are rescued in the Kaiser–Hogness system. The special role of the ends is also made clear by the discovery (Hershey, Burgi & Ingraham, 1963) that λ DNA molecules tend in some conditions to form linear polymers (or circles in dilute solution): the ends of the molecule are thus 'sticky'. The two ends are not identical, but complementary, since 'left' arms do not stick together, and 'right' arms do not either. However, when 'left' and 'right' arms are mixed in appropriate conditions they associate end-to-end (Hershey & Burgi, 1965).

The structure of the 'sticky ends' is now known. The 5'-terminal sequence (about 15 nucleotides) of each strand of the DNA molecule protrudes from the double helix as a short single-stranded chain (Wu & Kaiser, 1967). The two sequences are presumably complementary and

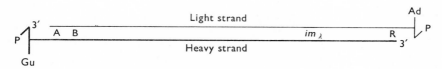

Fig. 2. The chromosome of bacteriophage λ. Ad stands for adenine, Gu for guanine; 3′ represents the 3′-OH terminal of the DNA strands; A, B, $im_λ$, R are λ genes. The two strands of λ DNA are designated 'heavy' and 'light', referring to their buoyant densities in alkaline CsCl. Modified from Wu & Kaiser (1967), with the permission of the authors and The National Academy of Sciences.

the formation of the rings is interpreted as a completion of the double-stranded structure by Watson–Crick base pairing between the ends. As far as one can judge, the sticky ends are the same in several related phages (434, 21, 424), which undergo recombination with λ, but they are different in two other (186, 299) apparently unrelated, temperate phages (Baldwin et al. 1966).

Circles are also formed in vivo following infection with λ (Bode & Kaiser, 1965a). However, in this case the ends become covalently bound. The circle is then maintained in a twisted form because free rotation is prevented. One single-strand break (induced enzymically or by ^{32}P decay: Ogawa & Tomizawa, 1967b) transforms the twisted circle into an untwisted circle. The formation of the covalently bound circles has recently been obtained in vitro (Gellert, 1967).

As will be seen below, the ability to form circles is a requirement of the Campbell model for the integration of the phage chromosome as a prophage. There are strong indications that the λ-DNA replicates vegetatively as a 'concatenate' made of a linear or closed association of several chromosomes (H. Ozeki, personal communication). This probably explains an important finding by Dove & Weigle (1965). When λ-DNA is extracted from infected cells at different times of the lytic cycle, it is found that soon after infection it loses its activity in the Kaiser–Hogness system. The infectivity is recovered only just before the appearance of the phage particles, apparently during the completion of the phage heads (Dove, 1966). This 'maturation' of the DNA undoubtedly involves the restitution of the monomeric linear structure with sticky ends, which is found in the DNA extracted from the virions.

The scission takes place at a well-defined place (between genes R and A). This appears clearly from the fact that copies of the λ chromosome are identical rather than circularly permutated as is the case for phage T4; the genetic map is linear rather than circular. From a chemical viewpoint, the scission implies the recognition by a highly specific nuclease of two sites about one and a half turns of the helix apart on the opposite strands of the DNA molecule.

In summary: the λ chromosome seems to replicate as structures in which the nucleotide sequences corresponding to the terminal genes (R and A) are covalently bound to each other, either in a closed monomeric structure, or in a (linear or closed) polymeric structure.

DNA maturation involves a scission of the strands between genes R and A. As a result the copies of the λ chromosomes have a unique gene sequence and the genetic map is linear rather than circular.

The points at which the two strands are interrupted are shifted by about fifteen nucleotides. This generates at the ends of the chromosome complementary single-stranded sequences, which can recognize each other and restore the probably original continuity.

3. IMMUNITY

As mentioned above, immunity is a quite specific character. A fundamental step towards understanding this specificity was reached when Kaiser & Jacob (1957) first crossed various mutants of λ with the heteroimmune phages 434, 82 and 21. They crossed, for instance, a clear (cII) mutant of λ with a normal (c⁺) 434. The progeny were plated on a strain lysogenic for λ (on which the phages with the immunity of 434 will form plaques) and on a strain lysogenic for 434 (on which the phages with the immunity of λ form plaques). It is found that the plaques of λ immunity are usually clear as in the parent λ. Some of them, however, are turbid like the 434 parent (and *vice versa* for the plaques of 434 specificity). This indicates that there is enough homology between these heteroimmune phages to allow, for instance, a c^+ marker to be rescued from a 434 phage into a λ cII phage. However, there is a region in which there appears to be a complete lack of homology between heteroimmune phages. The extent of this region is not characteristic of each phage but depends on the pair of phages crossed. For instance, the regions of λ which cannot be rescued from 434, 82 and 21 are different. There is a small segment which is common to these three regions of non-homology. This segment includes gene cI. It is thus not possible to recover λc^+ recombinants from a cross between λ cI and the heteroimmune phages tested: all the progeny phages with the cI^+ marker from the heteroimmune phage (say 434) have the specificity of immunity of 434, and all the progeny phages with the cI marker from λ have the specificity of immunity of λ. Moreover, the authors constructed a hybrid (434 hy, also called λim_{434}) whose chromosome is almost entirely derived from λ and has kept from 434 only a small segment bracketing gene cI. This hybrid behaves like 434 as regards immunity. It is thus clear that the genetic segment responsible for the specificity of immunity is not separable from segment cI in these heteroimmune crosses.

Before going further in the analysis of immunity one should look more carefully at the criteria used to determine its specificity. One can compare the specificity of immunity of a temperate phage with that of a known one, say λ, in two operationally distinct tests. A phage is said to have the same specificity of immunity as λ if: (a) the phage is unable to grow in a bacterial strain lysogenic for λ, although it can grow in the corresponding, non-lysogenic, strain; (b) a bacterial strain made lysogenic for the phage becomes immune to infection with λ.

It should be remarked that the specificity of immunity of the phages in a lysate can be checked on a large scale using *test a*: one just has to plate dilutions of the lysate on lawns of the proper lysogenic strains. In contrast, *test b* cannot be used on a large scale because it involves the isolation of a lysogenic strain from each phage to be tested. The first test tells us whether the phage in question is sensitive to the presence of a λ prophage in the cell it infects. The second test tells us whether, as a prophage, the phage in question prevents a superinfecting λ from growing. Thus, independently of any particular explanation of the phenomenon, one is led to attribute a dual character to the specificity of immunity: a *passive* one (sensitivity towards a product from the prophage) and an *active* one (production of something that prevents a superinfecting phage from growing).

These aspects are now well accounted for in terms of the general theory of the regulation of protein synthesis (Jacob & Campbell, 1959; Jacob & Monod, 1961 a). It has become clear that immunity is mediated by a diffusible product ('*immunity substance*': Bertani (1958); '*repressor*': Jacob & Monod (1961 a); Ptashne (1967 a)) specific for each type of temperate phage. According to the theory the repressor produced by a viral regulatory gene is able to combine specifically with a segment of the phage chromosome (the *operator*). This combination prevents the viral chromosome from expressing itself. In other words, each temperate phage has two genetic elements involved in immunity: a regulator gene and an operator. The product of the regulator gene (the repressor) recognizes the operator of the same phage but not of heteroimmune phages. In terms of this model *test a* gives information on the specificity of the *operator*, *test b*, on the specificity of the *regulator* gene.

We are now in a position to reconsider the meaning of the experiments of Kaiser & Jacob (1957). They show that in crosses between 434 and cI mutants of λ the progeny phages behave as λ or as 434 in *test a* according to whether their cI marker is derived from the λ or the 434 parent. This tells us that the *operator* of immunity is located in (or overlaps) the region of non-homology which includes cI. The relatively few progeny phage which have been scored by Kaiser & Jacob with *test b* in addition to *test a* derived their regulator gene from the same parent as their operator. But this is to be expected, since only particles which are capable of lysogenizing can be tested directly by method *b*, and a phage with an operator and regulator gene of different specificities would not be able to lysogenize by itself.

Experiments have thus been devised in order to check in more detail the possibility of separating the genetic determinants of immunity by recombination in heteroimmune crosses. The frequency of recombinants was artificially increased by selecting recombinants for two markers bracketing the immunity region. The frequency of recombinants (if any) between the operator and the regulator gene in these heteroimmune crosses is less than 4×10^{-6} (Thomas, 1964). Further experiments (R. Thomas, unpublished) considerably reduced this maximum value. If recombination were normal in this region, it would correspond to less than the distance between two pairs of nucleotides and the two structures would have to overlap. Alternatively, both structures are located in (or overlap) the region of non-homology, in which there is no recombination. Whatever the explanation, it is clear that the two genetic determinants of immunity are very closely linked. They behave, in fact, as absolutely linked in heteroimmune crosses. There may be an evolutionary reason for this (Thomas, 1964), since this linkage is a condition for the maintenance of the temperate character. It may be worthwhile to point out that in the lactose system as well, the operator and regulator genes are very closely linked, and this may be more than a coincidence.

One would now like to identify the genetic determinants of immunity in a more classical way, i.e. by finding mutants with the properties expected for alterations of these determinants.

As mentioned above (§2), the gene cI has been identified as the regulator gene of immunity. As a matter of fact, cI mutations behave in every respect as 'regulator constitutive' mutations, comparable to the i^- mutations in the lactose system. These mutants are unable to establish

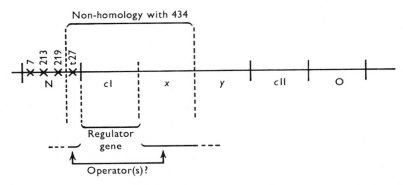

Fig. 3. The immunity region of bacteriophage λ. 7, 213, 219 are suppressor-sensitive (*sus*) mutations in gene N; t 27 is an 'absolute' defective mutation of this same gene.

or maintain the immune state. The character is recessive to the normal allele. In addition there is a so-called 'non-inducible' (*ind⁻*) mutation which makes immunity insensitive to inducing agents (Jacob & Campbell, 1959). This dominant mutation, which behaves like the i^s mutations of the lactose operon, has been located within the cI cistron.

An operator locus is defined by the so-called 'operator constitutive' (O^c) mutations. In contrast with i^-, these mutations are dominant (but only in *cis*). If such mutants exist in temperate bacteriophages, they are thus expected not only to be unable to lysogenize, but also to be immunity-insensitive. There are indeed immunity-insensitive (often called virulent) mutants. However, in the case of λ at least, one does not know any single mutation which confers immunity-insensitivity. The simplest known case (L. Pereira da Silva & F. Jacob, in preparation) is a *cis* combination of two clear mutations, the non-complementing mutation c_{17} and a cI mutation. This situation is not understood. There are other temperate phages (e.g. P2: Bertani, 1954) which yield one-step mutations to immunity-insensitivity, but their location relative to other elements involved in the specificity of immunity is not known precisely.

One can summarize the situation as follows. The regulator gene of λ, as identified by constitutive mutations, is identical with cistron cI. The operator has not been clearly identified by mutations, but it has nevertheless been located using heteroimmune crosses. Like cI, it is located in (or it overlaps) the small region of non-homology between λ and 434. Figure 3 shows the topography of this region.

The interference of immunity with viral replication

It is now clear that functions which involve protein synthesis are required for initiating vegetative replication (N, O, P in the case of λ: Brooks, 1965). If immunity blocks these functions (or only one of them) as it blocks most functions, replication will be thereby prevented. This indirect control of replication has been postulated by Jacob & Monod (1961 b) and it was later incorporated in a more general form in the hypothesis of the 'replicon' (Jacob & Brenner, 1963). According to this hypothesis each unit of replication (replicon) has a specific starting-point for replication (*replicator*) and produces a diffusible '*initiator*' which specifically interacts with the replicator, thereby starting replication. In lysogenic systems the immunity would prevent replication indirectly by blocking the synthesis of the initiator.

Although such an indirect control of replication undoubtedly exists, it would seem that it has no opportunity to operate in the normal conditions, since immunity also prevents replication by *directly* interacting with the phage chromosome (Thomas & Bertani, 1964; Thomas, 1965; Green, 1967). This conclusion is based on experiments involving mixed superinfection of a lysogenic bacterium with a homoimmune phage and with a closely related, immunity-insensitive phage. It is found that the block exerted by immunity on replication of the homoimmune chromosome cannot be by-passed by the immunity-insensitive chromosome. Nevertheless, the expression of the latter provides all the diffusible products necessary for replication: for instance, in spite of the fact that part of gene N overlaps the region of non-homology between λ and 434, the N products of λ and 434 are interchangeable (R. Thomas, unpublished results). It is suggested that the receptor site for the repressor coincides with the starting-point of replication. The presence of the repressor would thus prevent interaction between the chromosome and the enzyme or enzyme complex necessary to initiate replication.

Any explanation of the block exerted by immunity on the autonomous replication of the superinfecting viral chromosome must account for the fact that immunity does not prevent the prophage from replicating at the pace of the bacterial chromosome. As a part of the bacterial replicon, the prophage uses the bacterial replicating system and does not need the viral early enzymes or 'initiator'. This explains that the indirect effect of immunity does not operate in this case. The direct effect of immunity on the viral chromosome does not prevent the normal replication of the prophage either, indicating that the repressor does not operate by preventing progress of the wave of replication. Rather, as suggested above, it blocks initiation of the viral (but not of the bacterial) replication.

As noted by F. Jacob (personal communication) a direct block of replication by immunity is a strong argument in favour of the idea that the repressor operates at the genetic level rather than at the level of translation. This point will be considered in more detail below.

The interference of immunity with the expression of viral genes

It is known that immunity blocks the expression of most viral genes. A notable exception is the regulator gene itself and also the genes responsible for lysogenic conversion and for phage-mediated restriction and modification.

(1) Does the repressor operate at the level of transcription or of translation? The λ-specific messenger RNA is readily detected by hybridization (Hall & Spiegelman, 1961) of pulse-labelled RNA with denatured λ-DNA (Attardi *et al.* 1963; Isaacs, Echols & Sly, 1965; Green, 1966; Skalka, 1966; Naono & Gros, 1966). Recent improvements have made the method extremely sensitive (Szybalski, 1967). It is clear that immunity almost completely prevents λ transcription; the small amount of λ-specific mRNA produced in a lysogenic cell seems to correspond to the transcription of gene cI (the structural gene of the repressor). The simplest (but not the only: see Stent, 1964) explanation is that the repressor blocks transcription at the gene level. This view is supported by the elegant demonstration by Ptashne (1967*b*) that the λ repressor specifically interacts with λ DNA (and not with 434 hy DNA).

(2) Does the repressor block the whole phage chromosome directly? Alternatively, the repressor might act by preventing early functions whose expression is a prerequisite for the expression of other functions.

Information concerning this point has been provided by the finding that at least some prophage genes (notably F, P and R) are switched on upon superinfection with a closely related heteroimmune phage (complementation by prophage: Thomas (1966)). The system used consists of a lysogenic strain superinfected with heteroimmune defective mutants, which would fail to grow on the corresponding, non-lysogenic strain. For instance, a strain lysogenic for 434 hy liberates an average of 200 phage particles per bacterium on superinfection with a λR^-, whereas the corresponding, non-lysogenic strain yields only 0·2. The striking difference is due to the R^+ allele contributed by the prophage, since a strain lysogenic for a 434 hy R^- behaves like the non-lysogenic control. The effect is not due to the presence of non-defective λ recombinants. This was demonstrated in the cases of genes P and R, first by single burst experiments (Thomas, 1966), then by using combinations of strains which almost completely exclude prophage excision and drastically lower the frequency of recombinants (R. Thomas, S. Mousset & D. Wauters, in preparation). The possibility that the expression of the gene is due to normal leakage of the block exerted by immunity on prophage functions has been ruled out, in the case of gene R, by direct measurements of lysozyme activity (Dambly et al. 1968).

It is thus quite clear that at least some prophage genes can be switched on by heteroimmune superinfection. That the block exerted by immunity on these genes can be by-passed, shows that this block is an *indirect* one. That it can be by-passed in *trans* indicates that these genes are switched on by a diffusible product. The immunity mechanism thus presumably operates by preventing the expression of some early genes whose product is required for switching on later genes.

For obvious reasons of gene dosage an efficient complementation by prophage cannot be expected for 'stoichiometric' functions. Consequently, a negative result in heteroimmune superinfection with a defective mutant does not necessarily allow us to draw the conclusions that the corresponding prophage gene is not inducible. However, if the function is known to be catalytic, the negative result may be taken to mean that the gene is actually not induced following superinfection. It is suggested that genes N (Thomas, 1966) and O (L. Pereira da Silva & F. Jacob, personal communication; B. Egan, personal communication) are directly blocked by immunity. In addition, the results of Bode & Kaiser (1965b) show not only that genes cII and cIII are blocked in the immune state, but also that they are not induced following heteroimmune superinfection. There are thus two genetic segments located on each side of cI, which appear to be under direct immunity control. This fits well with the two possible locations assigned on entirely different grounds to the operator(s) of immunity (see Fig. 3).

4. THE ESTABLISHMENT AND BREAKDOWN OF THE LYSOGENIC STATE

The establishment and breakdown of the lysogenic state involve *structural* factors (the integration of the viral chromosome into the bacterial chromosome or its excision) and *regulatory* factors (immunity). As far as possible we will first consider these two aspects separately.

Integration and excision of the prophage

Prophages are attached at specific points of the bacterial chromosome (Wollman, 1953; Lederberg & Lederberg, 1953; Frédéricq, 1953; Lederberg, 1954; Appleyard, 1954a). In some cases, however, there are two to three (phage P2: Bertani & Six (1958)) or many (phage μ1: Taylor (1963)) possible locations. The phages which give generalized transduction (e.g. P1) do not seem to have a specific chromosomal location. There has been much discussion about the precise physical relation between prophage and bacterial chromosome (see, for instance, the reviews by Jacob & Wollman (1957) and by Bertani (1958)). Especially disturbing was the fact that the genetic map of the prophage behaves as a permutation of the vegetative map (Calef & Licciardello, 1960). To explain this paradox, as well as his own results concerning the peculiarities of the transducing phage λdg, Campbell (1962) formulated his well-known model (Fig. 4).

According to this model, the phage chromosome first circularizes, then pairs with a presumably homologous region of the bacterial chromosome. This homology region must be located on the viral chromosome between genes J and cIII. A genetic exchange between the homologous regions brings about integration of the viral chromosome into the bacterial chromosome. One easily realizes that if the model is correct the genetic map of the prophage must be a *circular* permutation of the vegetative map and is bounded by two copies of the homology region. Normal prophage excision is thought to take place by a reverse recombinational event. Transducing phages are formed as a result of rare illegitimate recombinational events. The genetic constitution of the various transducing particles should thus correspond to uninterrupted sections of genetic material spanning either of the homology regions and involving a partial phage genome together with some bacterial material. One expects therefore the availability of a bacterial gene for specialized transduction to depend only on its distance from the prophage. Finally, the insertion of a prophage should decrease the linkage between two bacterial markers which span the prophage, and inactivate the gene in which it is inserted.

All the predictions of the model have been verified: circularization (Hershey, Burgi & Ingraham, 1963; Kaiser & Inman, 1965), gene order (Rothman, 1965; Franklin, Dove & Yanofsky, 1965), genetic structure of the transducing phages (Campbell, 1963; E. R. Signer, unpublished experiments), distance dependence of the transducibility of bacterial markers, decrease of genetic linkage of markers following insertion of a prophage (Signer, 1966), inactivation of genes by insertion of phage $\mu 1$

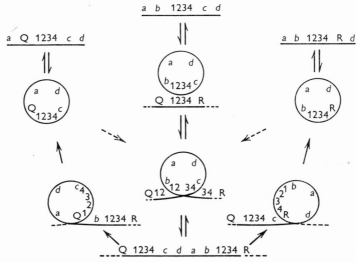

Fig. 4. The Campbell model. Q and R are bacterial markers; *a*, *b*, *c* and *d* are phage markers, and 1234 represents the nucleotide sequence of the special region of homology between the phage and its attachment site. The Figure shows normal attachment and detachment in the centre and, the formation of a transducing phage carrying Q (replacing *b*) and one carrying R (replacing *c*) at the left and right. The broken arrows indicate that the transducing phage genomes are expected to attach and detach by the same mechanism as the active phage genome in subsequent infection. From Signer & Beckwith (1966), with the permission of the authors and Academic Press.

(Taylor, 1963). There is little doubt indeed that the model is essentially correct. I would like to point out that there seems to be a very precise limit to the extent of the defectivity in specialized transduction. One knows a great number of different λdg, whose deletion covers a greater or lesser part of the left arm. The deletion frequently reaches gene A, or even spans all its known markers, but in no case does it penetrate into gene R. There seems to be an obvious explanation for this observation. To observe a specialized transduction one needs formation of an infective (although in general genetically defective) particle. A deletion spanning the region between A and R would eliminate the site for de-circularization of the phage chromosome, thus preventing the formation of an infective particle. In contrast, one may obtain defective strains by directly selecting deletions, some of which may extend into the prophage (Franklin *et al.* 1965). In this case the extent of the defectivity is not limited. Again in contrast with the defective transducing phages, the prophages originating from these deletions lack one of the *att* regions. This prevents phage detachment (W. F. Dove & N. C. Franklin, personal communication; Signer & Beckwith, 1966).

Starting from the Campbell model, studies on the mechanism of integration and excision are progressing rapidly. The phenomena are now understood in terms of two types of genetic elements: homology regions which are thought to pair, and structural genes which promote recombination between these homologous regions, presumably by an enzymic mechanism. A major system involves specific recognition of the Campbell homology region by a viral recombination-promoting system. However, integration and excision can also be mediated in special cases by viral or bacterial recombination systems which are not specific for a given genetic sequence.

Specific recombination in the Campbell homology region

It is well known that most prophages have each a characteristic location on the bacterial chromosome (Wollman & Jacob, 1957). For instance, λ normally attaches at a locus (att_λ) near the gal cluster, $\phi 80$ at a locus (att_{80}) near the try cluster. Contrary to early indications, this specificity of attachment is unrelated to the specificity of immunity. There are now numerous examples of hybrids between λ and 21 (Liedke-Kulke & Kaiser, 1967) or between λ and $\phi 80$ (Signer, 1965; Radding *et al.* 1967) which have derived their immunity from one parent and their attachment specificity from the other.

According to the Campbell model, the corresponding segment of the viral chromosome (which we will also symbolize as *att*) must be located between genes J and cIII. The density mutant $\lambda b2$ (Zichichi & Kellenberger, 1963), which is unable to attach to the bacterial chromosome, has the properties expected for a deletion of region *att*. The b2 marker is indeed located between J and cIII (Jordan, 1964). It is not clear yet, however, what exact relation exists between the (presumably large) region of homology and the (presumably small) sequence, in or adjacent to it, in which the recombination takes place.

Superinfection of a lysogenic bacterium with a related phage may result in prophage excision and curing of the lysogenic strain, but only if the attachment specificities of the prophage and superinfecting phage are the same (Signer, 1966; Gottesman & Yarmolinsky, 1967; Radding *et al.* 1967). The additional fact that curing does not occur following homoimmune superinfection indicates that the mechanism involved is repressed by immunity. These and other results (see Signer & Beckwith, 1966) are understandable if each temperate phage carries the information for a system specifically promoting recombination between its region *att* and the corresponding region of the bacterial chromosome. The genetic determinant (*int*) for this system is now identified in the *Salmonella* phage P22 (Smith & Levine, 1967) as well as in phage λ (Zissler, 1967; Gottesman & Yarmolinsky, 1967; R. Gingery & H. Echols, personal communication). *Int⁻* mutants normally do not lysogenize; however, they can be helped to lysogenize by simultaneous infection with an *int⁺* phage (which does not necessarily attach). A strain lysogenic for an *int⁻* mutant behaves as defective because the prophage cannot be excised properly. Finally, heteroimmune superinfection with an *int⁻* phage does not result in curing. Thus *int⁻* mutants are affected in the following abilities: as an infecting phage, to lysogenize; as a prophage, to detach; as a heteroimmune superinfecting phage, to excise the prophage. Clearly, a common mechanism is involved in integration and excision (following either induction or superinfection).

That *int* is distinct from *att* is suggested by the fact that b2 mutants complement with *int* mutants and are able to excise heteroimmune prophages of the proper specificity. It may be worth while to stress here the formal similarity between the genetic region involved in attachment and the region *im* responsible for immunity. In both cases the function and its specificity are ensured by a pair of complementary genetic elements, one passive (*att* in the integration system, the operator in the case of immunity), one active (*int*, in the integration system, the regulator gene in the case of immunity). As discussed above the elements of immunity are closely linked and they could not be separated in heteroimmune crosses. Although the situation has not been analysed carefully as yet in the case of the attachment specificity, it seems that here too the two elements are closely linked. In a cross between phages with different specificities of attachment (say λ and ϕ80), a recombination between *att* and *int* would presumably yield a hybrid with a strongly reduced ability to attach by itself. Such hybrids are at least not frequent.

The role of the non-specific bacterial recombination system (rec)

The role of the bacterial recombination system (*rec*) can now be analysed, thanks to the isolation of recombination-deficient (*rec⁻*) bacterial mutants (Clark & Margulies, 1965; Fuerst & Siminovitch, 1965; Van de Putte, Zwenk & Rörsch, 1966).

A normal phage lysogenizes a *rec⁻* strain at a normal rate, suggesting that the contribution of the *rec* system to the normal integration process is a minor one. Accordingly, in the absence of additional homology between the prophage and bacterial chromosome even *rec⁺* strains are not lysogenized (Gottesman & Yarmolinsky, 1967) or are very poorly lysogenized (Signer & Beckwith, 1966: see below) by *int⁻* phages. The efficiency of the *rec* mechanism should greatly increase if there is an additional region of homology between the prophage and bacterial chromosome. However, even then the relative contribution of this mechanism remains negligible, unless the *int* system is prevented from operating. For instance, when a bacterium diploid for att_λ and carrying a *heteroimmune* prophage at one of these sites, is superinfected with λ, the latter can ignore the presence in the bacterial chromosome of a large region of homology, i.e. the prophage, and attach at the vacant att_λ (J. Zissler, quoted by Signer & Beckwith, 1966).

The situation changes if *int* is inactive. The frequency of lysogenization is then greatly decreased, but what remains is due to the non-specific *rec* system. This is the case if experiments just mentioned are performed in *homoimmune* conditions, where immunity prevents the synthesis of the viral recombination systems. One finds indeed a low frequency of lysogenization, and in the rare double lysogenic survivors the second phage is always (38/38) integrated in *cis* (Campbell & Zissler, 1966). Under these conditions integration takes place within the pre-existing phage (Calef, 1967). Another well-documented case is that of the two defective, lactose-transducing phages φ80 *dlac* (Signer & Beckwith, 1966). Although their deletions do not overlap, these two phages are both deficient for the *int* system, as shown by the fact that they are completely unable to integrate in *rec⁻* (non-lysogenic) strains. They have two regions of homology with the bacterial chromosome (att_{80} and *lac*). In a *rec⁺* strain the φ80 *dlac* attach at *lac*, thus suggesting that this region of homology is considerably larger (or more homologous) than att_{80}. However, the viral genome can attach at a low frequency at att_{80}, as shown by experiments using bacterial strains with a complete deletion of the *lac* region. When a prophage is already present, it provides a homology region even greater than that offered by the bacterial *lac* cluster. Consequently, when a φ80 *dlac* infects a strain lysogenic for φ80 it usually inserts at the level of the prophage (Signer & Beckwith, 1966). Similarly, in lysogenization by *int⁻* phage, the helping effect of the bacterial recombinase becomes apparent in the presence of a (hetero-immune) prophage, whereas it is not detected in a non-lysogenic strain (Gottesman & Yarmolinsky, 1967).

As concerns a possible participation of the bacterial recombination system in prophage excision, the situation is complicated by the fact that in the most widely studied *rec⁻* strains the genetic block also prevents induction of the prophage by ultraviolet and similar agents. We will come back to this important point later (see p. 334). Suffice it to say here that the non-inducible character is not an unavoidable consequence of the *rec⁻* character, as would be the case, for instance, if excision were a prerequisite for the lifting of immunity; rather, both characters result from a common defect (Brooks & Clark, 1967; Hertman & Luria, 1967). There are also *rec⁻* strains—presumably blocked at a later stage— whose prophage is normally inducible (Van de Putte *et al.* 1966). In the *rec⁻* strains which are not u.v.-inducible, one can by-pass the difficulty by using prophages with a thermosensitive mutation in gene cI. These remain heat-inducible even in the *rec⁻*, non u.v.-inducible strains.

The fact that the spontaneous induction in these *rec⁻* is very low probably does not reflect an involvement of the bacterial 'recombinase' in spontaneous induction. Rather, the effect would mean that in spontaneous induction, as in u.v. induction, the inactivation of the repressor requires the operation of a mechanism blocked in those *rec⁻* strains.

The role of the non-specific viral recombination system

The existence of such a system is made clear by the occurrence of normal recombination between phages in *rec⁻* strains (Brooks & Clark, 1967; N. Lefèbvre, unpublished results) even if the phages are *int⁻* (Gottesman & Yarmolinsky, 1967). Mutants (*red*, for recombination-deficient) affected in this system have now been isolated (E. Signer, personal communication; H. Echols & R. Gingery, in preparation).

The role of the non-specific viral recombination mechanism in prophage integration and excision seems to be limited to rather special situations. *Rec⁻* strains made lysogenic (with a helper) for the thermoinducible phage λc_{857} belong to two types. All are induced by heat and lyse. Whereas a majority of these strains produce very little active phage (about 0·004 per bacterium) some of them have an almost normal production (Gottesman & Yarmolinsky, 1967). These authors have shown that these 'high-yielder' strains are doubly lysogenic. After thermal lifting of the immunity, a viral 'recombinase' is synthesized in any case, but only if the cell is doubly lysogenic can a phage be excised efficiently by pairing of the two prophages and recombination anywhere along them.

Likewise, a given strain, lysogenized with a given phage, may yield two types of isolates. A majority behave 'normally', i.e. liberate almost exclusively the superinfecting type upon heteroimmune superinfection. However, some strains liberate an abnormally high proportion of particles of the prophage type, even if the strain is rec^- and the superinfecting phage, int^- (R. Thomas, S. Mousset & D. Wauters, in preparation). These strains again are doubly lysogenic; excision must be mediated by a non-specific recombinase of the superinfecting, heteroimmune phage. The occurrence of such strains may be the reason for some apparent contradictions in the literature about whether (see Jacob *et al.* 1953; Ptashne, 1965) or not (Thomas & Bertani, 1964) superinfection with a heteroimmune or virulent phage breaks down immunity.

Regulatory aspects

Shortly after infection there is a decision as to whether the process will be productive (lytic) or reductive. In the case of the *coli* phages, very little is known about this early phase. In the case of the *Salmonella* phage, P22 (Smith & Levine, 1964), in conditions which favour lysogenization, there is a very early peak of viral DNA synthesis followed at about 6 min. by a transient and almost complete prevention of all DNA synthesis. This block is controlled by gene cI (the equivalent of the λ gene cII). Only later does the operation of gene c2 (the equivalent of the λ gene cI) establish the immune state, in which the bacterial DNA replicates normally, but the viral DNA does not replicate autonomously. There is a third gene (c3 for phage P22, cIII for λ) involved in lysogenization, but its role is unknown. In P22 as well as in λ, the regulator gene (c2 in P22, cI in λ) has to function continuously in order to maintain the lysogenic state whereas the other c genes operate only at a critical time during the establishment of the lysogenic state (Levine, 1957; Kaiser, 1957; Bode & Kaiser, 1965b).

Establishment and breakdown of lysogeny

There is no simple relation between the establishment of the immune state and prophage integration. Immunity can become established in such conditions that the phage chromosome cannot integrate (for instance, when a cell is infected with mutant b2). The viral material is then completely prevented from replicating and it is diluted out as cell divisions proceed (abortive lysogeny: Zichichi & Kellenberger, 1963; Ogawa & Tomizawa, 1967a). On the other hand, if integration takes place, then the viral material will be able to replicate as a part of the bacterial chromosome, and the lysogenic state will ensue.

One could provisionally visualize as follows the events which take place between infection and the establishment of lysogeny. A temperate phage entering a lysogenic bacterium expresses various functions. One of them, *int*, very efficiently promotes recombination between the *att* of the bacterial chromosome and the (circularized) phage chromosome. This may result in integration, although the reverse phenomenon (excision) may be just as frequent as long as there is *int* product present. In the meantime immunity has become established in a fraction of the bacterial population and no additional *int* product is synthesized. I suggest that the situation stabilizes as the *int* product disappears, thus leading to stable lysogeny if the phage chromosome was integrated at that time. This would perhaps explain an apparent paradox. Smith & Levine (1967) noted that integration in *Salmonella* phage P22 is a *late* function, which takes place around 100 min., well after the establishment of immunity. This result might be understood by assuming that in this case the *int* product (which is reversibly thermosensitive) decays only slowly (or has to be diluted out as cell divisions proceed). In λ, the integration process is also thermosensitive (Lieb, 1953; Zichichi & Kellenberger, 1963), but here the decision for or against integration seems to take place within 15 min. following infection.

Induction involves the inactivation of the immunity repressor. This may be a *direct* inactivation, as in thermal induction of thermoinducible lysogenic strains (Sussman & Jacob, 1962). In these cases the repressor is reversibly or irreversibly inactivated by heat.

Immunity can also be lifted by, for example, u.v.-irradiation, exposure to mitomycin, or thymine starvation, which have in common the property of blocking bacterial DNA synthesis. In these cases the repressor is inactivated by a more complex process, which requires the operation of bacterial genes and protein synthesis (Tomizawa & Ogawa, 1967). Many facts concerning the mechanism of u.v.-induction are accounted for by a scheme proposed by Goldthwait & Jacob (1964). In this scheme the lambda repressor is antagonized by a low molecular weight precursor of DNA. The inducing effect of u.v. and similar agents is explained by an accumulation of this product, as a consequence of the block of DNA synthesis. A major piece of evidence in favour of this theory was the isolation of a bacterial mutant (T44) in which any u.v.-inducible prophage is induced by heat. This mutation is thought of as a thermosensitive alteration of a system regulating the formation of the inducer. At intermediate temperatures the induction of this mutant is favoured by adding adenine derivatives and minimized by adding other nucleic acid precursors; it is suggested that the inducer is a derivative of adenine.

As mentioned above, several (but not all) *rec⁻* mutants are characterized, in addition to their deficiency in recombination, by their inability to lift immunity following u.v.-irradiation of their lysogenic derivatives (Hertman & Luria, 1967; Brooks & Clark, 1967). The authors interpret their results in the frame of Goldthwait & Jacob's hypothesis: either a common metabolic step is necessary for recombination and for the synthesis of the λ inducer (Brooks & Clark, 1967) or the inducer fails to act because the breakdown of DNA in rec⁻ strains causes accumulation of antagonists of the inducer (Hertman & Luria, 1967). The actual connection between the two above-mentioned types of bacterial mutations (rec⁻ and T44) is demonstrated by the recent finding (Kirby, Jacob & Goldthwait, 1967) that the thermosensitivity of the lysogenic derivatives of T44 can be lost either by reversion to wild-type or by a mutation which behaves in every respect like the rec⁻ mutations conferring non-inducibility.

Finally, as already mentioned, a lysogenic cell can be cured of its prophage, following heteroimmune superinfection. This normally results from excision of the prophage by the superinfecting *int* (or *red*) system. There is no lifting of immunity. The excised prophage does not replicate and, if the cell survives, it is diluted out as cell division proceeds.

5. THE DEVELOPMENT OF TEMPERATE PHAGES

Since this paper is devoted primarily to the lysogenic state, lytic development will be dealt with only briefly here.

Starting from a lysogenic cell, a first step is the breakdown of the immune state. Induction brings about the inactivation of the repressor by a low molecular weight inducer (or by heat in strains with a thermo-inducible prophage); or else immunity is lifted as the prophage is transferred into the cytoplasm of a sensitive cell (zygotic induction).

Early viral genes then begin to express themselves and, among them, genes involved in prophage excision and replication (Weisberg & Gallant, 1966). But, as mentioned above (§3), some functions are under direct immunity control, while other functions do not express themselves in the immune state because they are dependent on the previous expression of another gene (Thomas, 1966). There is reason to think that gene N operates as a primary switch (Protass & Korn, 1966; Thomas, 1966; Dambly *et al.* 1968) and that the operation of gene Q induces the late functions (Dove, 1966; Dambly *et al.* 1968). There are functions (*stoichiometric* functions) whose full expression requires the synthesis of a great number of protein molecules. This in turn requires the operation of many copies of the gene, and consequently these functions are replication-dependent.

A provisional scheme for these complex interactions (see Dambly *et al.* 1968) is given in Figure 5.

The operation of genes A to F culminates in the formation of phage heads (Weigle, 1966; L. Siminovitch, personal communication) containing 'mature' DNA (Dove, 1966). Genes G to J control tail synthesis. The association between heads and tails is spontaneous. Mixing a lysate from any A–F mutant with a lysate from any G–J mutant under appropriate conditions leads to the appearance of complete, infective particles. These particles harbour the genotype of the G–J 'parent', as expected from the fact that this phage is the 'head donor', the other supplying the tail (Weigle, 1966).

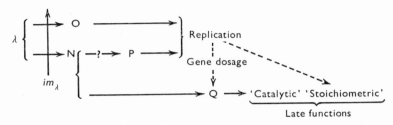

Fig. 5. A provisional scheme for the development of bacteriophage λ. Following this scheme, gene N directly or indirectly turns on the expression of gene P (which is required for replication) and of gene Q (which in turn is involved in the expression of late functions). In addition the full expression of some genes (including Q and a majority of the A–J genes) requires the synthesis of many protein molecules. These functions are thus replication-dependent for reasons of gene dosage. These factors (at least partly) account for the time schedule of gene expression.

Cell lysis requires the operation of gene R, the structural gene for lysozyme (Del Campillo-Campbell & Campbell, 1965) and of a second gene (Harris *et al.* 1967) whose product seems to be required in stoichiometric amounts. A very small amount of lysozyme suffices to bring about lysis in the presence of an excess of the second factor. However, if there is a shortage of this factor, lysozyme accumulates in the cell without being able to lyse it. The cells would lyse prematurely if lysozyme were the only factor required for lysis.

6. LYSOGENY AS A MODEL FOR POSSIBLE SIMILAR SITUATIONS IN HIGHER ORGANISMS

There is a widespread feeling that situations more or less similar to lysogeny occur in higher organisms. A major objection often made to this view is that lysogeny is quite easy to detect and to analyse: should a similar mechanism actually account for situations found in higher organisms, the various characteristics of lysogeny (immunity, production of virus, etc.) would have been demonstrated without difficulty.

Rather than try to analyse situations in a field which is not my own, I wish to terminate this paper: (1) by pointing out the real difficulties which may be encountered in identifying a strain as lysogenic, especially if it is defective; (2) by briefly indicating which methods may be used to detect lysogeny in marginal cases.

'Healthy' lysogeny results in the appearance of viral particles in the culture, and immunity towards the virus carried as a prophage. However, the detection of viral particles supposes that one has a sensitive host on which the particles can form plaques, or else, that there are enough virions to be detected with the electron microscope. One can also check whether the presumed lysogenic cells lyse following a treatment with an inducing agent; but this of course only holds for inducible strains. As for immunity, this can easily be tested, but only if one has the virus to which the cells are presumably immune.

Moreover, many lysogenic strains are defective. The defective strains usually produce extremely few viral particles, or none at all. If the strain is still immune there is no special problem, except for distinguishing between immunity and such other forms of resistance as non-adsorption of the virus, or restriction. This is usually not difficult in a well-explored system. The situation becomes much more delicate if one deals with a lysogenic strain which has lost the immunity region, for instance, the 'cryptic' strains (Fischer-Fantuzzi & Calef, 1964). One can rely neither on phage production nor on immunity to detect such a defective prophage. One solution is to detect viral genes present, either by complementation or by marker rescue, following superinfection with a related phage defective for one of the genes in question.

In summary, detection of a *healthy* provirus is easy if viral development can be induced by some agent or if there is a sensitive strain available on which the virus forms plaques. If one knows in advance which virus one is looking for and if an infective form of this virus is available, one can detect not only healthy, but also defective and even cryptic proviruses, by function- or marker-rescue following superinfection with a defective mutant of the virus.

ACKNOWLEDGEMENTS

I wish to thank Drs Yarmolinsky, Michael and Ann Roller who kindly read the manuscript and suggested useful modifications.

REFERENCES

APPLEYARD, R. K. (1954*a*). Segregation of lambda lysogenicity during bacterial recombination in *E. coli* K12. *Genetics*, **39**, 429.

APPLEYARD, R. K. (1954*b*). Segregation of new lysogenic types during growth of a doubly lysogenic strain derived from *Escherichia coli* K12. *Genetics*, **39**, 440.

APPLEYARD, R. K., McGREGOR, J. F. & BAIRD, K. M. (1956). Mutation to extended host range and the occurrence of phenotypic mixing in the temperate coliphage lambda. *Virology*, **2**, 565.

ARBER, W. & KELLENBERGER, G. (1958). Study of the properties of seven defective-lysogenic strains derived from *Escherichia coli* K12 (λ). *Virology*, **5**, 458.

ATTARDI, G., NAONO, S., ROUVIERE, J., JACOB, F. & GROS, F. (1963). Production of messenger RNA and regulation of protein synthesis. *Cold Spring Harb. Symp. quant. Biol.* **28**, 363.

BALDWIN, R. L., BARRAND, P., FRITSCH, A., GOLDTHWAIT, D. A. & JACOB, F. (1966). Cohesive sites on the deoxyribonucleic acids from several temperate coliphages. *J. molec. Biol.* **17**, 343.

BERTANI, G. (1954). Studies on lysogenesis. Superinfection of lysogenic *Shigella dysenteriae* with temperate mutants of the carried phage. *J. Bact.* **67**, 696.

BERTANI, G. (1958). Lysogeny. *Adv. Virus Res.* **5**, 151.

BERTANI, G. & SIX, E. (1958). Inheritance of prophage P2 in bacterial crosses. *Virology*, **6**, 357.

BODE, V. C. & KAISER, A. D. (1965*a*). Changes in the structure and activity of λ DNA in a superinfected immune bacterium. *J. molec. Biol.* **14**, 399.

BODE, V. C. & KAISER, A. D. (1965*b*). Repression of the cII and cIII cistrons of phage lambda in a lysogenic bacterium. *Virology*, **25**, 111.

BROOKS, K. (1965). Studies in the physiological genetics of some suppressor-sensitive mutants of bacteriophage λ. *Virology*, **26**, 489.

BROOKS, K. & CLARK, A. (1967). Behaviour of lambda bacteriophage in a recombination deficient strain of *E. coli*. *J. Virol.* **1**, 283.

BROWN, A. & ARBER, W. (1964). Temperature-sensitive mutants of coliphage lambda. *Virology*, **24**, 237.

CALEF, E. (1967). Mapping of integration and excision crossovers in superinfection double lysogens for phage lambda in *E. coli*. *Genetics*, **55**, 547.

CALEF, E. & LICCIARDELLO, G. (1960). Recombination experiments on prophage host relationships. *Virology*, **12**, 81.

CAMPBELL, A. (1959). Ordering of genetic sites in bacteriophage λ by the use of galactose-transducing defective phages. *Virology*, **9**, 293.

CAMPBELL, A. (1961). Sensitive mutants of bacteriophage λ. *Virology*, **14**, 22.

CAMPBELL, A. (1962). Episomes. *Adv. Genet.* **11**, 101.

CAMPBELL, A. (1963). Distribution of genetic types of tranducing lambda phages. *Genetics*, **48**, 409.

CAMPBELL, A. & ZISSLER, J. (1966). The steric effect in lysogenization by bacteriophage lambda. III. Superinfection of monolysogenic derivatives of a strain diploid for the prophage attachment site. *Virology*, **28**, 659.

CLARK, A. J. & MARGULIES, A. D. (1965). Isolation and characterization of recombination-deficient mutants of *E. coli* K12. *Proc. natn. Acad. Sci. U.S.A.* **53**, 451.

DAMBLY, C., COUTURIER, M. & THOMAS, R. (1968). Control of development in temperate bacteriophages. II. Control of lysozyme synthesis. *J. molec. Biol.* (in the Press).

DEL CAMPILLO-CAMPBELL, A. & CAMPBELL, A. (1965). Endolysin from mutants of bacteriophage lambda. *Biochem. Z.* **342**, 485.

Dove, W. F. (1966). The action of the lambda chromosome. I. The control of functions late in phage development. *J. molec. Biol.* **19**, 187.

Dove, W. F. & Weigle, J. J. (1965). Intracellular state of the chromosome of bacteriophage lambda. I. The eclipse of infectivity of the bacteriophage DNA. *J. molec. Biol.* **12**, 620.

Eisen, H. A., Fuerst, C. R., Siminovitch, L., Thomas, R., Lambert, L., Pereira da Silva, L. & Jacob, F. (1966). Genetics and physiology of defective lysogeny in K 12 (λ): studies of early mutants. *Virology*, **30**, 224.

Fischer-Fantuzzi, L. & Calef, E. (1964). A type of λ prophage unable to confer immunity. *Virology*, **23**, 209.

Franklin, N. C., Dove, W. F. & Yanofsky, C. (1965). The linear insertion of a prophage into the chromosome of *E. coli* shown by deletion mapping. *Biochem. biophys. Res. Commun.* **18**, 910.

Frédéricq, P. (1953). Transfert génétique des propriétés lysogènes chez *E. coli. C. r. Séanc. Soc. Biol.* **147**, 2046.

Fuerst, C. R. & Siminovitch, L. (1965). Characterization of an unusual defective lysogenic strain of *E. coli* K 12 (λ). *Virology*, **27**, 449.

Gellert, M. (1967). Formation of covalent circles of lambda DNA by *E. coli* extracts. *Proc. natn. Acad. Sci. U.S.A.* **57**, 148.

Goldthwait, D. & Jacob, F. (1964). Sur le mécanisme de l'induction du développement du prophage chez les bactéries lysogènes. *C. r. hebd. Séanc. Acad. Sci., Paris*, **259**, 661.

Gottesman, M. E. & Yarmolinsky, M. B. (1967). Integration-negative (*int*) mutants of lambda bacteriophage. *J. molec. Biol.* (submitted).

Green, M. H. (1966). Inactivation of the prophage lambda repressor without induction. *J. molec. Biol.* **16**, 134.

Green, M. H. (1967). Regulation of bacteriophage λ DNA replication by the λ repressor. *Fedn Proc.* **26**, 450.

Hall, B. D. & Spiegelman, S. (1961). Sequence complementarity of T2-DNA and T2-specific RNA. *Proc. natn. Acad. Sci. U.S.A.* **47**, 137.

Harris, A. W., Mount, D. W. A., Fuerst, C. R. & Siminovitch, L. (1967). Mutations in bacteriophage lambda affecting host cell lysis. *Virology*, **32**, 553.

Hershey, A. D. & Burgi, E. (1965). Complementary structure of interacting sites at the ends of lambda DNA molecules. *Proc. natn. Acad. Sci. U.S.A.* **53**, 325.

Hershey, A. D., Burgi, E., Frankel, F., Goldberg, E., Ingraham, L. & Mosig, G. (1963). In Annual report of Director of the Genetics Research Unit. *Yb. Carnegie Inst. Wash.* **62**, 481.

Hershey, A. D., Burgi, E. & Ingraham, L. (1963). Cohesion of DNA molecules isolated from phage lambda. *Proc. natn. Acad. Sci. U.S.A.* **49**, 748.

Hertman, I. & Luria, S. E. (1967). Transduction studies on the role of a rec$^+$ gene in the ultraviolet induction of prophage lambda. *J. molec. Biol.* **23**, 117.

Hogness, D. S. (1966). The structure and function of the DNA from bacteriophage lambda. *J. gen. Physiol.* **49**, 29.

Isaacs, N. L., Echols, H. & Sly, W. S. (1965). Control of lambda messenger RNA by the cI immunity region. *J. molec. Biol.* **13**, 963.

Jacob, F. (1954). Les bactéries lysogènes et la notion de provirus. *Monographies de l'Institut Pasteur.* Paris: Masson et Cie éd.

Jacob, F. & Brenner, S. (1963). Sur la régulation de la synthèse du DNA chez les bactéries: l'hypothèse du réplicon. *C. r. hebd. Séanc. Acad. Sci., Paris*, **256**, 298.

Jacob, F. & Campbell, A. (1959). Sur le système de répression assurant l'immunité chez les bactéries lysogènes. *C. r. hebd. Séanc. Acad. Sci., Paris*, **248**, 3219.

JACOB, F., FUERST, C. R. & WOLLMAN, E. L. (1957). Recherches sur les bactéries lysogènes défectives. II. Les types physiologiques liés aux mutations du prophage. *Ann. Inst. Pasteur*, **93**, 724.

JACOB, F. & MONOD, J. (1961*a*). Genetic regulatory mechanisms in the synthesis of proteins. *J. molec. Biol.* **3**, 318.

JACOB, F. & MONOD, J. (1961*b*). On the regulation of gene activity. *Cold Spring Harb. Symp. quant. Biol.* **23**, 193.

JACOB, F. & WOLLMAN, E. (1957). Genetic aspects of lysogeny. In *The Chemical Basis of Heredity*. Ed. McElroy and B. Glass. Baltimore: Johns Hopkins Press.

JACOB, F., WOLLMAN, E. & SIMINOVITCH, L. (1953). Propriétés inductrices des mutants virulents d'un phage tempéré. *C. r. hebd. Séanc. Acad. Sci., Paris*, **236**, 544.

JORDAN, E. (1964). The location of the b2 deletion of bacteriophage λ. *J. molec. Biol.* **10**, 341.

JOYNER, A., ISAACS, L. N., ECHOLS, H. & SLY, W. S. (1966). DNA replication and messenger RNA production after induction of wild-type λ bacteriophage and λ mutants. *J. molec. Biol.* **19**, 174.

KAISER, A. D. (1957). Mutations in a temperate bacteriophage affecting its ability to lysogenize *E. coli*. *Virology*, **3**, 42.

KAISER, A. D. & HOGNESS, D. S. (1960). The transformation of *E. coli* with DNA isolated from bacteriophage λ dg. *J. molec. Biol.* **2**, 392.

KAISER, A. D. & INMAN, R. B. (1965). Cohesion and the biological activity of bacteriophage lambda DNA. *J. molec. Biol.* **13**, 78.

KAISER, A. D. & JACOB, F. (1957). Recombination between related temperate bacteriophages and the genetic control of immunity and prophage localization. *Virology*, **4**, 509.

KIRBY, E. P., JACOB, F. & GOLDTHWAIT, D. A. (1967). Prophage induction in a thermo-inducible lysogenic bacterium. *Fedn Proc.* **26**, 449.

KORN, D. & WEISSBACH, A. (1963). The effect of lysogenic induction on the deoxyribonucleases of *E. coli* K12λ. I. Appearance of a new exonuclease activity. *J. biol. Chem.* **238**, 3390.

KORN, D. & WEISSBACH, A. (1964). The effects of lysogenic induction on the deoxyribonucleases of *E. coli* K12λ. II. The kinetics of formation of a new exonuclease and its relation to phage development. *Virology*, **22**, 91.

LEDERBERG, E. M. (1954). The inheritance of lysogenicity in interstrain crosses of *E. coli*. *Genetics*, **39**, 978.

LEDERBERG, E. M. & LEDERBERG, J. (1953). Genetic studies of lysogenicity in *E. coli*. *Genetics*, **38**, 51.

LEVINE, M. (1957). Mutations in the temperate phage P22 and lysogeny in *Salmonella*. *Virology*, **3**, 22.

LIEB, M. (1953). The establishment of lysogenicity in *E. coli*. *J. Bact.* **65**, 642.

LIEDKE-KULKE, M. & KAISER, A. D. (1967). Genetic control of prophage insertion specificity in bacteriophages λ and 21. *Virology*, **32**, 465.

LITTLE, J. W. (1967). An exonuclease induced by bacteriophage λ. II. Nature of the enzymatic reaction. *J. biol. Chem.* **242**, 679.

LWOFF, A. (1953). Lysogeny. *Bact. Rev.* **17**, 269.

NAONO, S. & GROS, F. (1966). Control and selectivity of λ DNA transcription in lysogenic bacteria. *Cold Spring Harb. Symp. quant. Biol.* **31**, 363.

OGAWA, T. & TOMIZAWA, J. I. (1967*a*). Abortive lysogenization of bacteriophage lambda b2 and residual immunity of non-lysogenic segregants. *J. molec. Biol.* **23**, 225.

OGAWA, H. & TOMIZAWA, J. I. (1967*b*). Bacteriophage lambda DNA with different structures found in infected cells. *J. molec. Biol.* **23**, 265.

PROTASS, J. J. & KORN, D. (1966). Function of the N cistron of bacteriophage lambda. *Proc. natn. Acad. Sci. U.S.A.* **55**, 1089.

PTASHNE, M. (1965). The detachment and maturation of conserved lambda prophage DNA. *J. molec. Biol.* **11**, 90.

PTASHNE, M. (1967a). Isolation of the λ phage repressor. *Proc. natn. Acad. Sci. U.S.A.* **57**, 306.

PTASHNE, M. (1967b). Specific binding of the λ phage repressor to λ DNA. *Nature, Lond.* **214**, 232.

RADDING, C. M. (1964). Nuclease activity in defective lysogens of phage λ. *Biochem. biophys. Res. Commun.* **15**, 8.

RADDING, C. M. & SHREFFLER, D. C. (1966). Regulation of λ exonuclease. II. Joint regulation of exonuclease and a new λ antigen. *J. molec. Biol.* **18**, 251.

RADDING, C. M., SZPIRER, J. & THOMAS, R. (1967). The structural gene for λ exonuclease. *Proc. natn. Acad. Sci. U.S.A.* **57**, 277.

ROTHMAN, J. L. (1965). Transduction studies on the relation between prophage and host chromosome. *J. molec. Biol.* **12**, 892.

SIGNER, E. R. (1965). Attachment specificity of prophage λ dg. *J. molec. Biol.* **14**, 582.

SIGNER, E. R. (1966). Interaction of prophages at the att_{80} site with the chromosome of *E. coli*. *J. molec. Biol.* **15**, 243.

SIGNER, E. R. & BECKWITH, J. R. (1966). Transposition of the *Lac* region of *E. coli*. III. The mechanism of attachment of bacteriophage $\phi 80$ to the bacterial chromosome. *J. molec. Biol.* **22**, 33.

SKALKA, A. (1966). Regional and temporal control of genetic transcription in phage lambda. *Proc. natn. Acad. Sci. U.S.A.* **55**, 1190.

SMITH, H. O. & LEVINE, M. (1964). Two sequential repressions of DNA synthesis in the establishment of lysogeny by phage P22 and its mutants. *Proc. natn. Acad. Sci. U.S.A.* **52**, 356.

SMITH, H. O. & LEVINE, M. (1967). A phage P22 gene controlling integration of prophage. *Virology*, **21**, 207.

STENT, G. (1964). The operon: on its third anniversary. *Science, N.Y.* **144**, 816.

SUSSMAN, R. & JACOB, F. (1962). Sur un système de répression thermosensible chez le bactériophage λ. *C. r. hebd. Séanc. Acad. Sci., Paris*, **254**, 1517.

SZYBALSKI, W. (1967). Asymmetric distribution of the transcribing regions on the complementary strands of coliphage λ DNA. *Proc. natn. Acad. Sci. U.S.A.* **57**, 1618.

TAYLOR, A. L. (1963). Bacteriophage-induced mutation in *E. coli*. *Proc. natn. Acad. Sci. U.S.A.* **50**, 1043.

THOMAS, R. (1964). On the structure of the genetic segment controlling immunity in temperate bacteriophages. *J. molec. Biol.* **8**, 247.

THOMAS, R. (1965). Le contrôle de la réplication génétique et de l'expression des fonctions chez les bactériophages tempérés. *Archs Biol.* **76**, 551.

THOMAS, R. (1966). Control of development in temperate bacteriophages. I. Induction of prophage genes following heteroimmune superinfection. *J. molec. Biol.* **22**, 79.

THOMAS, R. & BERTANI, L. E. (1964). On the control of the replication of temperate bacteriophages superinfecting immune hosts. *Virology*, **24**, 241.

THOMAS, R., LEURS, C., DAMBLY, C., PARMENTIER, D., LAMBERT, L., BRACHET, P., LEFÈBVRE, N., MOUSSET, S., PORCHERET, J., SZPIRER, J. & WAUTERS, D. (1967). Isolation and characterization of new sus (amber) mutants of bacteriophage λ. *Mutation Res.* **4**, 735

TOMIZAWA, J. I. & OWAGA, T. (1967). Effect of ultraviolet irradiation on bacteriophage lambda immunity. *J. molec. Biol.* **23**, 247.

VAN DE PUTTE, P., ZWENK, H. & RÖRSCH, A. (1966). Properties of four mutants of *E. coli* defective in genetic recombination. *Mutation Res.* **3**, 381.

WEIGLE, J. (1966). Assembly of phage lambda *in vitro*. *Proc. natn. Acad. Sci. U.S.A.* **55**, 1462.

WEISBERG, R. A. & GALLANT, J. A. (1966). Two functions under cI control in lambda lysogens. *Cold Spring Harb. Symp. quant. Biol.* **31**, 374.

WOLLMAN, E. L. (1953). Sur le déterminisme génétique de la lysogénie. *Annls Inst. Pasteur*, **84**, 281.

WOLLMAN, E. L. & JACOB, F. (1957). Sur le processus de conjugaison et de recombinaison chez *E. coli*. II. La localisation chromosomique du prophage λ et les conséquences génétiques de l'induction zygotique. *Annls Inst. Pasteur*, **93**, 323.

WU, R. & KAISER, A. D. (1967). Mapping the 5'-terminal nucleotides of the DNA of bacteriophage λ and related phages. *Proc. natn. Acad. Sci. U.S.A.* **57**, 170.

ZICHICHI, M. L. & KELLENBERGER, G. (1963). Two distinct functions in the lysogenization process: the repression of phage multiplication and the incorporation of the prophage in the bacterial genome. *Virology*, **19**, 450.

ZISSLER, J. (1967). Integration-negative (int) mutants of phage λ. *Virology*, **31**, 189.

BACTERIAL TRANSFER FACTORS AS VIRUSES

E. S. ANDERSON

Enteric Reference Laboratory, Public Health Laboratory Service, Colindale Avenue, London, N.W. 9

INTRODUCTION

The resemblance between a bacterial transfer factor and viruses was first pointed out by Hayes (1953*a,b*) who remarked on the similarities between the behaviour of the F factor of *Escherichia coli* K12 and that of temperate phages in the prophage state. The similarity was limited, however, by the fact that, in spite of the efficiency with which the F factor was transferred, cell-free filtrates of donor strains were inactive in this respect. Contact between donor and recipient cells was necessary for the transfer of the F factor to occur.

Speculation on the possible viral nature of transfer factors has been renewed recently (Hayes, 1964, 1965; Anderson, 1965*a*, 1966*a*), and in this article the evidence in support of this hypothesis will be reviewed.

THE DEFINITION OF VIRUSES

The discussion will be based on the definition of Lwoff (1959) who postulated three main criteria:

(1) Possession of only one type of nucleic acid.

(2) Dependence on the metabolic equipment of the host cell for the provision of energy systems.

(3) Infectivity by means of particles of organized morphology.

The question of pathogenicity is not included in this definition, and there seems to be no more reason for pathogenicity being a necessary property of viruses than of bacteria. As we know, therefore, that the majority of bacteria are non-pathogenic, we may conjecture that the majority of viruses are devoid of harmful effects on the host cell. However, in contrast to the ease with which non-pathogenic bacteria can be isolated because they can be cultivated on artificial media, the virologist is restricted to the study of the pathogenic viruses, because the presence of those free from pathogenicity may defy the usual isolation techniques. But if we accept the postulate that non-pathogenic viruses are probably numerous, and presumably widespread, we may also propose the existence of viruses that are indifferent to the welfare of the host, of viruses that are beneficial (Andrewes, 1966), and perhaps of viruses that are necessary for the survival of the host systems they inhabit. In the same way as life could not continue without the contribution of numerous non-pathogenic bacteria, therefore, it can be suggested that unidentified viruses may have a similar importance at all biological levels.

A truly non-pathogenic virus could not expect to be transmitted as a result of cell destruction produced by itself. If it were entirely dependent on its own resources, therefore, and were to enjoy anything better than strictly lineal transmission, it would either have to be extruded by its intact host cells in order to seek fresh hosts, or it would have to be transmitted directly from infected to non-infected cells by contact. It can be assumed that only extrusion would necessitate the existence of an organized particulate viral form, because direct transmission from cell to cell could be in the form of the viral nucleic acid, which comprises the whole viral genome.

Having postulated the existence of non-pathogenic viruses, and having suggested that they may have dispensed with the need of an externally infective form, let us seek for evidence of the existence of agents that might be regarded as fitting this description. The bacterial transfer factors, which are found in the Enterobacteriaceae, are our obvious choice. I shall describe some of them so that their properties may be compared with those of known viruses.

THE F FACTOR OF *ESCHERICHIA COLI* K 12

The first of the transfer factors to be discovered was the well-known F factor of *E. coli* K 12. This strain will be referred to hereafter as K 12.

Some years after the discovery of recombination in *E. coli* K 12 by Lederberg & Tatum (1946*a,b*), it was observed by Hayes (1952) that mating cells in K 12 crosses showed polarity, that is, that one strain acted as the male or donor and the other as the female or recipient partner. This led to the demonstration that the property of 'maleness' was infectious in these bacteria, and that the factor on which this property depended was transmitted to female cells at much higher frequency than the genetic characters whose transfer it was mediating (Lederberg, Cavalli & Lederberg, 1952; Hayes, 1953*a,b*; Cavalli-Sforza, Lederberg & Lederberg, 1953). This factor was designated the F or fertility factor, and has come to be regarded as the 'sex' factor of *E. coli* K 12. Cells possessing it are known as F + and act as donors of genetic characters. Cells lacking F are designated F−. As I have said, the F factor passes from F+ to F− cells with much higher frequency than the genetic characters it transfers. In fact it spreads in epidemic fashion in the recipient population, which then becomes F+. It is therefore evident that F multiplies for a time autonomously in freshly infected cells, so that each cell contains a number of copies of F (Lederberg *et al.* 1952; Cavalli-Sforza *et al.* 1953; de Haan & Stouthamer, 1963). This stage is succeeded by that in which the rate of multiplication of F is governed by that of its host, which then contains perhaps only one F copy per chromosome, although the F factor is extrachromosomal.

The F factor may become integrated into the bacterial chromosome, and when this occurs it mediates chromosomal transfer. During conjugation, a portion of the F factor is probably associated with the leading end of the chromosome and the remainder with the distal end. Cells containing F in this state transfer chromosomal characters with high frequency and are known as Hfr (High Frequency Recombinants). Because part of F is at the distal end of the chromosome during transfer, and because fracture of the transferring chromosome is common, the entire F factor is rarely transferred by Hfr strains, so that recipient cells usually remain F−.

A third state of F results from genetic exchange between it and the chromosome in regions adjacent to the site of integration of F in Hfr strains. The result is that F may pick up bacterial genes which are contiguous with its chromosomal location. One such gene is that for lactose fermentation (lac). These recombinant F factors are known as F′ factors (Adelberg & Burns, 1959, 1960). F′ factors will infect F− cells with high efficiency, but each cell receiving F also receives the bacterial gene associated with it. Thus, *E. coli* F− lac− recipients become simultaneously F+ and lac+ after infection with F-lac. Another property of F′ factors is brought about by the genetic exchange which results in their formation, presumably because of the region of the F factor left on the bacterial chromosome. This produces an oscillating relationship between F′ factors and the chromosome. Transfer of characters thus occurs in the same order as in the Hfr strain from which the F′ factor is derived, but is lower in frequency because of the transient nature of F′ integration. If original F′ strains are 'cured' of F with acridines the treated strain becomes F−, but reinfection with F produces, not ordinary F+, but an F′ strain, because the portion of F inserted into the chromosome is not eliminated by the acridines, and the newly introduced F factor assumes the same chromosomal relationship as its predecessor.

The F factor is composed of deoxyribonucleic acid (DNA), of which it contains about the same amount as a phage such as λ (Arber, 1960; Driskell-Zamenhoff & Adelberg, 1963), and the guanine + cytosine percentage of most of its DNA is similar to that of the DNA of K 12 (Falkow & Citarella, 1965). Although its properties suggest that it has some homology with the DNA of K 12, at least half its DNA appears to be peculiar to the F factor itself (Falkow & Citarella, 1965).

The autonomous replication of F in freshly infected cells is an indication that, until it submits to cellular control and settles down to the stable number of perhaps one per chromosome, the F factor behaves as an independent infective agent, using the shelter and metabolic provisions of the host cell to propagate, and able to invade fresh cells. Once an organism is infected with F it resists superinfection with the same factor, as shown by the experiments of Scaife & Gross (1962) in superinfection of F + cells with the F-lac factor.

Presumably the resident and the superinfecting factor compete in the host cell for a site which can accommodate only one of them. The result is that either the superinfecting or the resident factor is lost, so that the progeny ultimately contain either one or the other. This is reminiscent of interference between related viruses, and particularly of mutual exclusion between the temperate bacteriophages. It is, for example, difficult to make a bacterial line doubly lysogenic for closely related phages, and it is assumed that the phages concerned compete for the single or the few sites which the prophages can occupy.

TRANSFER FACTORS OTHER THAN F

Although the F factor is the most extensively studied transfer factor it is only one of a large group of factors which are widely distributed in the Enterobacteria. Transfer of chromosomal characters by factors other than F has been demonstrated to only a limited extent, but genes for other characters, perhaps not of chromosomal origin, are carried. The most striking among these are the genetic determinants which promote the synthesis of colicins, and those conferring antibiotic resistance. The latter group are of considerable medical and veterinary importance because they may spread in the Enterobacteria resistance to all currently available antibiotics, and to synthetic antibacterial drugs (reviews: Watanabe, 1963; Datta, 1965). The transfer factors carrying the determinants for drug resistance have a wide host range and will infect many so-called 'genera' in the Enterobacteriaceae. In practical terms their importance lies in the fact that a transmissible drug resistance, emerging in non-pathogens as the result of the use of antibiotics, may be transferred to pathogens, which then become resistant to the drugs concerned, and which in suitable conditions may cause epidemics difficult to control. Such an epidemic occurred in calves in Britain between 1964 and the end of 1966. It affected thousands of calves and resulted in hundreds of human infections. Attempts to control the calf disease were directed more towards the use of antibiotics than to the improvement of animal husbandry. The dominant infecting strain, which belonged to phage-type 29 of *S. typhimurium*, developed successively resistance to streptomycin and sulphonamides, tetracycline, ampicillin, neomycin and kanamycin, and furazolidone. Chloramphenicol resistance also appeared in a few instances. All these resistances were transmissible to *E. coli* and they may originally have arisen in non-pathogenic Enterobacteria in animals.

This incident, and others, showed the potential ecological importance of the transfer factors. The work it stimulated (Anderson, 1965*a,b*, 1966*a*, 1967; Anderson & Lewis, 1965*a,b*) gave us a considerable amount of information concerning the method of operation and the origin of the transmissible resistance factors or R-factors. It had been postulated (Watanabe, 1963) that the R-factors consisted of a linear linkage group of genetic determinants for drug resistance, the transmissibility of which was potentiated by a hypothetical resistance transfer factor or RTF. Our work established the existence of the transfer factors as independent entities, and showed that they could often be easily separated from the resistances they carried. It also showed that transfer factors originally associated with resistance transfer could carry colicinogeny and, conversely, that transfer factors originally associated with colicinogeny transfer could mediate the transfer of drug resistance. I believe that the transfer factors in general carry a wide range of characters that await identification. Our findings suggested that the resistance determinants were basically independent of each other and were independently attached to the transfer factor, that is, that the linear linkage between determinants suggested by Watanabe probably did not exist. Many wild, drug-sensitive strains of *S. typhimurium* were shown to carry transfer factors that could pick up resistance determinants, and wild drug-resistant *S. typhimurium* strains were identified which could transfer their resistance only after infection by a transfer factor. It was therefore suggested that R-factors arose in nature by contact between bacterial strains containing transfer factors and strains carrying drug-resistance determinants, in an environment in which selection pressure of antibiotics favoured the dominance of drug-resistant organisms and produced a high probability of contact between such organisms and those carrying the transfer factors which effected the mobilization (Anderson, 1965*a,b*).

In the system on which we concentrated, it was observed that some resistance determinants, those for ampicillin and streptomycin-sulphonamide resistance for example, were usually dissociated from the transfer factor carrying them, which has been designated the Δ factor (Anderson & Lewis, 1965*b*). This dissociation resulted in transmission of the Δ factor to recipient cells at a frequency of 5×10^{-1} or higher, but in a relatively low frequency of transfer of the drug resistance determinants (about 10^{-2}). Moreover, transfer of the drug resistance *without* Δ occasionally occurred. With the exception of the last feature, this activity of the Δ transfer system resembles the transfer of characters by F+ cells. The transfer of resistance determinants without Δ, which is easily demonstrated in crosses interrupted up to 2 hr., resembles the transfer of characters from Hfr to F− strains of K12. In Hfr × F− crosses the characters are transferred with high frequency, but the F factor, because of its distal position on the chromosome, rarely enters the recipient cells. On the other hand, it was shown that the determinant for tetracycline resistance was closely linked to the Δ factor, so that all cells receiving tetracycline resistance also received Δ. This aspect of Δ activity resembles that of F′ systems, and it has been suggested that the tetracycline resistance determinant is inserted into Δ in the same way as the lac gene, for example, is inserted into F in the F-lac factor (Anderson, 1966*a*, 1967). It has been observed, moreover, that Δ obstructs superinfection with further Δ complexes and with other related transfer factors, in the same way as F factors oppose superinfection with F (Anderson, 1966*a*). Thus although mediation of chromosomal transfer by Δ has not so far been demonstrated, the Δ transfer system identified in *S. typhimurium* and the F system in *E. coli* show close operational similarities. We have identified many other resistance transfer systems in *S. typhimurium*, and in other Enterobacteria, in which such similarities can be demonstrated, and we can conclude that the modes of operation of the transfer factors in general are the same throughout the Enterobacteria.

PHAGE INTERFERENCE BY TRANSFER FACTORS

A special property of the transfer factors is their effect on the phage sensitivity of their host cells. Watanabe *et al.* (1964) divided the transfer factors into two types. The first type inhibited the fertility of F+, F′ and

Hfr strains of K 12, and was called fi + ; the second was without effect on F fertility and was called fi − . Watanabe, Fukasawa & Takano (1962) also demonstrated that fi + factors reduced the sensitivity of F + strains to the so-called male-specific phages, and this property proved to be useful for testing the 'fi' property of transfer factors. It has been suggested (Datta, Lawn & Meynell, 1966) that the inhibitory effect of fi + transfer factors on F fertility is due to repression of the synthesis of the 'sex fimbriae', which are associated with the possession of F (Brinton, Gemski & Carnahan, 1964) and which have been postulated to play a part in bacterial conjugation. As these fimbriae are the receptors for the male-specific phage, their absence also removes the portal of entry of these phages into the cell.

There is, however, a phage inhibitory effect of the transfer factors that operates on a more intimate cellular plane. Watanabe *et al.* (1964) showed that certain fi − factors, when introduced into non-lysogenic K 12, inhibited the multiplication of externally invading phages λ and T 1. Watanabe, Takano, Arai & Sato (1966) later observed that phage restriction was also exerted by fi − factors on the multiplication of phage T 7 in K 12, and phage P 22 in *S. typhimurium* LT 2. The infecting phages were still adsorbed, but their DNA was degraded. Some invading phage particles could multiply but their progeny proved to have undergone host-controlled (phenotypic) modification; in these systems, therefore, the interference with phage multiplication by the transfer factors is similar to the interference with multiplication of invading phages by prophages already in the host cell (for review see Arber, 1965).

This similarity was observed in our work, at a time when we were unaware of Watanabe's observations (Anderson & Lewis, 1965*b*). We showed that the Δ transfer factor, which is fi −, had a restricting effect on the phage sensitivity of *S. typhimurium*. When Δ was introduced into type 36 of *S. typhimurium*, sensitive to all 30 of the *S. typhimurium* typing phages, the host strain was converted into type 6, sensitive to only nine of the phages. Δ also restricted the phage sensitivity of *S. paratyphi B* and *S. typhi* in a specific fashion, and it will probably produce characteristic phage restriction in any Salmonella it will infect (Anderson, 1966*a*). The significance of these observations lies in the fact that, in the case of *S. typhimurium*, *S. paratyphi B* and *S. typhi*, the phage-restricting effect of the transfer factor can be defined with precision because of the existence of the range of phages employed in the phage-typing scheme for each organism. The specificity of the phage types of the various Salmonellae has long been known to depend in part on the exclusion effects of temperate phages ('determining phages') carried by some phage types (for review see Anderson, 1962). We have not so far identified determining phages in type 6 of *S. typhimurium*, but the phage-restricting effect of a modified form of Δ, known as Δm, on the sensitivity of *S. typhi* to the typing adaptations of Vi-phage II (Craigie & Yen, 1938), is identical to that of a determining phage discovered some years ago and designated phage f2 (Anderson, 1951; Felix & Anderson, 1951; Anderson & Felix, 1953; Anderson, 1966a). (This phage is unrelated to the male-specific RNA phage also designated f2 (Loeb, 1960; Loeb & Zinder, 1961).)

Salmonella typhi carrying phage f2 (*S. typhi* (f 2)) is lysed by a specific host-range mutant of Vi-phage II, and it has been shown that *S. typhi* (Δm) is lysed by the same mutant, and will select that mutant from a population of the wild-type of Vi-phage II. Thus, Vi-phage II is unable to distinguish *S. typhi* carrying Δm from *S. typhi* carrying phage f2.

The demonstration of identity of pattern of the phage-restricting effect of Δm on the one hand, and phage f2 on the other, in *S. typhi*, suggests the possibility of similarity between the two agents. Vi-phage II unable to lyse *S. typhi* carrying either agent is nevertheless adsorbed, so that the restriction occurs at some stage in the phage multiplication beyond adsorption. Prophage f2 consists of DNA and it can be assumed from the work of Falkow & Citarella (1965) and Falkow, Citarella, Wohlhieter & Watanabe (1966) that Δm also consists of DNA. It is known that determining phages with closely similar restricting effects are closely related to each other (Anderson & Felix, 1953). If the same step in Vi-phage II multiplication is blocked by Δm and phage f2, therefore, the two agents may possess similar genes for phage restriction, so that they may have considerable regions of homology in their DNA. However, although the presence of Δm in *S. typhi* excludes phage f2, this phage, while exerting some exclusion effect on Δm, depresses its infectivity towards *S. typhi* by only about a log unit. More work is needed in this field and for the moment it can only be suggested that there may be a similarity between the DNA of Δm and that of phage f2.

The demonstration of the type-determining effects of Δ and Δm in *S. typhimurium*, *S. paratyphi* B and *S. typhi* has led us to use selected phage types of these three Salmonella serotypes as the 'standard' hosts in the evolution of a classification scheme for transfer factors. Because of the multiplicity of these agents, and because they come from a variety of hosts, Salmonellae, Shigellae, *E. coli* and other Enterobacteria, it occurred to us that the limited subdivision of transfer factors now available (fi+ and fi−) could be extended by examining the phage-restricting effects of transfer factors of different origin when introduced into the standard hosts (E. S. Anderson & J. S. Pitton, to be published). In general, fi − factors have more phage-restricting effect on Salmonellae than have fi + factors. Transfer factors having closely related restricting effects have been isolated from *E. coli*, *Shigella sonnei* and *S. typhimurium*. The most sensitive host for the demonstration of the phage-restricting effect of transfer factors seems to be *S. typhi*, although *S. typhimurium* and *S. paratyphi B* are also needed for the characterization of restriction.

There are determining phages in *S. typhi* which are so closely related that they can be distinguished only by the slight differences in their pattern of phage restriction (Anderson & Felix, 1953). This is true of the group of phages of which f2 is the prototype. If I were presented with a strain of *S. typhi* carrying Δm or other transfer factors resembling it, and were asked which determining phage it carried, I should identify Δm as a prophage of the f2 group. And I should not be greatly deterred by failure to isolate mature phage particles from the host strain, because the determining phages of *S. typhi* often become so defective as never to produce mature phage particles (Anderson, 1959; Bernstein & Wilson, 1963).

In any event, temperate determining phages can be identified by their phage-restricting effects on Salmonellae, and transfer factors are to some extent identifiable by the same method, sometimes by the same restriction patterns. The temptation to postulate a relationship between the two types of agent is very strong.

SUMMARY OF THE PROPERTIES OF TRANSFER FACTORS

(1) They are composed of one type of nucleic acid.

(2) They depend on the metabolic equipment of their host cells to provide energy for their survival and replication.

(3) They are infective, but not in the form of a particulate organized phase. Their infectivity depends on their ability to potentiate conjugation between donor and recipient bacterial cells.

(4) They associate with bacterial genetic determinants which they carry to new hosts.

(5) At least one transfer factor, the F factor, can insert itself into, and mediate the transfer of, the chromosome of its host.

(6) Transfer factors can recombine with the genetic material of the host so as to produce factors (F-lac for example) which confer at the same time, and with high efficiency, transfer factor infectivity—a property of the original transfer factor particle, and one or a few

genes coding for metabolic characters, independently of chromosomal transfer.

(7) They exert an exclusion effect against superinfection with themselves.

(8) They exert phage-restricting effects which may be closely similar to those of known phages.

If the form of infectivity is immaterial, the first three properties correspond to Lwoff's (1959) definition of viruses. The question that now arises is whether the invasion mechanism *is* a necessary part of viral make-up. Experiments which demonstrate the infectivity, albeit very low, of extracted viral nucleic acid suggest that it is not. However, for the transmission of an intact virus an invasion mechanism of some sort is clearly necessary, and the known viruses have developed this in various forms. The natural corollary is that they can be obtained in cell-free suspension, and that infection of susceptible host cells is effected by the adsorption of virus particles. The viral invasion mechanism does the rest. But if the nucleic acid carries all the information necessary for viral reproduction, the viruses that can exist in the form of infective particles capable of invading their hosts from without may constitute only one viral group. Alternative mechanisms that ensured transmission of their nucleic acid into fresh host cells could dispense with the need of an intrinsically viral invasion apparatus. What form would such a mechanism take? One answer to this question may be found in the bacterial transfer factors. The mechanism whereby the F factor gained entry into recipient cells was an obvious field for conjecture. Two main hypotheses have been proposed. The first is that a conjugation bridge is formed between the donor and recipient cells, and that the F factor passes through this bridge into the new host. Experimental evidence for this is sparse and, although bridges apparently connecting donors with recipients have been demonstrated in electron micrographs (Anderson, Wollman & Jacob, 1957) they are rare, and they may be artifacts. An alternative possibility arises from the work of Brinton *et al.* (1964) who demonstrated that possession of the F factor causes host cells to produce the special fimbriae which act as the receptors for the so-called male-specific bacteriophages. F − bacteria do not produce such fimbriae, and do not support multiplication of the male phages. The production of similar fimbriae has also been demonstrated in strains infected with transfer factors other than F (Datta *et al.* 1966). The suggestion that these fimbriae act as the channels through which transmission of the transfer factor takes place must obviously be considered.

As far as the availability of transfer factors for transmission is concerned, it is postulated that the F factor is attached to the cell membrane at a specific site (Jacob, Brenner & Cuzin, 1963). This site would coincide either with the local production of an antigen facilitating contact with recipient cells, followed by the synthesis of an intercellular bridge, or with the production of sex fimbriae through which transmission of the factor would take place. Whatever the structural details of the actual transmission apparatus, the present evidence favours the idea that it is synthesized by the host cells in response to a specific stimulus provided by the transfer factors. It therefore performs the functions normally carried out by, for example, the phage tail. It has been suggested (Hayes, 1965) that a curious group of phages which are of filamentous structure, not unlike fimbriae in appearance, and which may be released without bacterial lysis (Hofschneider & Preuss, 1963), may represent an intermediate stage between the phages of conventional morphology (that is, consisting of head and tail) and the transfer factors, in which transmission may be by way of fimbriae synthesized by the host.

TRANSFER OF BACTERIAL CHARACTERS BY BACTERIOPHAGES

The evidence adduced in favour of the hypothesis that transfer factors may be viral in nature is so far limited to pointing out that they conform to Lwoff's definition of viruses, which may of course be inadequate, and to describing their similarities to the phages, which may be coincidental. Let us now reverse the coin and seek an indication that viruses may act as factors which transfer genetic material between different hosts. Here again we must turn to the phages, for the evidence lies in the phenomenon of transduction, that is, the phage-mediated transmission of discrete or closely linked characters from one bacterial line to another (Zinder, 1953; Stocker, Zinder & Lederberg, 1953). The phage is grown on a bacterial strain possessing the characters concerned, and is then used to infect a recipient bacterial population lacking these characters. Recipient cells which have escaped destruction by the phage are then grown on a medium selecting for, or able to detect, the character under examination. Any detectable bacterial character may be acquired by transduction. Although the injection of phage DNA is necessary for transduction to occur, lysogenization is unnecessary, and transduction may also be carried out by virulent phages. The success of transduction is usually of a low order, 10^{-6} or less per phage particle.

One type of transduction, however, has a high success rate; that of galactose fermentation (gal) transfer by high-frequency transduction

(HFT) lysates of phage λ (Morse, Lederberg & Lederberg, 1956a,b). This may be summarized as follows: gal transduction is possible only with λ released by ultraviolet (u.v.) irradiated K 12 (λ) gal+ cells. This phage transduces the gal marker with the usual frequency of 10^{-6}. The resulting transductants, however, are diploid for the gal region, carrying both the parental gal− and the introduced gal+ allele. Ultraviolet induction of such cells results in the production of a λ lysate which transduces the gal marker with a frequency as high as 5×10^{-1}. This high frequency of transduction is probably due to the incorporation of the gal segment of the original bacterial host of λ into the phage genome, so that all cells of the first (restricted diploid) transductant series receive the λ-gal hybrid genome. Irradiation of these cells precipitates the release of a phage population which consists largely of λ-gal particles. Although such particles can infect, and protect against superinfection with λ, the cells they invade cannot produce mature phage particles (Arber, Kellenberger & Weigle, 1957). The λ-gal particles are thus defective, and it has been shown that such particles, belonging to the class designated λdg, have undergone a deletion of about 25 % of their chromosome (Arber *et al.* 1957; Arber, 1958). It is apparently this deleted region which is replaced by the gal marker.

The transduction of gal by phage λ belongs to the category of restricted transduction. It results from the fact that λ has a single site of chromosomal attachment, and can therefore pick up only genes related to that site. Some phages, such as P22, will transduce any character that can be detected although at a very low frequency ($\leqslant 10^{-6}$); this is known as generalized transduction. It has been suggested that the ability of these phages to transduce a wide range of characters is due to their capacity for attaching themselves to a large number of chromosomal sites, there to recombine with contiguous genes. This would group restricted and generalized transduction under the same heading, the difference of expression being governed by whether or not the prophage has a specific chromosomal attachment site.

Support for this hypothesis comes from the demonstration that HFT lysates can be prepared of some general transducing phages, such as P1 and P22, and that the production of such lysates is associated with a variable degree of defectivity in the phage particles concerned (Luria, Fraser, Adams & Burrous, 1958; Luria, Adams & Ting, 1960; Dubnau & Stocker, 1964).

There is at any rate a formal resemblance between the formation of λdg particles on the one hand, and the F-lac factor on the other, in that both carry a region of the bacterial chromosome. Although the tetracycline resistance determinant T may not be of chromosomal origin, the operational similarities between Δ-T and F-lac suggest that the Δ-T resistance factor may be included in the same category, and that its formation, as of that of any complex in which there is an intimate relationship between a transfer factor and the bacterial determinant it carries, is due to recombination between the bacterial genetic fragment and the transfer factor. Defective prophages may have a less stable relationship with the bacterial chromosome than normal prophages. And they can incorporate genetic fragments in their genome. When they cannot produce infective particles, although they may be transmitted lineally, they have no way of infecting fresh hosts. Whether it is justifiable to suggest that their next step was to persuade their bacterial hosts to participate in their transmission, I hesitate to say, but if such a step was taken, we have a possible origin for the transfer factors. Defective prophages could then be regarded as a transitional stage between viruses and transfer factors.

ANTIGENIC CONVERSION OF BACTERIA BY PHAGES AND TRANSFER FACTORS

I have mentioned earlier the similar properties of phage restriction bestowed on bacterial hosts by the genomes alone of the phages and the transfer factors. These characters, which appear in all cells receiving either type of agent, belong to the variety known as bacterial conversion. Another sort of conversion effected by phages is that of antigenic conversion (Uetake, Nagakawa & Akiba, 1955; Uetake, Luria & Burrous, 1958; Uetake & Hagiwara, 1960, 1961). In this, infection by the phage stimulates the host cell to produce a new antigen, and every phage-infected cell shows this change. The phenomenon has been best explored in the E group of Salmonellae, in which the production of two somatic antigens, 15 and 34, is dependent on the presence of phage ε15 and ε34 respectively. However, antigen 15 is necessary for the adsorption of phage ε34, so that prior infection with phage ε15 is necessary before ε34 can effect the conversion. In this system, infection by one phage (normally followed by lysogenization) is needed in order to promote the synthesis of the bacterial receptor to which another phage is adsorbed. If a parallel is to be found in the transfer factors, it may lie in the sex fimbriae, the synthesis of which depends on the presence of transfer factors in the cell. Apart from their possible role in conjugation, these fimbriae form the receptor for the adsorption of male-specific phages.

It was observed by Ørskov & Ørskov (1960) that possession of the

F factor stimulated the host cells to produce a specific surface antigen, which they called the f+ antigen. The fimbriae may consist of f+ antigen, so that the conjugational, phage adsorption and antigenic properties are different manifestations of the same conversion.

CONVERSION EFFECTS OF ANIMAL VIRUSES

I have limited my observations to a comparison of the properties of the transfer factors and the bacterial viruses, first, because they affect the same hosts in ways that are most susceptible to comparison; second, because transfer factors can be detected in the bacteria by a variety of techniques, but have not yet been identified in other life forms; and third, because genetic modification by viruses, other than that directly dependent on the virus genome (that is, conversion) has not yet been observed outside the bacteria. There are, however, examples of genetic conversion of host cells by animal viruses, that of oncogenesis being the most obvious. An oncogenic virus does not destroy its host cell, but modifies it in such a way that it ceases to submit to the regulatory processes of the macro-organism of which it forms a part. It thus becomes an independent growing unit, either a simple or malignant tumour. The actual mechanism of oncogenesis is not yet clear, but what is clear is that the effect of such viruses is modification of the genetic apparatus of the cells they infect so as to produce independent and aberrant replication. It has been suggested that tumour formation does not take place unless the virus nucleic acid is incorporated in the genome of the host cell. Support for this hypothesis comes from the fact that, once tumour formation is established, free particles of the causal oncogenic virus may not be detectable in the tumour tissue. For example, free virus particles may not be demonstrable in tumours formed in hamsters after infection by simian virus 40 (SV40), a DNA virus. If an indicator system of virus-sensitive monkey kidney cells is provided, however, free SV40 appears (Gerber & Kirschstein, 1962; Gerber, 1963, 1964, 1966). Gerber (1966) has established that contact between the tumour cells and the indicator cells is necessary for the production of free virus. He suggests that the synthesis of virus takes place in the indicator cells as the result of the transfer of the viral genome (presumably in the form of the viral DNA only) from the tumour cells to the indicator cells.

Similar observations were made by Svoboda, Chyle, Simkovic & Hilgert (1963) with the ribonucleic acid virus of Rous sarcoma. They were unable to demonstrate the presence of free virus in the XC tumour, produced in rats as the result of infection with Rous virus. Introduction of XC cells into chicks, however, produced Rous sarcoma which liberated virus. Intact XC tumour cells were needed to produce this effect. Simkovic, Valentova & Thurzo (1962) found that, when XC tumour cells were grown in tissue culture with chick fibroblasts, free Rous virus particles were released which were demonstrable by inoculation of chicks with culture fluids free from cells. Cell-free preparations of both tumour and indicator cells were inactive in promoting virus release (see Svoboda, this Symposium).

Huebner, Rowe, Turner & Lane (1963) found that virus antigens were produced in tumours induced in hamsters and rats by human adenoviruses types 12 and 18, although free virus could not be detected.

These observations establish that the genome of the oncogenic viruses is present in the respective tumour cells, and it seems probable that it is in the form of the viral nucleic acid, which is integrated into the genome of the host cells. Although this form does not normally liberate virus particles, its transfer to suitable indicator cells by contact with tumour cells results in the synthesis and liberation of mature virus particles. Gerber (1966) has suggested a similarity between this phenomenon and that of resistance transfer in the Enterobacteria, but this, while being an intriguing possibility, should be treated with caution until the similarities between the two phenomena are shown to be more than superficial.

The animal viruses mentioned possess a particulate infective form, but if analogues of the bacterial transfer factors exist which can produce similar effects, they could be genetically transmitted from either parent of a diploid system. They could then exercise their effect as a hereditary character, although that effect could have, from the point of view of oncogenesis, a tissue or organ specificity. This hypothesis need not be restricted to oncogenesis, but could form a basis for the re-examination of the pathogenesis of some diseases. At the moment, however, this is pure speculation and, until techniques are developed to test for the presence of transfer factors in other life forms than the protista, it must remain so.

It has been pointed out elsewhere (Anderson, 1966*b*, 1967) that a study of the spread of transferable drug resistance of recent years has shown the importance of transfer factors in bacterial ecology and that, knowing that they are widespread, and assuming that they can carry many characters other than drug resistance and colicinogeny, the conjecture that they have influenced and probably accelerated bacterial evolution is irresistible. If analogous agents exist in life forms other than the bacteria, the same arguments are applicable, and evolution in other biological fields may have been speedier than that envisaged purely in terms of mutation and natural selection.

CONCLUSIONS

This is a speculative paper, and I have accordingly extrapolated from what is known about bacterial transfer factors and bacterial viruses, in order to determine how good a case can be made for regarding the transfer factors as viruses. It is worthwhile now to summarize the known similarities and differences between the two types of agent.

(1) Both consist of only one sort of nucleic acid. They depend on their host's metabolism for provision of the energy systems for survival and replication, and they are infective. Whereas the viruses have an organized particulate form capable of external infection of new host cells, however, the transfer factors are dependent on transmission by cell-to-cell contact, their transmission mechanism being provided by donor cells under specific stimulation by the transfer factors.

(2) They both convert their host cells in similar ways, which are dependent on the genome of the transfer factors or the phages.

(3) They can both incorporate genetic determinants of their host cells into their genomes, and can transfer them to recipient cells.

(4) The transfer factors seem to have a mainly cytoplasmic location, whereas most prophages are believed to be attached to, or inserted into, the chromosome. However, the F factor can occupy a chromosomal site, although it is generally accepted that this is in the nature of an insertion rather than an attachment. The cytoplasmic location of the transfer factors may be optimal for their transmission, especially if, as has been suggested, they are attached to the cell membrane (Jacob *et al.* 1963). Their apparent general lack of a capacity for chromosomal attachment or insertion may be due to the fact that most transfer factors have lost, or have not acquired, a region of genetic homology with the bacterial chromosome.

(5) Because of the chromosomal attachment of the prophages, the bacterial genes transduced by phages are mainly of chromosomal origin, whereas those carried by transfer factors seem to be mainly extra-chromosomal. The F factor again forms a connecting link, however, because it transfers chromosomal characters.

Whether in the last analysis we should consider the transfer factors as viruses depends on the flexibility we are prepared to permit in our definition of viruses, the obvious differences between the two types of agent lying in the mode of infective spread. The decision is an important one, however, because, whatever the nature of the transfer factors, they are capable of autonomous replication and transmission to fresh hosts. If they cannot be classified as viruses, they must be regarded as yet another form of life.

REFERENCES

ADELBERG, E. A. & BURNS, S. N. (1959). A variant sex factor in *Escherichia coli*. *Genetics*, **44**, 497.

ADELBERG, E. A. & BURNS, S. N. (1960). Genetic variation in the sex factor of *Escherichia coli*. *J. Bact.* **79**, 321.

ANDERSON, E. S. (1951). The significance of Vi-phage types F1 and F2 of *Salmonella typhi*. *J. Hyg.*, *Camb.* **49**, 458.

ANDERSON, E. S. (1959). Mutation in the type-determining phages of *Salmonella typhi*. *Nature, Lond.* **184**, 1822.

ANDERSON, E. S. (1962). The genetic basis of bacteriophage typing. *Br. med. Bull.* **18**, 64.

ANDERSON, E. S. (1965a). Origin of transferable drug-resistance factors in the Enterobacteriaceae. *Br. med. J.* **2**, 1289.

ANDERSON, E. S. (1965b). A rapid screening test for transfer factors in drug-sensitive Enterobacteriaceae. *Nature, Lond.* **208**, 1016.

ANDERSON, E. S. (1966a). The influence of the Δ transfer factor on the phage sensitivity of Salmonellae. *Nature, Lond.* **212**, 795.

ANDERSON, E. S. (1966b). Possible importance of transfer factors in bacterial evolution. *Nature, Lond.* **209**, 637.

ANDERSON, E. S. (1967). Facteurs de transfert et résistance chez les Enterobactéries. *Annls Inst. Pasteur*, **112**, 547.

ANDERSON, E. S. & FELIX, A. (1953). The Vi-type determining phages carried by *Salmonella typhi*. *J. gen. Microbiol.* **9**, 65.

ANDERSON, E. S. & LEWIS, M. J. (1965a). Drug resistance and its transfer in *Salmonella typhimurium*. *Nature, Lond.* **206**, 579.

ANDERSON, E. S. & LEWIS, M. J. (1965b). Characterization of a transfer factor associated with drug resistance in *Salmonella typhimurium*. *Nature, Lond.* **208**, 843.

ANDERSON, T. F., WOLLMAN, E. L. & JACOB, F. (1957). Sur les processus de conjugaison et de recombinaison chez *E. coli*. III. Aspects morphologiques en microscopie électronique. *Annls Inst. Pasteur.* **93**, 450.

ANDREWES, C. H. (1966). Viruses and Evolution. *Huxley Lecture*, University of Birmingham.

ARBER, W. (1958). Transduction des caractères Gal par le bactériophage λ. *Archs Sci., Geneva*, **11**, 259.

ARBER, W. (1960). Transduction of chromosomal genes and episomes in *Escherichia coli*. *Virology*, **11**, 273.

ARBER, W. (1965). Host-controlled modification of bacteriophage. *A. Rev. Microbiol.* **19**, 365.

ARBER, W., KELLENBERGER, G. & WEIGLE, J. (1957). La défectuosité du phage lambda transducteur. *Schweiz. Z. Path. Bakt.* **20**, 659.

BERNSTEIN, A. & WILSON, E. M. J. (1963). An analysis of the Vi-phage typing scheme for *Salmonella typhi*. *J. gen. Microbiol.* **32**, 349.

BRINTON, C. C., JUN., GEMSKI, P., JUN. & CARNAHAN, J. (1964). A new type of bacterial pilus genetically controlled by the fertility factor of *E. coli* K12 and its role in chromosome transfer. *Proc. natn. Acad. Sci. U.S.A.* **52**, 776.

CAVALLI-SFORZA, L. L., LEDERBERG, J. & LEDERBERG, E. M. (1953). An infective factor controlling sex compatibility in *Bacterium coli. J. gen. Microbiol.* **8**, 89.

CRAIGIE, J. & YEN, C. H. (1938). The demonstration of types of *B. typhosus* by means of preparations of type II Vi-phage. *Can. publ. Hlth J.* **29**, 448.

DATTA, N. (1965). Infectious drug resistance. *Br. med. Bull.* **21**, 254.

DATTA, N., LAWN, A. M. & MEYNELL, E. W. (1966). The relationship of F type piliation and F phage sensitivity to drug resistance transfer in R+F− *Escherichia coli* K12. *J. gen. Microbiol.* **45**, 365.

DRISKELL-ZAMENHOFF, P. J. & ADELBERG, E. A. (1963). Studies on the chemical nature and size of sex factors of *Escherichia coli* K12. *J. molec. Biol.* **6**, 483.

DUBNAU, E. & STOCKER, B. A. D. (1964). Genetics of plasmids in *S. typhimurium. Nature, Lond.* **204**, 1112.

FALKOW, S. & CITARELLA, R. V. (1965). Molecular homology of F merogenote DNA. *J. molec. Biol.* **12**, 138.

FALKOW, S., CITARELLA, R. V., WOHLHIETER, J. A. & WATANABE, T. (1966). The molecular nature of R-factors. *J. molec. Biol.* **17**, 102.

FELIX, A. & ANDERSON, E. S. (1951). Bacteriophages carried by the Vi-phage types of *Salmonella typhi. Nature, Lond.* **167**, 603.

GERBER, P. (1963). Tumors induced in hamsters by simian virus 40: persistent subviral infection. *Science, N.Y.* **140**, 889.

GERBER, P. (1964). Virogenic hamster tumor cells: induction of virus synthesis. *Science, N.Y.* **145**, 833.

GERBER, P. (1966). Studies on the transfer of subviral infectivity from SV40-induced hamster tumor cells to indicator cells. *Virology*, **28**, 501.

GERBER, P. & KIRSCHSTEIN, R. L. (1962). SV$_{40}$-induced ependymomas in newborn hamsters. I. Virus-tumor relationships. *Virology*, **18**, 582.

DE HAAN, P. G. & STOUTHAMER, A. H. (1963). F-prime transfer and multiplication of sexduced cells. *Genet. Res., Camb.* **4**, 30.

HAYES, W. (1952). Recombination in *Bact. coli* K-12: unidirectional transfer of genetic material. *Nature, Lond.* **169**, 118.

HAYES, W. (1953a). Observations on a transmissible agent determining sexual differentiation in *Bact. coli. J. gen. Microbiol.* **8**, 72.

HAYES, W. (1953b). The mechanism of genetic recombination in *E. coli. Cold Spring Harb. Symp. quant. Biol.* **18**, 75.

HAYES, W. (1964). *The Genetics of Bacteria and their Viruses*, ch. 24. Oxford: Blackwell.

HAYES, W. (1965). Sex factors and viruses. *Proc. R. Soc.* B, **164**, 230.

HOFSCHNEIDER, P. H. & PREUSS, A. (1963). M13 bacteriophage liberation from intact bacteria as revealed by electron microscopy. *J. molec. Biol.* **7**, 450.

HUEBNER, R. J., ROWE, W. P., TURNER, H. C. & LANE, W. T. (1963). Specific adenovirus complement-fixing antigens in virus-free hamster and rat tumours. *Proc. natn. Acad. Sci. U.S.A.* **50**, 379.

JACOB, F., BRENNER, S. & CUZIN, F. (1963). On the regulation of DNA replication in bacteria. *Cold Spring Harb. Symp. quant. Biol.* **28**, 329.

LEDERBERG, J. & TATUM, E. L. (1946a). Novel genotypes in mixed cultures of biochemical mutants of bacteria. *Cold Spring Harb. Symp. quant. Biol.* **11**, 113.

LEDERBERG, J. & TATUM, E. L. (1946b). Gene recombination in *E. coli. Nature, Lond.* **158**, 558.

LEDERBERG, J., CAVALLI, L. L. & LEDERBERG, E. M. (1952). Sex compatibility in *E. coli. Genetics*, **37**, 720.

LOEB, T. (1960). Isolation of a bacteriophage specific for the F+ and Hfr mating types of *Escherichia coli* K12. *Science, N.Y.* **131**, 932.

LOEB, T. & ZINDER, N. D. (1961). A bacteriophage containing RNA. *Proc. natn. Acad. Sci. U.S.A.* **47**, 282.

LURIA, S. E., ADAMS, J. N. & TING, R. (1960). Transduction of lactose utilizing ability among strains of *E. coli* and *S. dysenteriae* and the properties of the transducing phage particles. *Virology*, **12**, 348.

LURIA, S. E., FRASER, D., ADAMS, J. N. & BURROUS, J. (1958). Lysogenization, transduction and genetic recombination in bacteria. *Cold Spring Harb. Symp. quant. Biol.* **23**, 71.

LWOFF, A. (1959). Bacteriophage as a model of host-virus relationship. In *The Viruses*, **2**, 187. Ed. F. M. Burnet and W. M. Stanley. New York: Academic Press Inc.

MORSE, M. L., LEDERBERG, J. & LEDERBERG, E. M. (1956a). Transduction in *E. coli* K12. *Genetics*, **41**, 142.

MORSE, M. L., LEDERBERG, J. & LEDERBERG, E. M. (1956b). Transductional heterogenotes in *Escherichia coli*. *Genetics*, **41**, 758.

ØRSKOV, I. & ØRSKOV, F. (1960). An antigen termed f+ occurring in F+ *E. coli* strains. *Acta path. microbiol. scand.* **48**, 37.

SCAIFE, J. & GROSS, J. D. (1962). Inhibition of multiplication of an F-lac factor in Hfr cells of *Escherichia coli* K12. *Biochem. biophys. Res. Commun.* **7**, 403.

SIMKOVIC, D., VALENTOVA, N. & THURZO, V. (1962). An *in vitro* system for the detection of the Rous sarcoma virus in the cells of the rat tumor XC. *Neoplasm*, **9**, 104.

STOCKER, B. A. D., ZINDER, N. D. & LEDERBERG, J. (1953). Transduction of flagellar characters. *J. gen. Microbiol.* **9**, 410.

SVOBODA, J., CHYLE, P., SIMKOVIC, D. & HILGERT, I. (1963). Demonstration of the absence of infectious Rous virus in rat tumor XC whose structurally intact cells produce Rous sarcoma when transferred to chick. *Folia biol.* **9**, 77.

UETAKE, H. & HAGIWARA, S. (1960). Somatic antigen 15 as a precursor of antigen 34 in *Salmonella*. *Nature, Lond.* **186**, 261.

UETAKE, H. & HAGIWARA, S. (1961). Genetic cooperation between unrelated phages. *Virology*, **13**, 500.

UETAKE, H., LURIA, S. E. & BURROUS, J. W. (1958). Conversion of somatic antigens in *Salmonella* by phage infection leading to lysis or lysogeny. *Virology*, **5**, 68.

UETAKE, H., NAGAKAWA, T. & AKIBA, T. (1955). The relationship of bacteriophage to antigenic changes in group E Salmonellas. *J. Bact.* **69**, 571.

WATANABE, T. (1963). Infectious heredity of multiple drug resistance in bacteria. *Bact. Rev.* **27**, 87.

WATANABE, T., FUKASAWA, T. & TAKANO, T. (1962). Conversion of male bacteria of *Escherichia coli* K12 to resistance to f phages by infection with the episome, 'resistance transfer factor'. *Virology*, **17**, 218.

WATANABE, T., NISHIDA, H., OGATA, C., ARAI, T. & SATO, S. (1964). Episome-mediated transfer of drug resistance in Enterobacteriaceae. VII. Two types of naturally occurring R factors. *J. Bact.* **88**, 716.

WATANABE, T., TAKANO, T., ARAI, H. & SATO, S. (1966). Episome mediated transfer of drug resistance in Enterobacteriaceae. X. Restriction and modification of phages by fi− R factors. *J. Bact.* **92**, 477.

ZINDER, N. D. (1953). Infective heredity in bacteria. *Cold Spring Harb. Symp. quant. Biol.* **18**, 261.

INDEX